The Mathematics Teacher Education Partnership: The Power of a Networked Improvement Community to Transform Secondary Mathematics Teacher Preparation

Volume 4

A Volume in:
The Association of Mathematics Teacher Educators (AMTE)
Professional Book Series

Series Editor

Babette M. Benken
California State University, Long Beach

The Association of Mathematics Teacher Educators (AMTE) Professional Book Series

Series Editor

Babette M. Benken
California State University, Long Beach

Building Support for Scholarly Practices in Mathematics Methods (2017)
Signe E. Kastberg, Andrew M. Tyminski, Alyson E. Lischka, and Wendy B. Sanchez

Elementary Mathematics Specialists: Developing, Refining, and Examining Programs That Support Mathematics Teaching and Learning (2017)
Maggie B. McGatha, and Nicole R. Rigelman

Cases for Mathematics Teacher Educators: Facilitating Conversations about Inequities in Mathematics Classrooms (2016)
Dorothy Y. White, Sandra Crespo, and Marta Civil

The Mathematics Teacher Education Partnership: The Power of a Networked Improvement Community to Transform Secondary Mathematics Teacher Preparation
Volume 4

Edited by

W. Gary Martin
Brian R. Lawler
Alyson E. Lischka
Wendy M. Smith

INFORMATION AGE PUBLISHING, INC.
Charlotte, NC • www.infoagepub.com

Library of Congress Cataloging-In-Publication Data

The CIP data for this book can be found on the Library of Congress website (loc.gov).

Paperback: 978-1-64113-931-1
Hardcover: 978-1-64113-932-8
E-Book: 978-1-64113-933-5

CONTENTS

SECTION I

IMPROVING THE PREPARATION OF
SECONDARY MATHEMATICS TEACHERS

v

SECTION II

OPPORTUNITIES TO LEARN MATHEMATICS

SECTION III

OPPORTUNITIES TO LEARN MATHEMATICS

SECTION IV

OPPORTUNITIES FOR RECRUITMENT AND RETENTION

SECTION V

THE POWER OF A NETWORKED IMPROVEMENT COMMUNITY

ACKNOWLEDGMENTS

The editors of this book would like to acknowledge the efforts of many people who have contributed to the work of the MTE-Partnership, which this book describes, and to the production of this book itself. Any list of this kind risks leaving out people who should be mentioned; please know that any omissions are unintentional.

Thanks to current and past members of the planning team who helped to build and nurture the MTE-Partnership. The MTE-Partnership would not exist without their creativity and dedication:

Emina Alibegovic	Diana Kasbaum	Margaret J. Mohr-Schroeder
Lisa Amick	Brian R. Lawler	Jennifer Oloff-Lewis
Laurie Cavey	W. J. "Jim" Lewis	Robert N. Ronau
Charles Coble	Alyson E. Lischka	Wendy M. Smith
Ed Dickey	Gary Martin	Eric Stade
Mark Ellis	James Martinez	Marilyn Strutchens
Maria L. Fernandez	Johannah Maynor	Diana Suddreth
Howard Gobstein	Mike Mays	David C. Webb
	Julie McNamara	

Thanks to the following organizations, which provided critical support during the formation and continued growth of the MTE-Partnership:

- The Association of Public and Land-grant Universities—Howard Gobstein;
- The Carnegie Foundation for the Advancement of Teaching—Paul LeMahieu; and
- The California State University Chancellor's Office—Joan Bissell.

Thanks to the following funders, which provided financial support for the MTE-Partnership:

- National Science Foundation—Jim Hamos and Sandra Richardson, program officers; and
- Helmsley Charitable Trust—Ryan Kelsey and Sue Cui, program officers.

Thanks to the following for their excellent administrative support:

- Katherine Hazelrigg;
- Mary Leskosky;
- Tia Freelove Kirk; and
- Nathalie Dwyer.

Particular thanks to the following for their support for the production of this book:

- Lindsay Augustyn—editing and production of the document as well as the cover design;
- Cathy Kessel—editing of early versions of some chapters in this book; and
- Babette Benken—series editor for the Association of Mathematics Teacher Educators.

Finally, thank you to all of the individuals on the local partnership teams, without your dedication and continued engagement, the MTE-Partnership would not exist.

FOREWORD

Mike Steele

The Association of Mathematics Teacher Educators (AMTE) is the largest professional organization dedicated to the improvement of mathematics teacher education. AMTE's roots trace back to the early 1990s, when mathematics educators interested in strengthening teacher preparation and professional development came together to help meet the recommendations of the standards released by the National Council of Teachers of Mathematics (NCTM; 1989, 1991, 1995). The AMTE community emerged out of a need for focused attention to the research and practice of developing mathematics teachers with strong content knowledge; excellent pedagogical skills; and attention to access, equity, and empowerment to strengthen mathematics learning for each and every student.

Nearly 30 years later, the work of AMTE and mathematics teacher educators has made a significant impact on the mathematics education landscape, highlighted most recently by the AMTE's 2017 release of the *Standards for the Preparation of Teachers of Mathematics* (AMTE Standards). Elementary mathematics achievement is on the rise, high-quality mathematics curricula are more available than they have ever been, and the field has clearer understandings of what effective teacher preparation and professional development look like (and how to measure its outcomes). Yet, persistent challenges remain. Most notably, little progress has been made in strengthening teaching and learning in middle and high school mathematics beyond localized successes (NCTM, 2018). Furthermore, new developments in recent years have made the mathematics teacher edu-

The Mathematics Teacher Education Partnership: The Power of a Networked Improvement Community to Transform Secondary Mathematics Teacher Preparation, pages xi–xii.
Copyright © 2020 by Information Age Publishing

cation landscape more rugged, such as increasingly disparate state requirements for teacher certification, the rise of quick-fix solutions to staffing shortages that include emergency licensure and low-quality preparation, removal of funding for sustained professional development, bounty-based salary contexts that increase teacher migration, and the rampant de-professionalization of teaching in the eyes of politicians and the public.

It is not enough for us to simply find ways to traverse this rugged landscape. Rather, we in mathematics teacher education seek to transform it, with the ultimate goal of meeting more students where they are to engage them in the joy, beauty, and wonder of mathematics. I see two grand challenges before our community as we seek that goal. First, we must re-professionalize the work of teaching by illustrating the ways in which effective teaching makes use of codified professional knowledge and practice. Particularly in secondary mathematics teaching, we as a field continue to push back against the myth that simply knowing mathematics well can qualify one to teach it effectively. Second, we must make a clear argument for why the work of mathematics teacher preparation and professional development matters. Arguing for the resources and investments needed for a high-quality preparation program means convincing stakeholders, districts, prospective teachers, and the community that such preparation will result in stronger mathematical outcomes for each and every student. Meeting these grand challenges requires sustained, networked, and articulated efforts across the mathematics teacher education community.

Reform efforts in teacher education in the 1980s and 1990s, guided largely by the Holmes Group (1986), were a story of small-scale innovation and experimentation with a common set of five goals: make teaching intellectually sound; recognize differences in teachers' knowledge, skill, and commitment; create defensible and intellectually relevant standards of entry into teaching; connect schools of education to the schools; and make schools better places for practicing teachers to work and learn. During that era of innovation, teacher preparation programs laid a strong practice-based, research-informed foundation for teacher preparation, in the context of wildly varying state licensure requirements and different university contexts and constraints. These efforts laid the groundwork for a second generation of teacher preparation reform, in which the efficacy of those common goals and practices across programs could be studied, iterated, and strengthened at scale.

The activities of the Mathematics Teacher Education Partnership (MTE-Partnership) described in this book provide a road map for this second generation of reform. The MTE-Partnership builds on the first-generation work that pulled together multiple stakeholders, mathematics educators, mathematicians, and schools, in local collaboration around secondary mathematics teacher preparation. The initial MTE-Partnership's *Guiding Principles for Secondary Mathematics Teacher Preparation Programs* (2014) are clear evolutions of the Holmes Group goals. Second-generation goals are further explicated in the AMTE Standards on which the MTE-Partnership is now focusing.

The work of the MTE-Partnership seeks to define wicked problems of practice that involve all those stakeholders and transcend the particularities of university preparation programs and state licensure requirements. These problems of practice create common sites for larger-scale scholarly inquiry and commonly designed innovations applied in diverse state and local contexts. By using the Networked Improvement Community model (Bryk, Gomez, Brunow, & LeMahieu, 2015; Martin & Gobstein, 2015), groups of mathematics teacher educators work on these problems of practice using defined structures and phases grounded in improvement science. Topics for innovation and inquiry are identified and situated within the broader problem space of secondary mathematics teacher education. They are scoped by teams across contexts and worked on through iterative cycles, with the work connected to the MTE-Partnership driver diagram. Some problems of practice are brought to a state of completion; some have a longer arc of work; and others defy the ability to be scoped and studied and are set aside or reworked. The outcomes of this work feed back into the definition of the problem space, driver diagram, and *Guiding Principles* of the MTE-Partnership. In this way, MTE-Partnership as a larger enterprise operates similarly to design-based implementation research.

The work of MTE-Partnership described in this book provides a response to the two grand challenges. It addresses the grand challenges of re-professionalizing the work of mathematics teaching by better defining the knowledge needed to teach, identifying successful recruitment and retention strategies, and creating stronger research-informed curricula for all facets of the preparation trajectory. The outcomes of the MTE-Partnership also have contributed to the second grand challenge, making an argument for why the work of mathematics teacher preparation matters. The research outcomes described here begin to build a comprehensive case for why the work that we do matters, both for teachers and for the students they will and do teach.

As you read this book, consider how the MTE-Partnership has transformed the landscape of secondary mathematics teacher preparation. What features of the terrain does their work help us better understand? How has their work shaped the hills, valleys, and towns that we see, and built roads to take us to new places? And what parts of the landscape remain unexplored, sparking your curiosity about what may be out there? The MTE-Partnership provides a model for second-generation reform of mathematics teacher preparation by charting this landscape. What does the future hold for the third generation? How will we take the work of the MTE-Partnership into the next decade and build larger, broader systems to understand the impact of our teacher preparation programs on secondary mathematics teachers and their students? What roles might technologies like big data, AI-assisted video analysis, and advances in neurological science and affect measurement play in an even larger-scale, faster, and more responsive improvement to mathematics teacher education? How can we pull in more partners across a next-generation Networked Improvement Community to build an unimpeachable argument for the value of mathematics teacher education as described in the AMTE Standards? And how can we build public awareness and support for why teachers, teacher education, and mathematics teacher education matters to each and every citizen of our country?

Let's hit the road.

—Mike Steele
President, Association of Mathematics Teacher Educators

REFERENCES

Association of Mathematics Teacher Educators. (2017). *Standards for preparing teachers of mathematics*. Raleigh, NC: Author. Retrieved from http://amte.net/standards

Bryk, A., Gomez, L. M., Grunow, A., & LeMahieu, P. (2015). *Learning to improve: How America's schools can get better at getting better.* Cambridge, MA: Harvard Education Press.

Holmes Group, Inc. (1986). *Tomorrow's teachers: A report of the Holmes Group.* East Lansing, MI: Author.

Martin, W. G., & Gobstein, H. (2015). Generating a networked improvement community to improve secondary mathematics teacher preparation: Network leadership, organization, and operation. *Journal of Teacher Education, 66*(5), 482–493.

Mathematics Teacher Education Partnership. (2014). *Guiding principles for secondary mathematics teacher preparation.* Washington, DC: Association of Public and Land-grant Universities. Retrieved from mtep.info/guidingprinciples

National Council of Teachers of Mathematics. (1989). *Curriculum and evaluation standards for school mathematics*. Reston, VA: Author.

National Council of Teachers of Mathematics. (1991). *Professional standards for teaching mathematics*. Reston, VA: Author.

National Council of Teachers of Mathematics. (1995). *Assessment standards for school mathematics*. Reston, VA: Author.

National Council of Teachers of Mathematics (2018). *Catalyzing change in high school mathematics: Initiating critical conversations.* Reston, VA: Author.

PREFACE

While the past decades have seen some progress in United States students' mathematical preparation, achievement at the secondary level has seen little improvement (National Council of Teachers of Mathematics, 2018). A major contributing factor to this situation is the shortage of well-prepared secondary mathematics teachers (Tatto & Senk, 2011). A number of significant national recommendations have been made to address this challenge, including *The Mathematical Education of Teachers II* (Conference Board of Mathematical Sciences, 2012), the *Statistical Education of Teachers* (Franklin et al., 2015), and more recently the Association of Mathematics Teacher Educators' *Standards for the Preparation of Teachers of Mathematics* (AMTE, 2017). This book reports on the work of the Mathematics Teacher Education Partnership (MTE-Partnership) to address this challenge. The MTE-Partnership is a consortium of over 90 universities, colleges, and their K–12 school partners, working collaboratively to improve secondary mathematics teacher preparation; it was organized by the Association of Public and Land-grant Universities (APLU), a national research and advocacy organization of public research universities, land-grant institutions, and state university systems.

The first section of the book describes the focus on improvement that undergirds the work of the MTE-Partnership. An introductory chapter provides an overview of how the MTE-Partnership was formed and its organization as a Networked Improvement Community (NIC; Bryk, Gomez, Brunow, & LeMahieu, 2015; Martin & Gobstein, 2015). Two following chapters discuss themes that undergird the work of the MTE-Partnership: program transformation and equity and social justice.

The following three sections describe research organized by the MTE-Partnership to address three particular challenges in the preparation of secondary mathematics candidates: their mathematical preparation, their clinical experiences, and their recruitment and retention. Each of these sections is introduced by a chapter that provides an overview of the literature in that area and the framing of MTE-Partnership's research, with particular reference to the AMTE Standards (2017). Subsequent chapters detail the work of Research Action Clusters (RACs) addressing particular problems of practice within those areas.

The final section of the book section provides a look back on the past seven years of work of the MTE-Partnership, as well as prospects for its continued development. Its first chapter outlines the major outcomes discussed across this book and their use in improvement efforts, followed by a discussion of how the MTE-Partnership will continue to develop in the coming years. Two additional chapters were written by representatives from the Carnegie Foundation for the Advancement of Teaching to reflect on the progress and prospects of the MTE-Partnership as an NIC.

It is our hope that this book will serve a number of purposes for a range of audiences, including (a) providing those engaged in mathematics education with an organized summary of the state of research related to specific aspects of secondary mathematics teacher preparation on which the MTE-Partnership is built; (b) outlining promising approaches to improving aspects of secondary mathematics teacher preparation from which mathematics teacher educators might learn as they strive to enact the vision of the AMTE Standards (2017), including specific products and approaches that have been developed by the RACs that programs might want to consider adopting; (c) discussing future directions in improving secondary mathematics teacher preparation that might be of importance to the field; (d) describing the NIC design, which might be emulated by those addressing other problem spaces in mathematics teacher education and beyond; and (e) suggesting ideas and strategies for how an interested stakeholder in mathematics teacher preparation might initiate improvement efforts for their local program.

DEFINITIONS

Terms in education frequently have different meanings. The following terminology is used throughout this book.

Clinical experiences. K–12 school-based experiences. Field experiences that occur early in preparation programs often include classroom observation or tutoring in an after-school program. Later experiences may include lesson planning and teaching small groups of students or single lessons, generally culminating in a full-time experience in a K–12 school. The synonymous term *field experiences* is sometimes used in the literature.

Driver diagram. A tool that visually represents a group's working theory of action to drive program improvement. A driver diagram organizes primary (and secondary, tertiary) potential *drivers* of change, that can contribute toward achieving an identified aim.

Fishbone diagram. A tool that visually represents a group's causal systems analysis. Sometimes known as a cause-and-effect diagram or Ishikawa diagram.

Induction. Formal or informal induction activities designed to support beginning teachers during the initial years of employment.

Mathematics and mathematics education university faculty. Designation follows the department in which a faculty member has an appointment. Note that some members of a mathematics department may have a primary interest in mathematics education, and some mathematics education faculty members may be in a department that has a more general mission, such as curriculum and instruction or STEM education.

Mentor teacher. A K–12 teacher who provides guidance to teacher candidates during their clinical experiences. Synonyms include cooperating, collaborating, practicing, or expert teacher.

Plan-Do-Study-Act Cycle. An iterative four-step process to plan and document the viability/efficacy of a proposed change to a system by planning a small change, implementing the change, observing the results, and then acting on what is learned.

Professional development. The experiences of a practicing teacher that are aimed at increasing teaching effectiveness. This is also known as in-service training, teacher training, or (at the higher education level) instructor training.

Prospective teacher. Someone who is in the process of being educated as a teacher but who is not yet licensed or certified as a teacher. Such persons may also be called a pre-service teacher. See related description of *teacher candidate* below.

Secondary teacher preparation programs. Programs preparing teachers for grades 6 through 12. These programs may focus on middle school level (which in some states may include grades 4–5), at the high school level, or across the secondary school grades.

Student teaching. Part of clinical experiences that includes full-time experience in a K–12 school, generally under the guidance of an experienced teacher. Its duration may range from eight weeks to a full school year, with some portion of that time spent teaching a full load of classes. Synonyms are *practice teaching* or *internship*.

Students. Refers to K–12 students. Individuals who are undergraduate or post-baccalaureate students are referred to as *prospective teachers* or *teacher candidates*, to not confuse them with K–12 students.

Teacher candidate. A prospective teacher who has advanced standing in the preparation process. They have typically been admitted to the teacher preparation program and are engaged in clinical experiences.

University supervisor. A university faculty or staff member who oversees clinical experiences (including student teaching) for teacher candidates or prospective teachers.

ABBREVIATIONS

The following abbreviations are commonly used throughout this report. Additional abbreviations specific to the topic of a chapter may be included.

ALM: Active Learning Mathematics
AMTE: Association of Mathematics Teacher Educators
APLU: Association of Public and Land-grant Universities
CCSS-M: Common Core State Standards for Mathematics
$MCOP^2$: Mathematics Classroom Observation Protocol for Practices
MKT: mathematical knowledge for teaching
MODULE(S^2): Mathematics of Doing, Understanding, and Learning for Secondary Schools
MTE-Partnership: Mathematics Teacher Education Partnership
NCTM: National Council of Teachers of Mathematics
NIC: Networked Improvement Community
PDSA Cycle: Plan-Do-Study-Act Cycle
PR^2: Program Recruitment and Retention
RAC: Research Action Cluster
STRIDES: Secondary Teacher Retention and Induction in Diverse Educational Settings
STEM: science, technology, engineering, and mathematics

REFERENCES

Association of Mathematics Teacher Educators. (2017). *Standards for preparing teachers of mathematics.* Raleigh, NC: Author. Retrieved from http://amte.net/standards

Bryk, A., Gomez, L. M., Grunow, A., & LeMahieu, P. (2015). *Learning to improve: How America's schools can get better at getting better.* Cambridge, MA: Harvard Education Press.

Conference Board of the Mathematical Sciences (CBMS). (2012). *The mathematical education of teachers II.* Providence, RI, and Washington, DC: American Mathematical Society and Mathematical Association of America.

Franklin, C., Bargagliotti, A., Case, C., Kader, G., Schaeffer, R., & Spangler, D. (2015). *Statistical education of teachers*. Arlington, VA: American Statistical Association. Retrieved from http://www.amstat.org/

Martin, W. G., & Gobstein, H. (2015). Generating a networked improvement community to improve secondary mathematics teacher preparation: Network leadership, organization, and operation. *Journal of Teacher Education, 66*(5), 482–493.

National Council of Teachers of Mathematics. (2018). *Catalyzing change in high school mathematics: Initiating critical conversations.* Reston, VA: Author.

Tatto, M. T., & Senk, S. (2011). The mathematics education of future primary and secondary teachers: Methods and findings from the Teacher Education and Development Study in Mathematics. *Journal of Teacher Education, 62*, 121–137.

SECTION I

IMPROVING THE PREPARATION OF SECONDARY MATHEMATICS TEACHERS

This section opens with an introduction of the Mathematics Teacher Education Partnership (MTE-Partnership), including its history, guiding principles, and connections to the Association of Mathematics Teacher Educators' (2017) *Standards for the Preparation of Teachers of Mathematics* (AMTE Standards). Chapters 2 and 3 focus on cross-cutting themes to the work of the MTE-Partnership: program transformation and equity and social justice. Chapter 2's focus on program transformation is particularly helpful for school-university partnerships trying to better align their secondary mathematics teacher preparation programs to the AMTE Standards. Educational outcomes in the United States remain inequitable; Chapter 3's focus on equity and social justice offers suggestions for how to transform programs with a commitment to equity and social justice as a central focus.

CHAPTER 1

OVERVIEW OF THE MATHEMATICS TEACHER EDUCATION PARTNERSHIP

W. Gary Martin and Howard Gobstein

Organized by the Association of Public and Land-Grant Universities (APLU) in 2012, the Mathematics Teacher Education Partnership (MTE-Partnership) encompasses more than 90 universities, colleges, and their K–12 school partners, working collaboratively to improve secondary mathematics teacher preparation. Increasing the quality and quantity of secondary mathematics teachers in the United States is complex, and engaging a large community of stakeholders across many different contexts is the MTE-Partnership's strategy to advance the field. This chapter provides an introduction to the MTE-Partnership, including its rationale, structure, and design.

RATIONALE

The initial impetus for the formation of the MTE-Partnership in 2012 was the release of the *Common Core State Standards for Mathematics* (*CCSS-M*; National Governors Association Center for Best Practices, Council of Chief State School Officers, 2010) two years earlier (Martin & Gobstein, 2015). This document, initially adopted by 45 states as their state standards, significantly raised the bar for student achievement, including an emphasis on mathematical understanding, de-

The Mathematics Teacher Education Partnership: The Power of a Networked Improvement Community to Transform Secondary Mathematics Teacher Preparation, pages 3–23.

fined as "the ability to justify, in a way appropriate to the student's mathematical maturity, why a particular mathematical statement is true or where a mathematical rule comes from" (National Governors Association Center for Best Practices, Council of Chief State School Officers, 2010, p. 4). In addition, the document's Standards for Mathematical Practice emphasize "varieties of expertise that mathematics educators at all levels should seek to develop in their students" (National Governors Association Center for Best Practices, Council of Chief State School Officers, 2010, p. 6), suggesting competence beyond content knowledge. These new, more-rigorous standards raised significant issues for the preparation of mathematics teachers, particularly at the secondary level.

Research on Secondary Mathematics Teacher Preparation

Indeed, the U.S. faces a continuing shortage of well-prepared secondary mathematics teachers. The percentages of high schools reporting difficulties in filling vacant mathematics teaching positions have been higher than almost every other subject since 1999 (Malkus, Hoyer, & Sparks, 2015). According to the National Center for Educational Statistics (Carver-Thomas & Darling-Hammond, 2017; National Center for Educational Statistics, NCES, n.d.), 1 in 14 secondary mathematics teachers leave the profession every year, and another 1 in 16 change schools. The attrition rate is particularly high for beginning mathematics teachers: nearly 1 in 7 leave teaching after their first year (Ingersoll, Merrill, & May, 2014). More than one-third of mathematics teachers report that they would not choose to become a teacher again (NCES, n.d.). Ingersoll, Merrill, and May (2014) report that quality of teacher preparation, particularly related to pedagogical practice, significantly impacts new teacher attrition. National studies note that problems of teacher attrition and turnover are not uniformly distributed within districts, but tend to be concentrated by school (Ingersoll & Perda, 2010; Ingersoll & Perda, 2010), affecting a disproportionate number of high-poverty and minority student-serving schools. Moreover, quality of mathematics instruction continues to be a concern, as seen in two national surveys of practicing secondary mathematics teachers: only half reported using instructional practices and goals aligned with the *CCSS-M* (Banilower et al., 2013; Markow, Macia, & Lee, 2013).

While research on teacher preparation is not robust, existing studies suggest that improving the quality of teacher preparation programs can do much to improve its graduates' instructional effectiveness and to retain them in their chosen profession. For example, analyses of nationally representative survey data find that in the first year of teaching, teachers with a mathematics baccalaureate but with little or no pedagogical preparation, left teaching at twice the rate of those with the same degree but with more comprehensive pedagogical preparation (Ingersoll, Merrill, & May, 2014). Such preparation tended to include a full semester of practice teaching. Pedagogical preparation is an opportunity for development of mathematical knowledge tailored to the work of teaching, a type of knowledge associated with instructional effectiveness (Hill, Rowan, & Ball, 2005).

Despite well-publicized alternative routes to certification, the overwhelming majority of teachers are prepared in undergraduate programs at postsecondary institutions. These programs have three main components: mathematics courses, methods courses, and clinical experiences in a K–12 school setting. Each of these components involves a different group of people: faculty members who teach mathematics, typically in mathematics departments; faculty members who teach methods courses, typically in schools of education; and mentor teachers, typically from nearby secondary schools. Traditionally, each of these groups have contributed separate components to teacher preparation—with unproductive results. As the American Association of Colleges for Teacher Preparation (American Association of Colleges for Teacher Preparation, AACTE, 2010) states:

> Teacher candidates generally completed course work on psychological principles, subject matter, and teaching methods before beginning student teaching—for about 8 weeks at the end of the program—with few connections to course content. School-based cooperating teachers were selected not necessarily on the basis of quality. Placements were idiosyncratic, with experiences ranging from primarily clerical work to solo teaching without assistance. (p. 1)

Not surprisingly, when entering their own classrooms, new teachers tended to revert to "what they knew best—the way they themselves had been taught" (p. 2).

Teaching is increasingly viewed as a clinical practice profession more akin to medicine, involving "observing, assessing, diagnosing, prescribing, and adjusting practice to reflect new knowledge" (AACTE, 2010, p. 1). This view has profound consequences for those who prepare teachers. Research, institutional examples, and national recommendations indicate that improving teacher education involves not only improving its three main components, but also the coordination among them. For example, mathematical knowledge specific to teaching could be viewed as the responsibility of mathematics instructors, as well as education instructors and mentor teachers. Because institutional structures, support, and leadership affect coordination among these three groups of people, the audience for this book includes administrators as well as faculty members, in institutions of higher education, in school districts, and at the state level.

Recommendations for Secondary Mathematics Teacher Preparation

A number of organizations have proposed recommendations for improving secondary mathematics teacher preparation. In 2012, the Conference Board of the Mathematical Sciences (CBMS) released *The Mathematical Education of Teachers II (MET II)*, a follow-up to an earlier set of recommendations addressing the mathematical content knowledge well-prepared beginning teachers of mathematics should have, including at the secondary level. In 2015, the American Statistical Association released *Statistical Education of Teachers* (Franklin et al., 2015) describing the statistical content knowledge needed by well-prepared beginning teachers of mathematics, including at the secondary level. The *NCTM*

CAEP Standards (National Council of Teachers of Mathematics and the Council for the Accreditation of Educator Preparation, 2012a, 2012b) provide standards for accreditation of secondary mathematics teacher preparation programs, including what prospective teachers of secondary mathematics should know and be able to do.

The Association of Mathematics Teacher Educators' (2017) *Standards for the Preparation of Teachers of Mathematics* (AMTE Standards) compiled these recommendations into a comprehensive vision for the nation, including standards for mathematics teacher candidates and for programs preparing mathematics teachers, with particular recommendations for Grades 6–8 and 9–12. The AMTE Standards authors argue that the standards put forth "an ambitious but achievable vision to prepare teachers who can effectively support the mathematics learning of each and every student in grades PK–12 education" (p. 163). The authors further point out, "Making the changes needed to achieve this vision will be challenging, requiring a significant investment of time and resources" (p. 163).

Although the MTE-Partnership predates the writing of the AMTE Standards, the MTE-Partnership builds on the same general theoretical and research base as the AMTE Standards, and so the MTE-Partnership vision generally aligns with that of the AMTE Standards. Indeed, multiple MTE-Partnership leaders were part of the writing team for the AMTE Standards, and a number of references are made to the MTE-Partnership and its activities throughout the document. The ultimate goal of the MTE-Partnership is to support secondary mathematics teacher preparation programs to transform their policies and practices to achieve the MTE-Partnership vision as well as that of the AMTE Standards.

GUIDING PRINCIPLES

The foundation of the MTE-Partnership is its *Guiding Principles for Secondary Mathematics Teacher Preparation Programs* (2014), which provides the framework on which subsequent analysis and activity were built. The *Guiding Principles* describe a shared vision for mathematics teacher preparation for MTE-Partnership teams and was developed through an intensive process of engagement with its members, composed of school-university partnerships. It is a living document intended to capture the MTE-Partnership's growing understandings of teacher preparation and the institutional structures needed to support it, with several revisions being made from the initial version. Since their release in 2017, the AMTE Standards have provided additional insights into particular areas of secondary mathematics teacher preparation; its recommendations are generally consistent with the *Guiding Principles*, as stated above. Thus, the *Guiding Principles* continue to serve as the foundation for the work of the MTE-Partnership, with additional support from the AMTE Standards. The current version of the Guiding Principles is posted at http://mtep.info/guiding-principles.

The *Guiding Principles* are organized into three sections, each of which includes three to four principles, as follows:

- Section I. Partnerships
  - o Guiding Principle 1. Partnerships as the Foundation
  - o Guiding Principle 2. Commitments by Institutions of Higher Education
  - o Guiding Principle 3. Commitments by School Districts and Schools
- Section II. Teacher Candidate Knowledge, Skills, and Dispositions
  - o Guiding Principle 4. Candidates' Knowledge and Use of Mathematics
  - o Guiding Principle 5. Candidates' Knowledge and Use of Educational Practices
  - o Guiding Principle 6. Professionalism, Advocacy, and Leadership
- Section III. Support Structures
  - o Guiding Principle 7. Clinical Experiences
  - o Guiding Principle 8. Student Recruitment, Selection, and Support
  - o Guiding Principle 9. Beginning and Inservice Teacher Support
  - o Guiding Principle 10. Tracking Student Success

Each guiding principle includes a number of more specific indicators. For example, Guiding Principle 4: Candidates' Knowledge and Use of Mathematics, has the following indicators: Mathematical Habits of Mind, Knowledge of the Discipline, Specialized Knowledge of Mathematics for Teaching, and Nature of Mathematics.

A discussion of each section of the *Guiding Principles* follows, along with its connections with the AMTE Standards.

Partnerships as the Foundation

The first section of the *Guiding Principles* establishes that partnerships are foundational to the enterprise of preparing secondary mathematics teachers, with key players including institutions of higher education (including departments and colleges of mathematics and education), K–12 partner schools districts, and other stakeholders. The first principle emphasizes the importance of the partners having shared goals and vision for preparing teachers, a commitment to mutual learning about how to better prepare teachers, and a sense of shared responsibility for the success of the program. Second, institutions of higher education should maintain focus on and provide support for secondary mathematics teacher preparation, with an emphasis on building partnerships across departments and with their K–12 partners. Finally, K–12 school districts should interact with teacher preparation programs across the teacher development continuum from prospective to induction in the profession and beyond, with a particular emphasis on mentoring. Note that developing partnerships is also central to the AMTE Standards; Assumption 4 states, "Multiple stakeholders must be responsible for and invested in preparing teachers of mathematics" (AMTE, 2017, p. 2). Further discussion can be found in Standard P.1, which states, "An effective mathematics teacher preparation program has significant input and participation from all appropriate stakeholders" (p. 27).

Teacher Candidate Knowledge, Skills, and Dispositions

This section includes three guiding principles. First, programs must ensure that "teacher candidates have the knowledge and understanding of mathematics necessary to promote student success in mathematics as described in the *MET II* (CBMS, 2012), *CCSS-M*, and other college- and career-ready standards" (MTE-Partnership, 2014, p. 3). This includes mathematical habits of mind, disciplinary knowledge, specialized knowledge for teaching, and an appreciation of the nature of mathematics. These emphases overlap substantially with AMTE (2017) Standards C.1 and P.2.

Second, candidates' knowledge and use of educational practices should include being able to design, implement, and assess instruction that engages students in developing mathematical practices and processes, which engaged in learning important mathematics. Particular emphasis is placed on effective use of technology and attention to diversity, which directly relate to AMTE (2017) Standards C.2 and P.3, as well as at least indirectly to Standards C.3 and C.4. The final guiding principle focuses on professionalism, advocacy, and leadership; these emphases also overlaps substantially with aspects of Standards C.3 and C.4. However, other aspects of the *Guiding Principles*, such as an emphasis on developing an intellectual spirit of learning and growth, are less present in the AMTE Standards, while the AMTE Standards contain a more-detailed discussion of issues related to equity and social justice.

Support Structures

This section focuses on particular elements of the secondary mathematics teacher preparation program that support candidates' growth across the professional development continuum. The first guiding principle focuses on recruitment, selection, and support of candidates, with particular emphasis on maintaining quality while promoting diversity of candidates and supporting candidates as they progress through the program; this guiding principle corresponds to AMTE (2017) Standard P.5. Second, candidates need to engage in clinical experiences that are aligned with program goals and extend across the program, mirroring the recommendations in AMTE Standard P. 4. The need for programs to engage with their school partners in continued mentoring as they enter the profession is also emphasized; Assumption 2 in the AMTE Standards states, "Teaching mathematics effectively requires career-long learning" (p. 1), although this is less of an emphasis in the standards themselves. Finally, programs need to track their success in preparing successful beginning teachers, with an emphasis on mutual improvement. This perspective can also be found in Assumption 5 of the AMTE Standards, which states, "Those involved in mathematics teacher preparation must be committed to improving their effectiveness in preparing future teachers of mathematics" (p. 2.). Chapter 8 of the AMTE Standards provides further recommendations about assessment of candidates and programs.

Development of the *Guiding Principles* was one of the initial accomplishments of the MTE-Partnership. They were foundational to the identification of particular problem areas within secondary mathematics teacher preparation that were selected for attention by the MTE-Partnership and have guided the ongoing work of the MTE-Partnership, with additional guidance provided by the AMTE Standards (2017).

STRUCTURE OF THE MTE-PARTNERSHIP

The MTE-Partnership consists of teams of institutions that are collaborating to improve secondary mathematics teacher preparation. Each MTE-Partnership team is headed by an APLU member institution and includes at least one K–12 partner school or district and at least one other partner engaged in secondary mathematics teacher preparation. To be accepted into the MTE-Partnership, a team needs to demonstrate commitment to transforming secondary mathematics teacher preparation, including collaboration across team institutions; active involvement of stakeholders, including mathematicians, mathematics educators, and K–12 personnel; and institutional support for the effort.

Since its inception in 2012, the MTE-Partnership has grown to include 40 teams in 31 states, including more than 90 universities, university systems, and community colleges; numerous K–12 schools and school districts; and several state departments of education, education consortia, and other education-focused organizations. Teams vary in size and organization, depending on what they determined would best meet their needs, ranging from a single university and its partners to

FIGURE 1.1. Participation in the MTE-Partnership as of 2019. Stars Represent Lead Institutions for a Team, and Circles Represent Other Participating Universities and Colleges.

a collaboration of multiple campuses within a state and their partners. Figure 1.1 provides a map of the MTE-Partnership membership as of 2018. A current list of teams and member institutions can be found at http://mtep.info/team-list.

MTE-PARTNERSHIP DESIGN

To meets its established goal of transforming secondary mathematics teacher preparation, the MTE-Partnership adopted the Networked Improvement Community (NIC) model, which includes use of improvement science methods (Bryk, Gomez, Brunow, & LeMahieu, 2015; Martin & Gobstein, 2015) to develop products and approaches to address challenges in secondary mathematics teacher preparation through iterative cycles of design, testing, and redesign. The NIC structure used in educational reform (e.g., Dolle, Gomez, Russell, & Bryk, 2013) adapts ideas and organizational structures used in health care reform (e.g., Plsek, 2003). These organizational structures are designed to change a complex adaptive system, that is, a system which is adaptive: its elements have the ability to respond to stimuli in different and sometimes unpredictable ways that change the context for other elements of the system (Plsek, 2003). Examples of such systems range from antibiotic-resistant bacteria, a colony of social insects, or almost any collection of human beings (e.g., the U.S. health care system). Iteration is central to this design of the organizational structures used to improve such systems: beginning with a shared hypothesis about the problem structure that affords rapid-cycle experimentation on subproblems in priority areas, measuring and comparing experimental outcomes, spreading successful innovations, refining the hypothesis, and recalibrating priorities.

Development of the NIC Design

When it launched in 2012, the MTE-Partnership faced a number of design challenges identified by the project leaders (Martin & Gobstein, 2015). A primary challenge was how to fully engage all of the teams in the effort to develop a true partnership, rather than following a more traditional model in which several lead teams are designated demonstration sites, with less participation from the other teams. An additional challenge was how to maintain a strong scholarly focus in the work rather than undertaking more ad hoc efforts that lacked rigor in their design. A final challenge was how to keep the *Guiding Principles*, established as the foundation for the MTE-Partnership as described in the previous section, as a central focus of its work. The NIC design, then being developed for additional replication by the Carnegie Foundation for the Advancement of Teaching, was identified by the planning team as a promising answer to these challenges. The decision to organize the MTE-Partnership as an NIC was ratified by the membership in an online survey conducted in spring 2013. Subsequently, the NIC design was identified in the AMTE Standards (2017) as a particularly useful means of building collaboration across a system that supports the improvement of mathematics teacher preparation.

NICs are distinguished by four essential characteristics (Bryk, Gomez, Brunow, & LeMahieu, 2015); each characteristic is described below, along with a discussion of how the MTE-Partnership NIC addresses that characteristic.

Focused on a Well-Specified Common Aim. As stated by Bryk et al. (2015), "an improvement aim articulates the specific problem to be solved and the measures of accomplishment to which the community will hold itself accountable. It imbues the community with purpose" (p. 150). The MTE-Partnership is focused on the aim of increasing by 40% the quantity of teacher candidates who meet a gold standard of preparedness by 2020. A collaborative process of collecting data from the individual teams and programs was used to establish this aim. An annual survey of teams provides information on their production of candidates, and two measures track the quality of programs in alignment with the *Guiding Principles*—a survey of how well-prepared candidates believe they are and a program self-study of their effectiveness. See further discussion of measures of this aim in a later section of this chapter. NICs need to have a common aim: although many solutions might be proposed by its members, the aim serves as a litmus test of whether those solutions should be pursued by the MTE-Partnership.

Guided by a Deep Understanding of the Problem and the System that Produces It. The MTE-Partnership makes the hypothesis that teacher preparation is an outcome of a complex adaptive system. Within each team, the parts of this system typically include a university—both a mathematics department and a teacher preparation unit (e.g., a school of education)—and a school district, as well as other institutions. The system is adaptive in that its outcomes change in response to changes in its elements (e.g., changes in school district or mathematics department policy) and external changes (e.g., state mandates for teacher certification). Selected outcomes of teacher preparation programs—the quality and quantity of graduates—can be improved by changing elements of the system. The *Guiding Principles*, described in an earlier section, outline major components of the system of secondary mathematics teacher preparation.

The *Guiding Principles* subsequently served as the framework for an intensive analysis of significant challenges in secondary mathematics teacher preparation. A fishbone diagram in Figure 1.2—used to visually represent a group's causal systems analysis (Bryk et al., 2015)—includes four priority problem areas identified using a rigorous process that included garnering input from teams about areas where they felt work was needed along with areas where they felt that had the capacity to participate. A combination of both high need and high capacity for participation guided the selection of problem areas. See Martin and Strutchens (2014) for a detailed description of the process.

Working groups were established in August 2012 to better understand each of four major challenges: Developing a common vision across stakeholders, improving candidates' mathematical preparation, preparing and supporting mentor

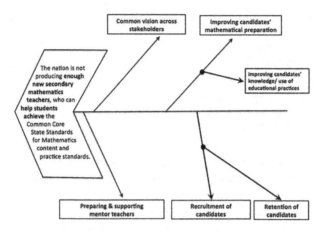

FIGURE 1.2. Fishbone diagram representing the problem space for secondary mathematics teacher preparation.

teachers, and recruitment and retention of candidates. More than 130 representatives from MTE-Partnership teams were involved in these working groups, whose work culminated in a series of white papers that explicated the research in each area. Revised versions of these white papers appear in Chapters 4, 7, and 13 of this book, introducing Sections II, III, and IV.

Their work resulted in an initial plan of action, captured in the driver diagram in Figure 1.3. As described by the Carnegie Foundation, "primary drivers are the major causal explanations hypothesized to produce currently observed results. Secondary drivers, in contrast, are interventions in the system aimed at advancing

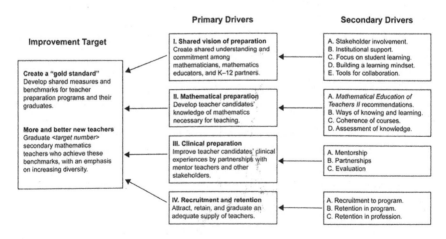

FIGURE 1.3. Initial driver diagram for the MTE-Partnership.

improvement toward targets" (Bryk, Gomez, & Grunow, 2011, p. 142). Driver diagrams are primary tools for analyzing problems and explicating ideas for specific work to address them; they are not comprehensive descriptions of the system. Instead, they depict the drivers hypothesized to change the state of the system, moving its outcomes toward the desired aim.

As a part of their agenda, the working groups described 13 specific change ideas on which the MTE-Partnership might begin work to address the secondary drivers. Through an extended period of input, these change ideas were narrowed down to a set of five areas, and a Research Action Cluster (RAC) was formed to begin work in each area. One RAC was subsequently retired, and a new RAC was added, resulting in the following list of active RACs in the MTE-Partnership:

- Program Recruitment and Retention, PR^2 RAC, actively recruiting high-quality and diverse teacher candidates and monitoring and providing support to ensure program completion;
- Active Learning Mathematics, ALM RAC, focusing on improving the content preparation of candidates in introductory mathematics classes (precalculus to calculus 2) using active learning strategies (Freeman et al., 2014);
- Mathematics of Doing, Understanding, Learning, and Educating for Secondary Schools, MODULE(S^2) RAC, producing modules or courses specifically aimed at developing mathematical knowledge for teaching (cf. Ball, Thames, & Phelps, 2008);
- Clinical Experiences RAC, working to improve clinical experiences, including new models for student teaching (Leatham & Peterson, 2010) as well as professional development for mentor teachers; and
- Secondary Teacher Retention and Induction in Diverse Educational Settings, STRIDES RAC, developing strategies to retain new secondary mathematics teachers in the profession.

Two additional working groups have been formed to address equity and social justice, along with strategies supporting overall program transformation. These working groups may eventually evolve into full-fledged RACs. In addition, other RACs may be retired as they achieve their primary aim. The RACs and working groups are discussed in more detail in following chapters.

Disciplined by the Rigor of Improvement Science. The use of evidence to guide the development of interventions ensures that the changes being proposed are actual improvements. The RACs employ the Plan-Do-Study-Act (PDSA) Cycle (see Figure 1.4) to iteratively prototype, test, and refine interventions; use of the PDSA Cycle has the potential to lead to timely solutions to important problems (Bryk et al., 2015). The plan stage of the cycle involves defining a change, predicting its impact, and designing a way to test its impact. The do stage involves carrying out the change and collecting relevant evidence. The study stage involves analyzing the evidence and developing insights about what happened. Finally, the

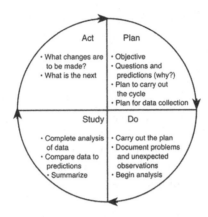

FIGURE 1.4.   The Plan-Do-Study-Act (PDSA) Cycle.

act stage involves deciding what to do next—perhaps refining the idea or extending it to further testing, or perhaps abandoning it if it does not prove useful.

Note that each RAC has developed its own aim statement and driver diagram that relates to the overall MTE-Partnership aim and driver diagram, and it then conducts PDSA Cycles in alignment with its driver diagram. In some sense, the RACs may be considered sub-NICs. A reporting form is commonly used across the RACs to document their PDSA Cycles (see Appendix A). Further details about the work of the RACs are included in Chapters 5–6, 9–11, and 14–16 of this book.

Networked to Accelerate Progress. NICs are designed to marry precepts of improvement science with precepts of networked improvement, so that the improvement cycles can be carried out across a range of contexts (Bryk et al., 2011; Bryk et al., 2015). Thus, partners are mobilized to work in a parallel and coordinated manner to address critical sub-problems hindering the transformation of mathematics teacher preparation. Rather than trying to control variation, as typical in traditional educational research, the NIC design embraces variation to study how interventions need to be adapted to respond to the differing conditions under which they are used. As they are tested and refined, interventions can gradually spread across the network, supporting scale up (Bryk et al., 2015). Thus, rather than developing a treatment that is tested against a control group, the initial development and testing of an intervention begins in a small number of settings. As its efficacy is demonstrated, it is tested in an increasing number of settings, noting adaptations that are needed due to differences in the context. Eventually, the interventions designed should be useful by teams across the MTE-Partnership.

The networked organization further allows a divide and conquer approach in which subsets of teams can address different problem areas. Thus, while each local partnership team typically participates in one or perhaps two RACs, their participation in the network provides them access to a wider range of interventions as the work of the RACs progresses, including those in which they are not engaged. Attending to more interventions, however, may provide teams additional challenges, as teams may lack the resources needed to integrate work across multiple areas. Additionally, scaling up the work of individual RACs so that it is accessible across the network presents a significant challenge for the MTE-Partnership leadership. These issues are being addressed by the Transformations Working Group, whose work is discussed in Chapter 2.

Summary

In addition to serving as the overall design for the MTE-Partnership, each RAC draws upon the design in its work. Each RAC essentially operates as a sub-NIC, with its own aim and driver diagram, and running improvement cycles based on the driver diagram. While each RAC has its own adaptation of the model based on its particular circumstances, the need to have clear objectives, a theory of change, and a dedication to the use of evidence in making decisions is foundational to the success of any group focusing on improvement.

MEASURING CANDIDATE AND PROGRAM QUALITY

Collection of evidence about candidate and problem quality can be undertaken for multiple purposes. It may be summative, designed to certify a candidate's readiness to begin teaching or a program's compliance with accreditation standards. Or, it may be formative, designed to provide useful information about a candidate's or program's progress, with the intent of guiding efforts to improve (AMTE, 2017). Not surprisingly, an NIC's use of evidence focuses on formative assessments for the purpose of improvement. Given this purpose, a further distinction must be drawn between measures for improvement and measures for research. While research measures often focus on developing general theory about some class of phenomena, measures for improvement tend to be much narrower in scope, focusing on "a relatively small number of impactful concepts that can be used reliably to guide action" (Bryk et al., 2015, p. 100). The focus of MTE-Partnership data collection is on practical, improvement-focused measures. Data collection occurs at multiple levels, including on a Partnership-wide basis to track progress toward the overall aim, at the RAC level to track progress toward the RAC aim, and at an even finer grain size to inform particular PDSA Cycles.

An Overview of Measures Used by MTE-Partnership Programs

Identifying common measures is an important aspect of being able to track progress across an NIC and to establish a common identity across its research efforts (Bryk et al., 2015; Martin & Gobstein, 2015). In the case of the MTE-Partnership, identifying measures that could be used to track teacher candidate quality proved particularly challenging, given that each program already had an assessment system in place. To better understand the landscape of assessment practices that were in place, a section of a baseline survey completed by all teams in the spring of 2013 focused on their assessment practices. This section included an item that asked them to identify the types of assessments used to assess teacher candidates and follow-up questions asked for details of the particular assessments used.

General Findings

Table 1.1 summarizes responses to a multiple-choice item in which respondents were asked to mention all types of measures used in their teacher preparation programs. In addition, the survey asked programs to list the particular assessments used, describe how they were developed, and when they were administered to teacher candidates. Over 100 examples were provided, for an average of about 3.5 per program. In light of the larger number of assessments reported in the previous section, respondents clearly gave only a sampling of the assessments they use. Table 1.1 includes frequent responses for this item, shown in italics.

TABLE 1.1   Types of Measures Used in Partnership Programs

	N (32)	%
Which of the following types of measures does your program use in assessing the ability of your teacher candidates to effectively teach mathematics?		
Written examination of mathematical knowledge (including mathematical knowledge specific to teaching)	24	75%
Praxis II[1]	18	56%
State-developed tests	12	38%
Locally-developed tests	2	6%
Written examination of mathematics-specific pedagogical knowledge	16	50%
Written examination of general pedagogical knowledge	23	72%
Praxis Principles of Teaching and Learning	9	39%
State-developed tests	4	13%
Evaluations of pedagogical practice[2]		
Observations of teaching using a fixed protocol	27	84%
General evaluation of characteristics and abilities	23	72%
Portfolio of evidence collected throughout program	19	59%
Portfolio of evidence collected throughout student teaching	14	44%
Extended performance assessment, such as a teacher work sample	25	78%
Locally-produced rubric	17	53%
edTPA	4	13%
Measures of student achievement	8	25%

[1] Italicized items summarize common examples provided in a follow-up item.
[2] Heading added to help with organization; this heading was not included in the survey.

Programs reported using as few as 4 or as many as 8 different types of measures. The median number of types of measures used was 6, with 57% using between 5 and 7 types, 28% using less than 5 types, and 16% using more than 7 types. Most programs (84%) reported using at least one written examination of candidate knowledge, and more than two-thirds use more than one. Most commonly, programs use a written examination of mathematical knowledge—three-quarters reported doing so. Although over two-thirds administer a written examination of general pedagogical knowledge, only half include a test of mathematics-specific pedagogical knowledge.

Every program reported using at least one assessment of teaching proficiency, with nearly 90% using more than one such assessment. Observation of teaching rated with a fixed protocol was the most commonly reported assessment overall, used by nearly 85% of the programs. Performance assessments (such as a work sample) are used by three-quarters of the programs. Over half of the programs use portfolios assembled throughout a candidate's progress in the program, and less than half use a portfolio assembled during the student teaching experience. Over 70% used a general evaluation of candidate characteristics and abilities. Finally, a fourth of the programs reported using a measure of student achievement in assessing candidate proficiency.

These data informed the early development of the MTE-Partnership measures system. In addition, a shortened version of this survey was included in the 2018 MTE-Partnership Program Progress Survey discussed below. The level of use of mathematical content measures was generally consistent, with 28 of 34 responding programs (82%) requiring some such assessment, and 19 of 34 (56%) specifically requiring the *Praxis II*. However, major changes were seen in the use of extended performance assessments. While 27 or 34 responding programs (79%) required some sort of extended performance assessment, 15 of 34 responding programs (44%) require the edTPA and 9 of 34 (26%) require a state-designed performance assessment. In contrast, only 5 of 34 (15%) used a locally designed assessment, which was the case for more than half of the programs in 2013.

Measures Used by the MTE-Partnership

The MTE-Partnership established a Measures Working Group in the early stages of development of the MTE-Partnership to identify and design common measures to be used across the partnership. These measures are used to track the overall progress of the MTE-Partnership, progress within specific RACs, and progress by local partnership teams. A brief discussion of the primary MTE-Partnership measures follows.

MTE-Partnership Candidate Production Survey. The MTE-Partnership annually collects data on the number of candidates produced by programs to track the progress of the MTE-Partnership as a whole in reaching its aim of increasing the number of teacher candidates. See Figure 1.5. Local teams can also use this data to track their progress.

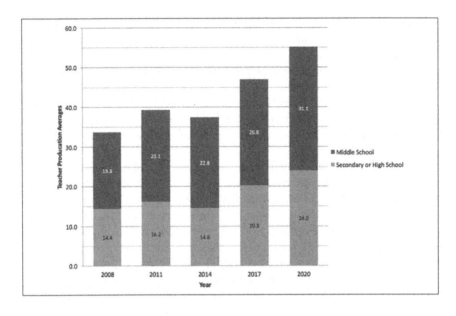

FIGURE 1.5. Average number of candidates produced by MTE-Partnership programs through 2014, along with targets for 2017 and 2020.

MTE-Partnership Program Completer Survey. The Measures Working Group also designed a survey of candidates completing programs offered by local Partnership teams, asking how well prepared they feel they are, as well as how effective they feel the program they are completing is. Questions are again aligned with the *Guiding Principles* to provide a point of triangulation with the MTE-Partnership Program Progress Survey. These data are primarily used to inform local teams, although RACs may also consider particular areas of the survey aligned with their missions.

Mathematics Classroom Observation Protocol for Practices (MCOP²). After reviewing multiple classroom observation protocols, the Measures Working Group identified the MCOP² (Gleason, Livers, & Zelkowski, 2015) as most closely aligned with the *Guiding Principles*. Although programs may not be able to replace current observation protocols, they may elect to use this protocol with a sample of teacher candidates at the conclusion of their culminating student teaching experience to better understand their effectiveness in reaching the *Guiding Principles*. A number of the RACs also use this protocol as a part of their data collection. A training session was developed to help ensure reliability in its use, and minor modifications were suggested to better reflect an emphasis on candidates' ability to promote equity in their instructional practices.

MTE-Partnership Program Progress Survey. This survey, designed by the Measures Working Group, primarily serves as a self-assessment of local partnership teams progress in reaching the MTE-Partnership *Guiding Principles*. A set of multiple-choice questions ask participants to rate their progress on the guiding principles and their specific indicators. Additional open-ended questions invite additional reflection on their progress and provide feedback on the progress of the MTE-Partnership as a whole. These data are used by the MTE-Partnership planning team to track areas in which additional work may be needed, by RACs to get input from teams that are not in their RAC, and by local teams to track their progress in producing candidates that meet the vision of the MTE-Partnership.

These common measures help the MTE-Partnership to track its progress toward its aim of increasing the number and quality of secondary mathematics teacher candidates. In some cases, these measures are used to inform the work of the RACs, and RACs have identified or designed additional measures that inform particular aspects of their work. The identification of useful measures has been a major challenge for the MTE-Partnership and is essential to its progress as an NIC.

CONTINUED DEVELOPMENT OF THE MTE-PARTNERSHIP

The MTE-Partnership has grown and evolved since its inception in 2012. Five RACs were launched in 2013. One was subsequently retired in 2016 as it was unable to establish a clear aim. Another was formed to address an additional area of concern related to retention of graduates from secondary mathematics teacher preparation programs in the profession.

Two additional areas of concerns have emerged, which led to the formation of working groups to begin to address these concerns. Both working groups developed a preliminary driver diagram, possible targets and measures, and promising approaches, which could lead to their formation as RACs. The Transformations Working Group was charged with establishing a foundation for the MTE-Partnership's strategic focus on overall transformation of secondary mathematics teacher preparation programs; the discussion in Chapter 9 of the AMTE Standards (2017) reinforces the critical nature of this work. The Equity and Social Justice Working Group is charged with advising the MTE-Partnership community on how to better incorporate attention to equity and social justice throughout the work of the MTE-Partnership. This area of work reflects the primacy of equity within the AMTE (2017) Standards; as stated in Assumption 1, "Ensuring the success of each and every learner requires a deep, integrated focus on equity in every program that prepares teachers of mathematics" (p. 1).

The current structure of the MTE-Partnership is depicted in Figure 1.6. This simplified driver diagram shows how the RACs and working groups address the four primary drivers initially identified by the MTE-Partnership, along with two new, cross-cutting drivers addressed by working groups. Progress by these two working groups is described in Chapters 2 and 3 in this section of the book. The

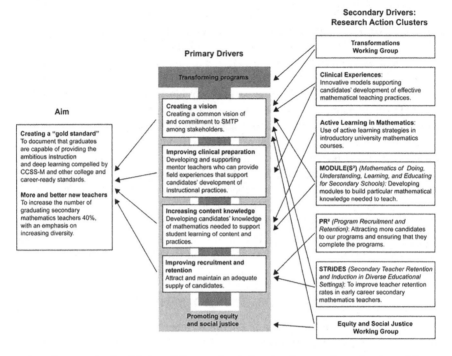

FIGURE 1.6.   Revised Driver Diagram Depicting the Current Organization of the MTE-Partnership.

following three sections of the book describe work of the RACs addressing three of the primary drivers related to content knowledge, clinical experience, and recruitment and retention. Note that the primary driver, creating a vision, is not directly addressed by any of the RACs but is indirectly addressed by all of them. The final section of the book provides commentary of the work of MTE-Partnership as an NIC and how it may continue to grow and evolve.

ENDNOTES

1.   Work on this chapter was supported in part by a grant from the National Science Foundation (DRL #1147987) and The Helmsley Charitable Trust. All findings and opinions are those of the authors and not necessarily those of the funding agencies.

2.   Nancee Garcia contributed to an earlier draft of the discussion on measurement. W. James Lewis provided useful input to an earlier version of this chapter. Special acknowledgment is due to Wendy M. Smith, corresponding editor, who provided invaluable guidance.

APPENDIX A: PDSA FORM WITH PROMPTS

Test Title:	A general title of the change concept being used		Date:	
Tester:		Cycle#:		Driver:
What change idea is being tested?	SPECIFIC change idea to test (e.g. Using a "request to retest" form)			
What is the goal of the test?*	What are we trying to learn by performing this test (e.g. Is the change feasible? Do the tweaks work well? Does it work in another classroom?)			

*Identify your overall goal: To make something work better? Learn now a new innovation works? Learn how to text in a new context? Learn how to spread or implement?

1) PLAN			3) STUDY	
Questions: Questions you have about what will happen. What do you want to learn?	Predictions: Make a prediction for each question. Not optional.	Data: Data you'll collect to test predictions	What were the results? Comment on your predictions in the rows below.	
e.g. Did the change idea happen and as planned?	You may have multiple questions about why the change idea may or may not have happened	What data will you collect to test your predictions?	→	You may have collected data from multiple sources, compile here.
e.g. How did the change idea go?	You may have predictions about what went well or badly.		→	
e.g. Do I/others like the change idea?	Sometimes qualitative data of whether people like the new idea can help our learning.		→	
How many...? How much time...?	Add other learning questions based on other questions or predictions you may have.		→	
Details: Describe the who/what/when/where of the test. Include your data collection plan.			What did you learn?	
This space is important because it helps us plan the execution of the change idea. If the test fails, then we need to think about two possibilities: the change wasn't executed well or it wasn't a good idea. This section helps us look back to see if we executed the change properly. If we didn't then we may want to try again.			What did you learn about this change idea? If the goal is the get change to happen reliably, then what did this test teach us about how to get change to happen in our context? What have we learned about our system or our context? (e.g. I learned that I was right about my students liking the activity, but the way I phrased the instructions could have been better and the worksheets were unclear...)	
2) DO (Briefly describe what happened during the test, surprises, difficulty getting data, obstacles, successes, etc.)			4) ACT (Describe modifications and/or decisions for the next cycle; what will you do next?)	
This section helps us note down our observations during or immediately after the test happens. What did you notice about how it went? What surprised you? What barriers existed that you did or did not expect? This section is a reflection on the test itself, before we think about what we learned from it.			Can be an adapt (continue tweaking the change to make it better), adopt (start to think about trying it in new contexts), or abandon (the idea isn't working out, let's try something new).	

Morales, L. & Grunow, A. (C) 2016 by the Carnegie Foundation for the Advancement of Teaching. Licensed for public use under the Creative Commons license CC-BY-NC-SA 2.0.

REFERENCES

American Association of Colleges for Teacher Preparation. (2010). *The clinical preparation of teachers: A policy brief.* Washington, DC: Author.

Association of Mathematics Teacher Educators. (2017). *Standards for preparing teachers of mathematics.* Raleigh, NC: Author. Retrieved from http://amte.net/standards

Ball, D. L., Thames, M. H., & Phelps, G. (2008). Content knowledge for teaching: What makes it special? *Journal of Teacher Education, 59*(5), 389–407.

Banilower, E. R., Smith, P. S., Weiss, I. R., Malzahn, K. A., Campbell, K. M., & Weis, A. M. (2013). *Report of the 2012 National Survey of Science and Mathematics Education.* Chapel Hill, NC: Horizon Research.

Bryk, A., Gomez, L. M., Grunow, A., & LeMahieu, P. (2015). *Learning to improve: How America's schools can get better at getting better.* Cambridge, MA: Harvard Education Press.

Bryk, A. S., Gomez, L. M., & Grunow, A. (2011). Getting ideas into action: Building networked improvement communities in education. In M. T. Hallinan (Ed.), *Frontiers in Sociology of Education* (pp. 127–162). Dordrecht, The Netherlands: Springer.

Carver-Thomas, D., & Darling-Hammond, L. (2017). *Teacher turnover: Why it matters and what we can do about it.* Palo Alto, CA: Learning Policy Institute. Retrieved from: https://learningpolicyinstitute.org/product/teacher-turnover

Conference Board of the Mathematical Sciences (CBMS). (2012). *The mathematical education of teachers II*. Providence, RI, and Washington, DC: American Mathematical Society and Mathematical Association of America.

Dolle, J. R., Gomez, L. M., Russell, J. L., & Bryk, A. S. (2013). More than a network: Building professional communities for educational improvement. *National Society for the Study of Education Yearbook, 112*(2), 443–463.

Franklin, C., Bargagliotti, A., Case, C., Kader, G., Schaeffer, R., & Spangler, D. (2015). *Statistical education of teachers*. Arlington, VA: American Statistical Association. Retrieved from http://www.amstat.org/

Freeman, S., Eddy, S., McDonough, M., Smith, M., Okoroafor, N., Jordt, H., & Wenderoth, M. P. (2014). Active learning increases student performance in science, engineering, and mathematics. *Proceedings of the National Academy of Sciences, 111*(23), 8410–8415.

Gleason, J., Livers, S. D., & Zelkowski, J. (2015). *Mathematics classroom observation protocol for practices: Descriptors manual*. Retrieved from http://jgleason.people.ua.edu/mcop2.html

Hill, H. C., Rowan, B., & Ball, D. (2005). Effects of teachers' mathematical knowledge for teaching on student achievement. *American Educational Research Journal, 42*(2), 371–406.

Ingersoll, R., Merrill, L., & May, H. (2014). *What are the effects of teacher education and preparation on beginning teacher attrition?* Research Report (#RR-82). Philadelphia, PA: Consortium for Policy Research in Education, University of Pennsylvania.

Ingersoll, R. M., & Perda, D. (2010). Is the supply of mathematics and science teachers sufficient? *American Educational Research Journal, 43*(3), 563–594.

Leatham, K. R., & Peterson, B. E. (2010). Purposefully designing student teaching to focus on students' mathematical thinking. In J. W. Lott & J. Luebeck (Eds.), *Mathematics teaching: Putting research into practice at all levels* (pp. 225–239). San Diego, CA: Association of Mathematics Teacher Educators.

Malkus, N., Hoyer, K. M., & Sparks, D. (2015, November). Teaching vacancies and difficult-to-staff teaching positions in public schools. *Stats in Brief* (NCES 2015-065). Retrieved from http://nces.ed.gov/pubs2015/2015065.pdf

Markow, D., Macia, L., & Lee, H. (2013). *The MetLife Survey of the American Teacher: Challenges for school leadership*. New York, NY: MetLife.

Martin, W. G., & Gobstein, H. (2015). Generating a networked improvement community to improve secondary mathematics teacher preparation: Network leadership, organization, and operation. *Journal of Teacher Education, 66*(5), 482–493.

Martin, W. G., & Strutchens, M. E. (2014, April). *Priorities for the improvement of secondary mathematics teacher preparation for the Common Core era*. Presentation to the annual meeting of the American Education Research Association, Philadelphia, PA.

Mathematics Teacher Education Partnership. (2014). *Guiding principles for secondary mathematics teacher preparation*. Washington, DC: Association of Public and Land-grant Universities. Retrieved from mtep.info/guidingprinciples

National Center for Educational Statistics. (n.d.). *Schools and staffing survey (SASS)*. Retrieved from https://nces.ed.gov/surveys/sass/

National Council of Teachers of Mathematics and the Council for the Accreditation of Educator Preparation (NCTM & CAEP). (2012a). *NCTM CAEP Standards—Middle grades (Initial preparation)*. Retrieved from http://www.nctm.org/uploadedFiles/

Standards_and_Positions/CAEP_Standards/NCTM%20CAEP%20Sta ndards%20 2012%20-%20Middle%20Grades.pdf

National Council of Teachers of Mathematics and the Council for the Accreditation of Educator Preparation (NCTM & CAEP). (2012b). *NCTM CAEP Standards—Secondary (Initial Preparation)*. Retrieved from http://www.nctm.org/uploadedFiles/ Standards_and_Positions/CAEP_Standards/NCTM%20CAEP%20Sta ndards%20 2012%20-%20Secondary.pdf

National Governors Association Center for Best Practices, Council of Chief State School Officers. (2010). Common core state standards: Mathematics. Washington, DC: Author.

Plsek, P. (2003). *Complexity and the adoption of innovation in health care*. Position statement prepared for the conference Accelerating Quality Improvement in Health Care: Strategies to Speed the Diffusion of Evidence-Based Innovations. Retrieved from https://www.nihcm.org/complexity-and-the-adoption-of-innovation-in-health-care

CHAPTER 2

TRANSFORMING SECONDARY MATHEMATICS TEACHER PREPARATION PROGRAMS[1]

W. Gary Martin, Wendy M. Smith, and Margaret J. Mohr-Schroeder

From its inception, the Mathematics Teacher Education Partnership (MTE-Partnership) has had as its goal of transforming secondary mathematics teacher preparation in alignment with the *Common Core State Standards for Mathematics* (National Governors Association Center for Best Practices, Council of Chief State School Officers, 2010) and other rigorous standards. More recently, the goal has expanded to encompass the Association of Mathematics Teacher Educators' (2017) *Standards for the Preparation of Teachers of Mathematics* (AMTE Standards). As the MTE-Partnership adapted the Networked Improvement Community (NIC) design (Bryk, Gomez, Brunow, & LeMahieu, 2015), two aims were set: (a) increase the supply and (b) increase the quality of secondary mathematics candidates. The MTE-Partnership also identified a set of four primary drivers (see Chapter 1) and disaggregated its work into five Research Action Clusters (RACs) addressing various aspects of the primary drivers, thus allowing the MTE-Partnership to "accelerate learning" through the power of the network (Bryk et al., 2015, p. 141). This approach, however, results in a conundrum: Each partnership team, generally, is only involved in one (or perhaps two) of these RACs—meaning that they are addressing only some of the areas of critical need. To fully meet the aim

The Mathematics Teacher Education Partnership: The Power of a Networked Improvement Community to Transform Secondary Mathematics Teacher Preparation, pages 25–55.

of the MTE-Partnership, teams must shift toward more holistic program transformation and integrate the work of the MTE-Partnership across multiple RACs into their local improvement efforts. However, making such a shift will, in many cases, raise a number of significant challenges including capacity and human capital, issues with the will to improve mathematics teacher preparation across stakeholder groups, and issues with institutional resources and support structures.

An additional challenge lies in how to effectively scale up the work of the RACs to members of the MTE-Partnership who are not a part of a given RAC, and ultimately to institutions beyond the MTE-Partnership. Fully benefiting from the work of the RACs in transforming their programs requires easy access to the resources being developed. In the literature, such systems are often referred to as knowledge management systems (Bukowitz & Williams, 2000). However, given the NIC design, attention needs to extend beyond merely creating a library of resources to creating a dynamic system to support the generation and propagation of knowledge, as additional teams use the resources of the RACs.

The Transformations Working Group was formed in the spring of 2016 with the following charge: To establish a foundation for the MTE-Partnership's strategic focus on overall transformation of secondary mathematics teacher preparation programs. The approach proposed by the MTE-Partnership Planning Committee was that the working group design ways to support teams in creating "strategic pathways" to scale up incorporation of the MTE-Partnership's improvements. The ultimate aim was comprehensive program transformation, with a focus on building capacity and infrastructure, collaboration with K–12 and other stakeholders, and cross-team collaboration. The group has explored the literature on institutional change (e.g., Corbo, Reinholz, Dancy, Deetz, & Finkelstein, 2016; Elrod & Kezar, 2016), conducted several surveys of the membership, and done extensive brainstorming about how to best support transformational change across the MTE-Partnership teams, leading to initial attempts at supporting program transformation at several institutions supported by a grant from the National Science Foundation.

In this chapter, we provide a review of literature relevant to attaining program transformation, discuss the analysis of the problem undertaken by the Transformations Working Group, present several cases of the kind of work initially being done by the group, and discuss what others might learn from these efforts.

REVIEW OF LITERATURE

Chapter 1 outlined key research related to the need for the improvement of secondary mathematics teacher preparation programs. In this section, we build on that review of literature to connect the transformational imperative to the AMTE Standards (2017), explore literature on institutional change, and consider existing work with knowledge management.

Improving Mathematics Teacher Preparation

The AMTE Standards (2017) posit that meeting the aspirational vision they present requires a commitment to improving mathematics teacher preparation. The standards further suggest that improvement is "not a linear process but is more cyclic in nature while efforts are made to improve various aspects of the program" (p. 164) and propose the diagram in Figure 2.1 depicting the process of improvement.

The cycle of change in Figure 2.1 is similar to the NIC model of the Plan-Do-Study-Act (PDSA) Cycle, with the addition of building a common vision at the front end and ensuring sustainability at the back end. However, if this process were easy, everyone would already be engaged in systemic improvements to their secondary mathematics teacher preparation programs. Further, this cycle necessitates multiple stakeholders collaborating, as well as cooperation at the institutional level.

Institutional Change

Institutional change involves multiple complex, interconnected systems, with the professional networks of multiple stakeholders interwoven (e.g., Corbo et al., 2016). Undergraduate institutions are, by design, resistant to change (e.g., Birnbaum, 1991/1999). From the organization of majors, to course syllabi, to expected classroom practices, changes are challenging to implement and sustain. K–12 systems also tend to resist change (e.g., Bryk at al., 2015; Fullan, 2016), while at the same time suffer from innovation fatigue (e.g., Fullan, 2016). However, research

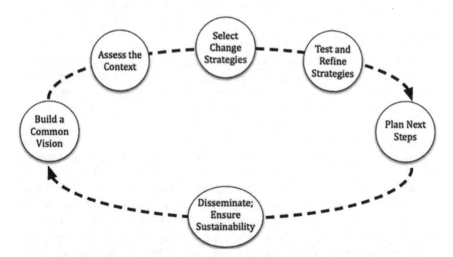

FIGURE 2.1. The Ongoing and Cyclic Nature of Improving Mathematics Teacher Preparation Programs (AMTE, 2017, p. 165). Used by permission.

on institutional change (Fullan, 2016; Hayward, Kogan, & Laursen, 2015; Kezar, 2014) has shown that with well-articulated goals, leadership, and proper incentives, traditions can be disrupted so that changes to the status quo can be enacted and sustained through key institutional supports that retain innovations as the new normal.

Serious research on institutional change is fairly recent in education; focusing on contexts and cultural change as levers for institutional change only became a theory in the 1980s. Yet the broader field knows quite a bit about effective and ineffective change strategies: "Research demonstrates that change strategies are successful if they are culturally coherent or aligned with the culture" (Kezar, 2014, pp. 32–33). Whereas more of the change literature is related to businesses than strictly to educational settings (e.g., Burnes, 2011), educational institutions can learn quite a bit from that research. Kezar (2014) and Corbo et al. (2016) discuss reasons why education change efforts fail—either by never achieving their goals or failing to be sustained.

One cause of failure is ignoring or leaving implicit theories of change: "Most change agents believe that once they have a vision or idea for change the major work is done, that implementation is nothing but an afterthought" (Kezar, 2014, p. x). Yet implementation is incredibly important; oversimplifying change models and implementation strategies is another common cause of failed change efforts. Successful change efforts do not just consider, but are built around, local contexts (e.g., Corbo et al., 2016; Hughes, 2011; McClellan, 2011). Change efforts also fail when they ignore the relative power and authority of the change efforts (e.g., Corbo et al., 2016; Raelin & Cataldo, 2011). Finally, failed efforts often are characterized by ignoring what research has shown to be effective about institutional change (Corbo et al., 2016; Kezar, 2014). Thus, successful change efforts need to be built on plans that take into account these common reasons for failure.

Knowledge Management

One of the primary strategies to support the success of a NIC is to develop a system that cultivates generation of knowledge and manages it in a dynamic and interactive manner; for the MTE-Partnership, the knowledge management and generation supports the improvement of secondary mathematics teacher preparation. "By formalizing the identification, capture, and organization of practical knowledge, a hub can accelerate the spread and use of the products of past improvement research" (Bryk et al., 2015, p. 158). While knowledge management can be defined in several ways, there is consensus that its central purpose is to capture knowledge in a way that adds value to an organization (Dalkir, 2017). In this sense of the term, knowledge is understood to encompass more than data or information; it must be based on experience and be highly contextual. Moreover, it encompasses both explicit knowledge, which has been more formally expressed and defined, as well as tacit knowledge, which may be less tangible (Dalkir, 2017). A NIC needs to capture the knowledge generated in the improvement process, addressing important contextual factors that impact the success of that product or

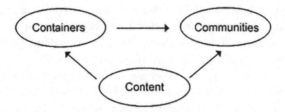

FIGURE 2.2. The Three Components of Knowledge Management: Containers, Communities, and Content.

strategy, in order to guide successful adaptation of the improvement by additional members of the NIC. Moreover, support for ongoing knowledge refinement must be provided as testing expands through succeeding improvement cycles (Bryk et al., 2015). Accordingly, Dalkir (2017) states knowledge management can "significantly contribute to all phases of the innovation cycle" (p. 25).

Dalkir (2017) identifies three major interactive components of knowledge management (Figure 2.2). Historically, focus has been put on containers, which can be roughly construed as the database function of storing and retrieving information. Containers, however, are not sufficient. As Mejia (2014) suggests, the existence of knowledge does not automatically lead to propagation of that knowledge. Thus, attention must also be paid to the development of a community or network that interacts with knowledge and generates new content to be included in the system. For example, Bukowitz and Williams (2000) proposed a Knowledge Management Process Framework (see Figure 2.3) that shows this interaction

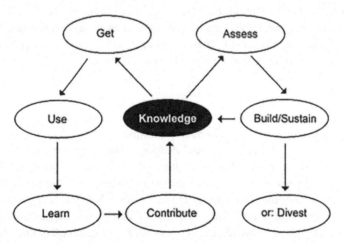

FIGURE 2.3. The Knowledge Management Process Framework shows the Interaction of Knowledge within a Community or Network.

of knowledge with the network. Effective knowledge networks are "collections of individuals and teams who come together across organizational, spatial, and disciplinary boundaries to invent and share a body of knowledge" (Pugh & Prusak, 2013, p. 79). Indeed, a NIC incorporates the norms and attributes of a community of practice—joint enterprise, mutual engagement, and shared repertoire—as described by Wenger (1999), with an additional focus on structured inquiry.

Some research addresses systems designed to manage the generation and sharing of knowledge. While the research literature often refers to such systems as knowledge management systems, in common usage, this term tends to connote a focus on creating a library of resources rather than on creating a dynamic system to support the generation and propagation of knowledge. The KGMS (Knowledge Generation and Management System) term better emphasizes a broader perspective. Research suggests a number of factors influence the success of such a system. In a study in a health-care setting, Sho-Fang, Ping-Jung, and Hui-Fang (2016) the importance of knowledge quality to both user satisfaction and system use. System quality issues, such as perceived ease of use, also affected satisfaction and use. Ali, Tretiakov, Whiddett, and Hunter (2017) found similar factors influencing use. Perceived usefulness is a major determinant of its use for both retrieval and sharing, which is impacted mostly by content quality but also by system quality. Dalkir (2017) describes four barriers to collaboration: "not-invented-here" (p. 87) in which people are hesitant to accept knowledge from other sources; "hoarding" (p. 87) of information in which people are reluctant to share their knowledge; ineffective searching infrastructure; and the reluctance to collaborate with people who are not familiar. Wang, Meister, and Gray (2013) found that bottom-up support among users influences system use the most.

Dalkir (2017) emphasized that many tools and platforms can be useful in supporting knowledge generation. He posits a number of factors that should be considered, including the blend between virtual and in-person collaboration, how content is stored, and issues of security and access. The research suggests that a KGMS should have the characteristics listed in Figure 2.4.

Summary

Program improvement efforts are complex and require complex, multi-dimensional plans for improvement, enacted by multiple stakeholders working toward common goals, which makes the NIC design an attractive approach to support program transformation. Furthermore, although much is known about knowledge management, the research tends to be related to businesses rather than education systems and focuses more on aggregating knowledge than generating knowledge. Working as a NIC (or interrelated sets of NICs) requires an efficient and effective way to not only share knowledge being generated, but also for that knowledge to be dynamic as the knowledge is applied and adapted in different contexts by different stakeholders.

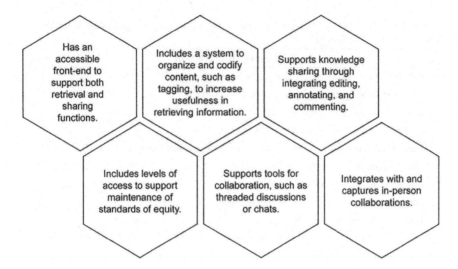

FIGURE 2.4. Necessary Characteristics of a KGMS for MTE-Partnership.

INITIAL TRANSFORMATIONS WORKING GROUP EFFORTS AT DEFINING AN ACTION PLAN

Successful change efforts are founded with explicit plans; research on change in education (e.g., Bryk et al., 2015) identifies key principles to guide improvement efforts. The work of the Transformations Working Group was built on the Carnegie Foundation for the Advancement of Teaching's (n.d.) six principles for change (see Table 2.1); these same principles are embodied in the discussion of the characteristics of a NIC in Chapter 1. These improvement principles are embedded in our overall theory of the cyclic nature of improving mathematics teacher preparation programs (see Figure 2.1) and highlight key issues, particularly around anticipated barriers to change or common causes of failure in improvement efforts in education. This section describes the efforts of Transformations Working Group members to understand the complexity of the issues around secondary mathematics teacher preparation program improvement in order to develop a plan of action.

Understanding the Problem

To better understand the current system that produces the current outcomes (as in the third item of Table 2.1), the Transformations Working Group developed a fishbone diagram, which illustrates and organizes the key challenges (see Figure 2.5). This development was built on an extended process of gathering input from different groups of stakeholders, including multiple brainstorming sessions and a survey of MTE-Partnership institutions that asked them about their key challenges.

TABLE 2.1.   Six Principles of Improvement from the Carnegie Foundation for the Advancement of Teaching

Principles of Improvement
(1) Make the work problem-specific and user-centered
• What specifically is the problem we are trying to solve?
• Engage key participants early and often
(2) Variation in performance is the core problem to address
• Critical issue is what works, for whom, and under what set of conditions
• Aim for equitable outcomes
(3) See the system that produces the current outcomes
• Need to understand what led to the current situation/problem
• Need a wide view of the whole system in order to solve a systemic problem
(4) We cannot improve at scale what we cannot measure
• Measure progress on key outcomes
• Not all change is an improvement
(5) Anchor practice in disciplined inquiry
• Rapid plan, do, study, act cycles allow for learning quickly, failing quickly, and improving quickly
• Learning from failures is key
(6) Accelerate improvements through networked communities
• Embrace the wisdom of crowds
• We can accomplish more together

Establishing an Aim

An effective NIC has an explicit and measurable aim. Based on its analyses of the problem space, the Transformations Working Group proposed the following aim to guide the emerging work in this area:

> In order to attain the overall MTE-Partnership aim ("gold standard" as expressed in its 2014 *Guiding Principles for Secondary Mathematics Teacher Preparation Programs* and number of candidates produced), *N* teams will be engaged in an explicitly defined continuous improvement process of overall transformation of their secondary mathematics teacher preparation programs by June 2019, in collaboration with other teams engaged in that process.

Several notes are made to better understand this statement:

- *Programs* as used here include the continuum from recruitment of prospective teachers of mathematics, undergraduate content coursework, early fieldwork experiences, methods coursework, clinical experiences with

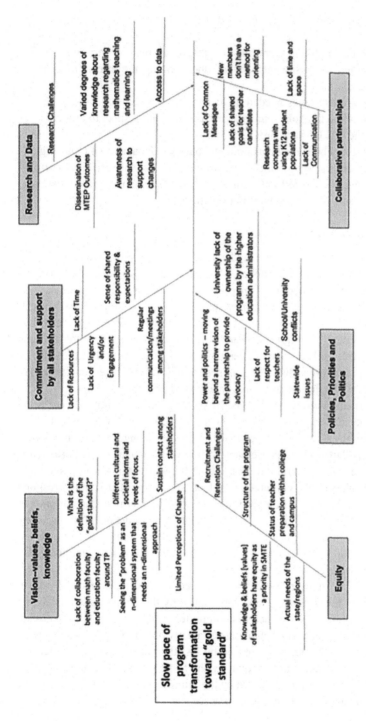

FIGURE 2.5. The Transformations Working Group's Fishbone Diagram Explores the Systemic Challenges Underlying Program Transformation Efforts.

mentor teachers in partner school districts, to early career induction support.

- To meet the condition, there must be an explicit plan for improvement for the program, including methods of documenting the effects of changes made.
- Continued attention is needed as to whether the *Guiding Principles* sufficiently define the gold standard, particularly with respect to induction, in light of the AMTE Standards (2017).
- *N* will initially be somewhat small (5), but then expand to be more aggressive (perhaps up to 80), and then ultimately encompass all MTE-Partnership teams. The funded NIC-Transform proposal involves five local partnership teams in pilot transformation efforts.

Building a Plan of Action

Following this attempt to better understand the scope of the challenges, the working group turned to thinking about solutions. In keeping with the NIC design, the working group identified a set of primary drivers (key levers necessary to achieve the aim), secondary drivers (key steps in achieving the primary drivers), and change strategies (see Table 2.1) that align with the challenges identified in the fishbone diagram (see Figure 2.5) and would comprise the early transformation efforts. Whereas all teams will eventually need to address all of the drivers, the NIC design means that different teams start in different places, and with differing priorities of the drivers to improve their local programs.

Survey of Transformational Needs. In keeping with the improvement principle that one cannot improve what is not measured (as outlined in Table 2.1), the Transformations Working Group is also involved in ongoing data collection across the MTE-Partnership. Participants at the 2018 annual MTE-Partnership Conference were asked several questions about transformations progress at their institutions as a part of the conference evaluation. Of the 34 individual respondents, 82% stated that a focus on program transformation at their institution was important, with 49% rating it as very important. A subsequent question asked about their team's progress in transformations efforts at their institutions. Only 32% reported making "some progress," while none felt their transformation goals were fully met (see Figure 2.7).

Participants were further asked about barriers or challenges faced by their team, drawn from the fishbone analysis produced by the Transformations Working Group (see Figure 2.5). The lack of commitment and support by all stakeholders was identified as a barrier or challenge by nearly half of the participants, with only lack of attention to equity identified by less than one-quarter of the teams. Based on the level of agreement shown by the respondents, there appears to be strong support for the work undertaken by the Transformation Working Group, as well as the initial analysis of the problem space.

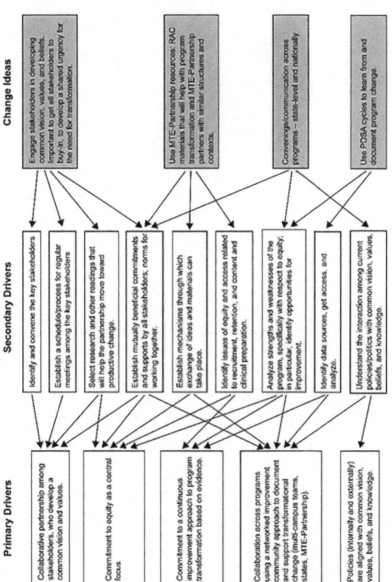

FIGURE 2.6. Transformations Working Group's Driver Diagram, Current as of 2019.

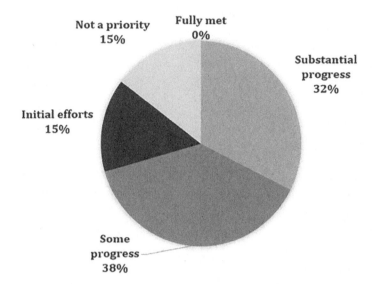

FIGURE 2.7. Ratings of Team Progress in Transforming Their Secondary Mathematics Teacher Preparation Programs (n = 34) on Survey Given at the 2018 MTE-Partnership Conference.

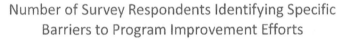

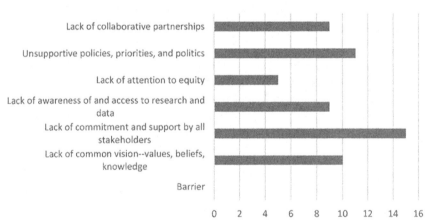

FIGURE 2.8. Biggest Challenges to Program Transformation Faced by Secondary Mathematics Teacher Preparation Programs, as Identified by MTE-Partnership Survey Respondents (n = 31).

CURRENT TRANSFORMATIONS
WORKING GROUP EFFORTS

The work of exploring literature related to institutional change, defining the problem space of the difficulty of making progress toward a gold standard of secondary mathematics teacher preparation, and devising a general plan of action extended over nearly two years. The Transformations Working Group had numerous discussions about transitioning to become an MTE-Partnership RAC; however, the work of RACs typically is undertaken by individuals from different local partnership teams, and the work of program transformation requires multiple individuals working together from specific local teams. Further, transformation efforts depend on the work of the five RACs, seeking to pull together the related efforts to improve local programs. Thus, the Transformations Working Group remains a working group focused on cross-cutting efforts in parallel to the RACs.

Initial Steps Toward Program Transformation

Within the fishbone and driver diagrams, the Transformations Working Group sought reasonable starting points on which further work could build. An obvious starting point seemed to be the first change idea from Figure 2.6: Engage stakeholders in developing common vision, values, and beliefs—important to get all stakeholders to buy-in, to develop a shared urgency for the need for transformation. This is consistent with the early stages of the change process (see Figure 2.1). Members of the Working Group initiated PDSA Cycles with their local teams, to track progress in addressing this change idea. The regular Transformations Working Group meetings provided accountability: members were more likely to make time to attend to group efforts between meetings, knowing they would need to report on progress and submit the PDSA Cycle reports (see Appendix A from Chapter 1). The ongoing meetings also supported learning from one another's efforts. Large-scale transformation of programs is daunting, making the Working Group's focus on starting small particularly helpful: even holding a meeting can be progress, such as sharing the local team's vision and progress with a new dean or superintendent. Having stakeholder meetings is an important part of the transformation process, particularly as stakeholders engage in productive dialogue. A common vision shared by the key stakeholders to change efforts provides the necessary foundation for transformations, so initial meetings constitute a small but crucial step.

Next Steps Toward Program Transformation

As the Working Group's work progressed, it was less clear how the other change ideas beyond stakeholder engagement might be addressed, as they would require more resources than could be reasonably expected of its members. As a result, a subgroup began developing a proposal to the National Science Foundation in the spring of 2018, subsequently funded in fall of 2018, to address several of the other change ideas. The funded project, *Using Networked Improvement Communities to Design and Implement Program Transformation Tools for Secondary Mathematics Teacher Preparation (*NIC-Transform; NSF 1834539 and 1834551, 2018–2020*)*, is a collaborative research project among Auburn University, the University of Nebraska-Lincoln, the University of Kentucky, California State University-Fullerton, and Mississippi State University, in association with the Association of Public and Land-grant Universities. NIC-Transform is an exploration and design tier for an institution and community transformation project to support and study the challenging work of transformation of secondary mathematics teacher preparation programs, which meets the national vision put forward by AMTE (2017) and other organizations (Conference Board of the Mathematical Sciences [CBMS], 2012; Franklin et al., 2015). The work of NIC-Transform builds on the initial efforts of the Transformations Working Group; the collaborative partners on the funded grant represent those local partnership teams who could commit to advancing local program transformation efforts and the related data collection beginning in fall 2018.

The specific overarching goals of NIC-Transform are two-fold. The first goal is to build a networked improvement community (NIC) of institutions focused on collaboratively developing and sharing tools and strategies for program transformation, incorporating attention to institutional change. This goal addresses a change idea from the driver diagram (see Figure 2.6): Using PDSA Cycles to learn from and document change. Teacher preparation programs are situated in local contexts and local education systems; complex systems cannot be improved via individual, uncoordinated change efforts. Improving a system requires stakeholders to see the system and plan for changes that will address multiple aspects of the system. One cannot expect a 1- or 2-dimensional solution to solve an n-dimensional problem ($n > 2$); NIC work attempts to keep the system in mind for effecting improvement.

Thus, a new adaptation of the NIC model was incorporated to support efforts by local teams to transform their programs. NIC-Transform is part of the MTE-Partnership NIC (see Chapter 1) but also is a NIC of its own. Each local team (e.g., university and school district partners) is also a NIC. These NICs are not independent, but rather interrelated. Figure 2.9 further illustrates the connections among the different NICs involved in transformation work.

The second goal of NIC-Transform is to create a knowledge generation and management system (KGMS) that facilitates the generation and capture of validated products and approaches useful in transforming secondary mathematics teacher preparation, as depicted in the center of Figure 2.9. This goal is also in alignment with the second change idea in the driver diagram (see Figure 2.6): making RAC materials available to teams as they work on their local transformations. Although research literature exists on knowledge management systems (as discussed earlier in this chapter), such systems typically create libraries of

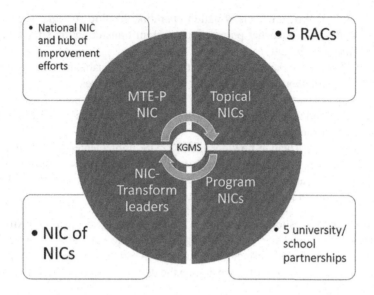

FIGURE 2.9.   Interrelated NICs Engaged in the Transformations Working Group; the Work Centers on an Effective KGMS.

resources. NIC-Transform's vision is to create a dynamic system that is based on a repository of collective knowledge and supports the generation, propagation, and evolution of that knowledge over time and on a large scale. In such a system, users would access a strategy and then contribute to further refinements by sharing written summaries of their own contexts and how they adapted the strategy to these local contexts and goals. Such a process then enriches the information known about a strategy for all future users.

As stated above, a KGMS would profitably include the features listed in Figure 2.4, although no single platform will likely meet the needs of all NICs. If a NIC already has an online collaborative system, there would be little benefit to asking them to migrate their information unless and until a new system is significantly better. However, an effective KGMS is critical in supporting transformation efforts, and the five NIC-Transform sites will participate in developing a KGMS to support their local efforts, to share the knowledge they are generating about program transformation, and to support the MTE-Partnership as a whole.

Finally, NIC-Transform also supports two annual workshops to support communications across MTE-Partnership team focused on program transformation, which is aligned with another change idea from the driver diagram (convenings/communication across program; see Figure 2.6).

In support of these goals, the specific research questions for NIC-Transform include:

1.   How is program transformation operationalized to structure secondary mathematics teacher preparation program transformation efforts across multiple contexts?

2.   What are the benefits and limitations of operationalizing a common program transformation framework across multiple sites?

   a.   In particular, how does the NIC structure, implemented within sites and across sites, impact efforts at operationalizing program transformation?

   b.   In particular, how does the KGMS, implemented within sites and across sites, enhance the generation of new knowledge and the application of that knowledge to support efforts at operationalizing program transformation?

To answer these questions, the project will analyze progress toward transformation at each of the local sites—considering documentation of PDSA Cycles, driver diagrams, meeting notes, periodic self-studies—as well as the progress of the collective set of sites, particularly focusing on the development of a KGMS that will address multiple use cases.

Thus, the NIC-Transform research questions advance the work of the Transformations Working Group. The resulting research should provide a foundation from which to help scale the work from a pilot phase to the larger MTE-Partnership. The objective of the NIC-Transform project is to provide infrastructure that will support program transformation efforts across MTE-Partnership teams and beyond.

CASES OF TRANSFORMATION EFFORTS

Each of the five sites collaborating as part of NIC-Transform (and as active members of the Transformations Working Group) has an initial focus for transformation efforts. All five are focused on institutionalizing recruitment strategies, including underrepresented populations. Tables 2.2 and 2.3 provide an overview of the transformation focus and some basic demographics about each of the five teams, respectively.

The undergraduate certification for traditional students is a four-year bachelor's degree in a college of education. The programs where the teacher candidates major in mathematics or double major in mathematics and education are noted in Table 2.3. All programs include graduate level certification; four of the sites only offer graduate certification in conjunction with a master's degree. At Fullerton (as with most California universities), teacher candidates complete an undergraduate major in mathematics, and then certification is a fifth-year program completed the following year, as a graduate student (which does not result in a master's degree). Table 2.3 also lists the main RAC participation by team members at each of the five sites. While all five teams have unique stories, three vignettes illustrate these change efforts.[3]

TABLE 2.2.   Information About Local Teams' Transformation Foci

Institution	Transformation Focus
Auburn University	Increase administrator support and buy-in; institutionalize courses in mathematics content relevant to teaching, as well as use of the paired placement model for student teaching.
University of Nebraska	Infuse a focus on equity throughout teacher preparation program.
University of Kentucky	Infuse integrated STEM curriculum throughout the education coursework; establish formal partnerships with the local school district to teach embedded, clinically-based methods courses in the schools.
California State University, Fullerton	Infuse a focus on equity throughout teacher preparation program; strengthen district partnership role in identifying mathematics teachers whose practice is aligned with the MTE-Partnership vision.
Mississippi State University	Infuse integrated STEM curriculum throughout the education coursework; expand pathways to certification in multiple STEM areas to expand options in rural and small districts.

TABLE 2.3.   Information About NIC-Transform Teams

	Auburn	Nebraska	Kentucky	Fullerton	Mississippi State
Enrollment	25,000	25,000	30,000	35,000	22,000
Math Teachers per year	10–15	25–30	10–15	30–35	10–15
Certification (Undergraduate)	Traditional (90%)	Traditional (85–90%)	Double major (85–90%)	Math major (pre-certification)	Traditional (90–95%)
Certification (Graduate)	1-year master's (10%)	1-year master's (10–15%)	1-year master's (10–15%)	2-semester post-baccalaureate (100%)	2-year master's (5–10%)
MTE-P RAC membership	MTE-P Hub Clinical Experiences ALM MODULE(S^2)	ALM STRIDES MODULE(S^2)	STRIDES PR^2	Clinical Experiences ALM PR^2	PR^2

Transformation Efforts at California State University, Fullerton
Contributed by Mark Ellis

California State University at Fullerton (CSUF) is part of the 22-campus teacher preparation network of the CSU system, which also comprises the CSU MTE-Partnership team. CSUF has a rich local context within the partnership. The local work we have done has helped in a bidirectional sense—we have been able to secure additional funding and then use that funding (e.g., S.D. Bechtel, Jr. Foundation) to develop more partnerships and collaborations.

Importantly, though the MTE-Partnership is focused on the preparation of secondary teachers of mathematics, we have expanded that to include the work of preparing and supporting teachers of mathematics K–12. Through this work, we have been able to envision teacher preparation development as a continuum over time instead of isolated incidences of learning. Through the Bechtel Foundation grant, CSUF developed a shared vision within their own institution and in partnership with three local school districts that allowed for more explicit conversations across the local partnership, especially focused on the qualities of a well-prepared teacher of mathematics and the practices of mathematics teachers who support *all* K–12 students with learning mathematics.

One important tool that is now used across credential-program methods coursework and field observations, as well as within district partner professional learning work, is the Mathematics Classroom Observation Protocol for Practices (MCOP^2; Gleason, Livers, & Zelkowski, 2015), a rubric grounded in the Standards for Mathematical Practice (NCTM, 2014). This work will be further enhanced through a NSF Noyce Master Teaching Fellowship grant received in 2017 to work with 20 secondary teachers of mathematics (half of whom come from CSUF MTE-Partnership partner districts) to strength practice within their own classrooms. These 20 MTFs in the Advancing Teachers of Mathematics to Advance Learning for All (see atmala.weebly.com) will serve as mentors to CSUF teacher candidates using the co-plan, co-teach model and will support district efforts to transform mathematics teaching by developing and facilitating microcredential modules aimed at building proficiency with specific instructional skills related to culturally responsive mathematics teaching.

New faculty have been another important contributor to the success with program transformation. Since 2014, CSUF has hired four tenure-track mathematics education faculty within the departments of Mathematics and Secondary Education (with plans for a search for an Elementary Mathematics Teacher Education faculty in 2018–2019). From their start at CSUF, these faculty have been part of the MTE-Partnership conversations about elements of program transformation (e.g., co-plan/co-teach; district partnerships); these became "normalized" conversations. Collaboration among mathematicians and mathematics educators across departments and colleges as well as between university and school district partners came to be seen as the norm. A recent example of this came when discussing how to generate more enrollment in optional (but valuable) mathematics courses for future elementary and secondary teachers; in less than an hour the departments of Mathematics, Elementary & Bilingual Education, and Secondary Education agreed to leverage existing funds to create a scholarship program for students who complete with a B- or better at least three courses in this four-course sequence. Having new deans in both colleges at the start of the 2016–2017 academic year aided in solidifying the program transformation as they understood the transformed program as something that was typical.

One set of challenges everyone has faced involves time: the time for planning; the time for implementing; the time for collecting data about the implementation; and the analysis of the data and revisions to that model. For example, Plan-Do-Study-

Act (PDSA) Cycles take time. However, as this work becomes more routine and less novel, these actions are less burdensome. Another challenge remains the time required for some long-time faculty to embrace the transformed program as the new normal, but this will be helped in part through the efforts of two of the recently hired faculty in Mathematics, Alison Marzocchi and Roberto Soto, who received a grant from SEMINAL, a project of the Active Learning Mathematics RAC funded by the National Science Foundation, to support their work on bringing active learning strategies into Calculus courses over the next several years.

Zooming out to think about the CSU MTE-Partnership group, having the support of the statewide network of the 22-campus CSU team has helped address the challenge of time through leveraging collective expertise and realizing there is strength in numbers. Through annual convenings at the CSU Chancellor's Office, faculty from throughout the CSU system have had opportunities to share strategies, resources, and form new collaborations that support teacher preparation efforts. Among the insights gained from conversations across the 22-campus network was the realization that some campus credential programs did not require content specialists to do supervision of teacher candidates. When this surfaced in a survey generated by the CSU MTE-Partnership team, it provided faculty with substantive data to bring to local campus administrators to advocate for program changes. Collectively, it is essential that program quality not be negatively impacted by different contexts.

These convenings also have allowed us to have a louder voice within the CSU system. One example is the success with advocating for the addition of 16 mathematics-specific items to a statewide exit survey given to all credential program completers; hearing that through CSU MTE-Partnership mathematics teacher preparation faculty would be routinely examining such data for program improvement, the committee charged with updating the survey agreed to include these additional items. Data from this survey will allow faculty to look deeper into their programs both locally and across institutions using a recently launched data dashboard on the CSU's Educator Quality Center's website: https://www2.calstate.edu/impact-of-the-csu/teacher-education/educator-quality-center/edq-dataview-dashboards/Pages/default.aspx

NebraskaMATH STEP: Secondary Teacher Education Partnership
Contributed by Wendy M. Smith and Lorraine M. Males

Nebraska joined the MTE-Partnership as a partnership among the three University of Nebraska campuses that have teacher preparation programs, along with the public school systems in those cities (Lincoln, Omaha, and Kearney). Prior to the MTE-Partnership, the Center for Science, Mathematics and Computer Education (est. 1990) had built a statewide partnership with public school districts and intermediary education agencies (called Education Service Units [ESUs]), focused on mathematics teacher professional development.

When Research Action Clusters (RACs) first formed, the Nebraska partnership joined the Active Learning Mathematics RAC (ALM RAC). Subsequently, members of the Nebraska partnership have also worked with the MODULE(S²) RAC, STRIDES RAC, Equity & Social Justice Working Group, and Transformations Working Group. Additionally, Nebraska has been engaged in local efforts to form a statewide Networked Improvement Community (NIC), to translate the work of the MTE-Partnership into local efforts to align programs with the new *Standards for the Preparation of Teachers of Mathematics* (AMTE, 2017).

Active Learning Mathematics RAC: The University of Nebraska-Lincoln (UNL) and University of Nebraska at Omaha (UNO) were founding members of the ALM RAC and received some funds from the Helmsley Charitable Trust as part of MTE-Partnership. Funds on both campuses were used to launch a Learning Assistants (LAs) program (hiring undergraduates as assistants in freshmen-level mathematics courses to help facilitate active learning). Prior to ALM RAC, UNL was already working to improve freshmen-level courses below Calculus (Intermediate Algebra, College Algebra, Trigonometry, and College Algebra & Trigonometry). Prior to reforms, UNL had some measures of coordination, including: common syllabus, common exams, and common grading of common exams. Since 2011, changes at UNL have been extensive, including: hiring a full-time director of first-year mathematics to coordinate courses below Calculus; teaching courses in renovated rooms with movable tables/chairs and wrap-around whiteboards; hiring LAs; creating common lesson plans that incorporate active learning and group-work structures; adding time to courses without changing the credit hours (75-minute classes instead of 50-minute classes); and graduate student professional development (before-semester workshop and course during first year as an instructor of record).

In conjunction with these reforms, the Department of Mathematics began collecting (and gaining access to) extensive student data in order to measure student success, including passing rates, course-taking trajectories, and attitudes. Efforts have been very successful, raising passing rates (C or better) from around 60% to consistently around 80%. Additionally, efforts have expanded from Precalculus courses to Business Calculus, Calculus 1 and Calculus 2, and a second professor of practice will begin in Fall of 2018 to share in the coordination and mentoring duties. Future efforts include ongoing refinement of courses, exploring online exam options (not just multiple-choice items), and expansion of active learning structures into additional courses.

UNL is also a collaborative partner in the research grant SEMINAL: Student Engagement in Mathematics through an Institutional Network for Active Learning. Through this grant, the research team is seeking to understand the contextual and leadership factors that initiate and sustain institutional change in ways that increase student engagement and success in freshmen-level mathematics.

MODULE(S²): UNL also became active in MODULE(S²), with a faculty member as a core member of the subset of the RAC that received a grant from NSF. This faculty member has led the efforts to create the algebra module, and has piloted and assisted with other modules as well.

STRIDES: UNL has been involved with STRIDES, particularly focused on retention. Initial efforts were made to understand the support received and needed by teachers within the first 3 to 5 years of teaching. The UNL group has worked to develop measures and collect data related to these supports in order to develop induction programs.

Recent Efforts: Equity & Social Justice Working Group (ESJWG): UNL has recently been active in the ESJWG. This work has focused on identifying the problem space related to preparing secondary mathematics teachers to work in diverse settings. Specifically, the working group has developed definitions for diversity, equity, and social justice, with UNL higher education and K–12 partners contributing to the definitions of diversity and social justice. In addition, this work has spurred more intentional conversations across UNL secondary math education faculty and faculty in multicultural education, and it has spurred more intentional work across our secondary mathematics education professional coursework (two methods courses, associated practicum, and student teaching and student teaching seminar). For example, NCTM (2014)'s *Principles to Actions* Access and Equity principle has become a focus in the first of two methods courses, is emphasized in the second methods course when students are in an associated practicum in a diverse school and is the main focus in the student teaching seminar. This seminar primarily focuses on access, equity, and identity. In addition to *Principles to Actions*, the course texts include Fernandes, Crespo, and Civil's (2017) *Access and Equity Promoting High Quality Mathematics Instruction Grades 6–8* and Aguirre, Mayfield-Ingram, and Martin's (2013) *Impact of Identity in K–8 Classrooms: Rethinking Equity-Based Practice.* Throughout the professional sequence, prospective teachers are required to reflect on their own experiences learning mathematics and interrogate their assumptions about what it means to do mathematics and what that looks like for their students.

Recent Efforts: Statewide NIC: UNL has been part of MTE-Partnership's Transformations Working Group. In 2017–2018, the NebraskaMATH partnership has grown in two directions: encompassing most of the 16 colleges and universities that prepare teachers in Nebraska and expanding to consider pre-service preparation of elementary teachers in addition to secondary teachers. These two changes were closely related; only UNL has a large enough student and faculty population to have separate elementary and secondary mathematics teacher preparation; in the other institutions, the same faculty may teach elementary or secondary mathematics and/ or methods courses.

At the September 2017 statewide NCTM affiliate meeting, the existing Nebraska-MATH Secondary Teacher Education Partnership (STEP--the local MTE-Partnership team) group convened to discuss this potential expansion. At a first meeting in October 2017, representatives from 11 of the 16 teacher preparation institutions, the Nebraska Department of Education, and five of the largest school districts met. In addition to mathematicians, mathematics teachers, and mathematics teacher educators, there were also special education and English-language learner faculty and district personnel. The purpose of this meeting was to try to form a statewide NIC to work toward the new AMTE Standards (SPTM; AMTE, 2017). The 34 attendees were excited to work together toward these aspirational standards, with UNL providing leadership and logistical support.

After an overview of the NIC process, the group brainstormed the problem space and then had discussions to determine priorities. Working groups formed around each priority: math dispositions, teacher preparation programs, clinical experiences/ cooperating teachers, and partnerships. The overall mood of the group was excitement to be collaborating across institutions, particularly on the part of K–12 personnel, who in the past have not been a large part of the conversation about improving teacher preparation. The colleges and universities have a fair amount of rivalry, since most are recruiting from the same pool of in-state students, and the state schools are often forced to compete with one another for state resources. Faculty and district personnel were glad to be working directly with one another, without waiting for official institutional collaboration. While we discussed possibilities such as joint programs where students could take classes from different campuses, such a possibility is far in the future after the working groups make some progress.

The working group focused on mathematics dispositions wanted to better understand the attitudes of future teachers in our programs, so developed and piloted a survey in winter 2017–2018. The results of the open-ended items were used to develop some word-clouds to represent the responses. These word clouds then were used for program representatives to discuss how programs might be revised to develop more positive attitudes toward mathematics, particularly for prospective elementary teachers.

The teacher preparation programs group was designed to survey the preparation programs in the state to determine the status quo. Depending on many factors, programs at different institutions are quite different in terms of number and sequence of courses and field experiences. As a first step toward borrowing the best from each institution, there was a desire to know and understand what each program does. Conversations included people sharing strengths of their programs. For most people, this was the first time there was an organized way to learn about what other Nebraska programs are doing. Most programs deal with small numbers (1 to 5 secondary mathematics teachers graduating in a year), which leads to particular challenges in offering courses specific for future mathematics teachers.

The clinical experiences/cooperating teachers working group was focused mostly on developing some type of shared cooperating teacher training (likely online); at present, the most any cooperating teacher is required to do in terms of training is to attend a district-sponsored orientation put on by the human resources department and is more about nuts-and-bolts (and things like sexual harassment) and not about how to mentor a novice mathematics teacher. The working group's first steps are to gather data from stakeholders to develop a shared vision for high-quality mentor teachers in mathematics. Following that, school personnel and university faculty plan to work together to design workshops for mentor teachers.

The partnerships working group decided to work on leveraging expertise on both sides of the K–12 – higher education partners, and develop more opportunities to converse, including at the Nebraska Association of Teachers of Mathematics, the state administrator days, and other local conferences and workshops. A first step is to develop "conversation starters" to help get partnerships started. This group also discussed partnerships within and across higher education, particularly how to connect math and education departments, as well as special education and English language learning. The working group noted that while the bureaucratic processes are prohibitive, establishing strong partnerships is worth doing in order to better prepare and support novice mathematics teachers.

A subset of the overall state group met in conjunction with the M4 conference (a mathematics education conference for Nebraska, Iowa, Kansas, and Missouri) in March 2018. Working groups reported on progress and next steps. The group next plans to meet in September 2018, at the state's Nebraska Association of Teachers of Mathematics meeting.

The Central Alabama MTE-Partnership (CAMTEP)
Contributed by Marilyn Strutchens and W. Gary Martin

Auburn University is the lead institution for the Central Alabama MTE-Partnership (CAMTEP), which also includes Tuskegee University, Alabama State University, the Auburn University site of the state-funded Alabama Mathematics, Science, and Technology Initiative, and local school districts. This partnership built on the 15-year track record of Transforming East Alabama Mathematics (TEAM-Math), a collaboration of Auburn University, Tuskegee University, and local districts to improve mathematics education in east Alabama. TEAM-Math has had nearly $15 million in funding from the NSFs Math and Science Partnership Program and the Noyce Scholarship Program, as well as from state and local funding.

In addition, Auburn University was a founding member of the Association of Public and Land-grant Universities' Science and Mathematics Teaching Imperative, which launched the MTE-Partnership. Faculty from Auburn University served on the initial planning committee for MTE-Partnership and continue to hold leadership positions; Martin serves as the co-director for the MTE-Partnership, and Strutchens leads the Clinical Experiences RAC. The team has been involved in several of the RACs in various capacities, including Clinical Experiences, Active Learning Mathematics, and MODULE(S^2), as well as the Equity Working Group and the Transformations Working Group.

Clinical Experiences RAC: Auburn University and its school partners have been involved in the MTE-Partnership work associated with clinical experiences from its inception. Marilyn Strutchens led the initial work and planning in the area, including the problem analysis and resulting white paper, as well as the work of the RAC as a whole and the development of the paired placement model for student teaching in particular. Auburn University faculty and graduate assistants have served as supervisors for the paired placement model during student teaching, members of school districts participating in CAMTEP or TEAM-Math have served as mentor teachers, and a mathematician serves as an advisor and has observed several of the paired placements. The responses to the use of the model have been overwhelmingly positive from all involved, including the teacher candidates who participated in the model, accelerating the teacher candidates' progress toward enacting effective instruction in alignment with the Mathematics Teaching Practices found in *Principles to Actions: Ensuring Success for All* (NCTM, 2014). Auburn University is one of the lead institutions for an NSF grant supporting the work of the Clinical Experiences RAC, and faculty members continue to work to refine their use of the paired placement model. In addition, Auburn is beginning to explore using materials developed by the Clinical Experiences sub-RAC focusing on field experiences in the methods courses in order to provide a continuum of experiences.

Active Learning Mathematics RAC: Auburn University was also a founding member of the Active Learning Mathematics RAC and was a recipient of funding from the Helmsley Foundation that supported use of active learning techniques in Calculus II, including learning assistants. A senior mathematician led the effort, which also included a mathematics educator and graduate teaching assistants in mathematics. While some signs of progress were evident in over two years of work, the effort failed to gain traction with members of the mathematics department beyond those involved in the funded project. The mathematician retired with no obvious successor in place to take over his leadership role. Thus, participation in this RAC is currently dormant. Tuskegee University has also participated in this RAC, with a focus on precalculus.

MODULE(S²) RAC: Through the work of TEAM-Math, Auburn University had begun work on the mathematical preparation of preservice teachers K–12. This led to the redesign of the content courses for the elementary mathematics major and the addition of a capstone course for the secondary mathematics major in alignment with the recommendations of the *Mathematical Preparation of Teachers* (CBMS, 2010). The capstone course is co-taught by a mathematics educator and a mathematician, and they began to pilot units being developed by MODULE(S²) in Spring 2016. The last iteration of the course included units from the geometry, algebra, and statistics materials. While Auburn University is not a member of the MODULE(S²) RAC, it has been providing feedback to the RAC over the past years, and Martin serves as the chair of the advisory board of the NSF project supporting their work.

Equity and Social Justice Working Group: Marilyn Strutchens has been an active member of this group, based on her research interests in this area, and has helped to develop the initial framing of the work for the group. In addition, equity is a major focus of the ongoing work of the NSF-funded work of the Clinical Experiences RAC.

Moving toward Program Transformation: While Auburn University has been deeply involved in many facets of the MTE-Partnership work, much remains to be done to move toward program transformation. Both Martin and Strutchens have been involved in the initial work of the Transformations Working Group. To initiate this new focus, meetings were held with key stakeholders at Auburn University over the past year, beginning with the dean of the College of Education and the chair of the Department of Curriculum and Teaching. With their support, a meeting was held with representatives of the mathematics department to discuss improving the mathematics preparation of secondary mathematics teachers to better align with the AMTE Standards (2017) recommendations. Initial work was made in laying out a sequence of three mathematics content courses particularly designed for prospective secondary mathematics teachers, although the departure of the chair of the mathematics department will require further work when a new chair is identified. An additional area of concern lies in program recruitment and retention, given falling enrollments in the secondary mathematics program.

The MTE-Partnership has played a central role in our efforts to improve the secondary mathematics teacher preparation program, and it will help accelerate our continuing progress toward program transformation. Auburn University is a lead institution in NIC-Transform project funded by NSF and one of the five participating sites. As a part of this project we will be focusing on how we can garner additional administrator support and buy-in for our efforts, including institutionalizing our current work. Our involvement with the MTE-Partnership Clinical Experiences and MODULE(S^2) RACs will support these efforts, and we will explore how we might incorporate the work of additional RACs, such as the work PR^2 to focus on recruitment and retention. NIC-Transform will provide an opportunity to reflect on our overall progress and learn how we can more effectively work toward overall program transformation.

Across the three cases described here, each one outlines different initial efforts by local (sometimes statewide) partnership teams. For all of the teams, their transformation efforts build directly on the work of individuals as part of different RACs and working groups in the MTE-Partnership. In each case, local leaders are also leaders within the MTE-Partnership and active members of the Transformations Working Group. In keeping with a central tenet of NICs, change plans have common features, but all change is local, and strategies are adapted to fit local contexts and priorities.

Although the work of each local partnership team is connected to the driver diagram of the MTE-Partnership (see Figure 1.6) and the Transformations Working Group (see Figure 2.6), the local partnership teams have developed their own driver diagrams in order to prioritize local transformation efforts. All three of these cases feature local teams working on all four of the Transformation Working Group's change ideas (Figure 2.6): engage stakeholders in a common vision; use MTE-Partnership resources developed by RACs; convene stakeholders; and use PDSA Cycles to learn from and document program changes. Along with engaging and convening stakeholders, the local sites each have an intentional focus on building a NIC among stakeholders to support program transformation efforts. The change efforts and transformation processes are closely aligned to AMTE Standards (2017) in all three locations, as well as the MTE-Partnership's *Guiding Principles* (2014). Creating unique local driver diagrams within local NICs follows recommendations of transformation efforts, since all change ultimately occurs at the local level (e.g., Bryk et al., 2015).

In line with the primary drivers of the Transformation Working Group (Figure 2.6), all three of the cases feature efforts to build a collaborative partnership among stakeholders, with a common vision for program improvement toward the aspirational AMTE Standards (2017). Although not necessarily explicitly described in each of the vignettes in this chapter, each of the local partnership teams is centering improvement efforts on a commitment to equity and takes a stance of continuous improvement supported by evidence (PDSA Cycles). In Nebraska and California, the NIC efforts have a statewide reach. Thus, in keeping with improvement principles (Figure 2.4) and cyclical change efforts (Figure 2.1), each local partnership team is using the lenses of the MTE-Partnership, AMTE Standards, and Transformations Working Group to enact change ideas and drivers within their local contexts.

The Alabama and Fullerton teams both have leaders from the Clinical Experiences RAC represented on their local teams and thus have more of a focus on improving teacher candidates' clinical experiences than does the Nebraska team. In both Fullerton and Nebraska, since the teacher educators involved largely have K–12 responsibilities, the change efforts have expanded to encompass elementary in addition to secondary teacher preparation in mathematics. However, the efforts in California and Nebraska look quite different; most Nebraska mathematics teacher educators work alone or with few colleagues, in geographically isolated locations. In Alabama, in particular, extensive personnel turnover has caused the cyclical change process to re-focus on getting (new) stakeholders on board with a common vision for program improvement. Thus, each of the local partnerships is making progress, situated in their own contexts.

NIC-Transform is developing cases with rich descriptions of the contexts, to better help other local teams understand the contexts and how those contexts and priorities influenced local teams' progress on chosen change strategies. Following the NIC model, the investigators seek to understand five cases deeply, then propagate efforts to a larger number of universities in the near future. Section II: Opportunities to Learn Mathematics of this book provides some specific suggestions for how stakeholders may seek to initiate their own program transformation efforts.

FOCUS ON IMPROVEMENT

The Transformations Working Group has the central aim of propagating transformation efforts nationwide. Thus, the working group welcomes additional involvement; see http://mtep.info/TWG for information about how to connect with the NIC and its ongoing work. This website will also contain information about the lessons the Transformations Working Group and NIC-Transform are learning and contain developing cases and more refined recommendations, based on the most current work.

Programs seeking to initiate transformation efforts may want to start by convening a group of stakeholders; the three cases provided in this chapter describe different ways such efforts might begin. University mathematics educators and mathematicians should be part of the conversation, as well as K–12 mathematics supervisors, field placement supervisors, and department of education personnel. Additionally, stakeholders at the district, state department, and university levels involved with special education and English language learners should be included, since teachers in these areas are tasked with teaching mathematics (AMTE, 2017). The MTE-Partnership and NIC-Transform efforts are directed at secondary mathematics teacher preparation improvement; however, the principles of program transformation can be adapted for K–6 and for other content areas (as Fullerton and Nebraska have done with their local efforts). Partnerships may be focused around a single university and one local school district, or may encompass a much larger group of universities and districts.

The change cycle outlined in Figure 2.1 will likely be helpful to programs seeking to transform teacher preparation programs. Stakeholders may want to familiarize themselves with key documents such as the AMTE Standards (2017), and Elrod and Kezar's (2016) guide to systemic institutional change; see the recommended reading list at the end of this chapter. When stakeholders are convened (in-person if feasible), an initial focus is to determine a common vision. Creating a fishbone diagram to organize the challenges is a potential stepping stone toward building a common vision (see Figure 2.5). The stakeholders should then create an aim that reflects the common vision, and an accompanying driver diagram (see Figure 2.6) to elaborate change strategies. Note that this process of gathering stakeholders and getting to a common vision and driver diagram takes a considerable amount of time, often six months to a year (e.g., Elrod & Kezar, 2016).

Part of successful change involves measuring intended outcomes for growth. As mentioned in the principles of change (see Table 2.1), one cannot improve what cannot be measured. Teams of stakeholders should discuss how they plan to measure intended outcomes; MTE-Partnership resources for measurement are also available to support transformation efforts (see mtep.info/nic-transform). When working on change strategies, it is important to remember that not all changes are improvements, and that one should expect some failures through trying new strategies; the key is to learn (quickly) from those failures to keep the overall transformation efforts moving forward.

While selecting, testing, and refining change strategies, it is important to keep the end goal in mind. Successful improvement strategies plan for sustainability from the beginning (Kezar, 2014). In any large system, personnel turnover is inevitable. A team of stakeholders should make explicit plans for how to handle such turnover and to bring the new people on board, while giving them a chance to influence the process. Plans for personnel turnover need to consider how to engender feelings of shared ownership when someone is brought into transformation efforts that are already underway.

Transformation efforts are not easy, but they are crucial to making progress toward the AMTE's aspirational standards for preparing teachers of mathematics. Schools in the United States have not yet solved the problems of inequitable opportunities for students to learn, directly causing subsequent differential student outcomes. Having more and better qualified mathematics teachers is one step toward improving our education system.

Reading List

The following is a list of recommended readings for stakeholders interested in program transformation.

1. Association of Mathematics Teacher Educators. (2017). *Standards for the preparation of teachers of mathematics.* Raleigh, NC: Author. Retrieved from http://amte.net/standards
   - This publication outlines aspirational standards for the preparation of all teachers of mathematics (including special education, English language learners, etc.) with a focus on equity and on university-school partnerships.

2. Elrod, S., & Kezar, A. (2016). *Increasing student success in STEM: A guide to systemic institutional change.* Washington, DC: Association of American Colleges and Universities. Retrieved from https://www. aacu.org/publications-research/publications/increasing-student-success-stem-guide-systemic-institutional
   - This handbook includes worksheets a program might use to work through the change process. The chapters are designed to be used like a handbook, and provide practical suggestions, grounded in research literature.

3. Conference Board of the Mathematical Sciences. (2012). *The mathematical education of teachers II.* Providence, RI and Washington, DC: American Mathematical Society and Mathematical Association of America. https://www.cbmsweb.org/archive/MET2/met2.pdf
   - This publication recommends what mathematics teachers should learn and understand in order to be effect teachers of mathematics (see Chapter 4 of this book, for more information).

4. Bryk, A. S., Gomez, L., Grunow, Al., & LeMahieu, P. (2015). *Learning to improve: How America's schools can get better at getting better.* Boston: Harvard Education Publishing. Retrieved from https://www.carnegiefoundation.org/resources/publications/learning-to-improve/
   - This book takes a deep look at NICs and how networked communities can be used to improve complex educational systems.

5. Strutchens, M. E., Huang, R., Potari, D., & Losano, L. (Eds.). (2018). *Educating prospective secondary mathematics teachers: Knowledge, identity, and pedagogical practices.* Cham, Switzerland: Springer. Retrieved from https://www.springer.com/us/book/9783319910581
   - This book provides an international review of literature on secondary mathematics teacher preparation, including sections on clinical

experiences, technologies, tools and resources, teacher knowledge, and teacher professional identities.

6. Martin, W. G., & Gobstein, H. (2015). Generating a networked improvement community to improve secondary mathematics teacher preparation: Network leadership, organization, and operation. *Journal of Teacher Education, 66*(5), 482–493.
    – This publication describes the organization of the MTE-Partnership as a NIC, including lessons learned and challenges faced.

For current information on the Transformations Working Group, please visit mtep.info/TWG. For more   information on NIC-Transform, please visit mtep. info/NIC-Transform.

ENDNOTES

1. Work on this chapter was supported in part by grants from the National Science Foundation Directorate for Education and Human Resources, Division of Undergraduate Education (DUE)—Improving Undergraduate STEM Education (IUSE), Exploration and Design: Institutional and Community Transformation, Grant ID #s 1834539 and 1834551 and from the Helmsley Charitable Trust. Any opinions, findings, and conclusions or recommendations expressed in this material are those of the author(s) and do not necessarily reflect the views of the National Science Foundation or the Helmsley Charitable Trust.
2. Other contributors to this chapter include: Mark Ellis, California State University, Fullerton and Dana Franz, Mississippi State University. The active members of the Transformation Working Group also provided the foundation for this chapter, by engaging in the hard work of the group. In addition to the authors and contributors, this group includes Pier Junor Clarke, Georgia State University; Judy Kysh, San Francisco State University; Jennifer Oloff-Lewis, California State University, Chico; Robert Ronau, University of Louisville; Marilyn Strutchens, Auburn University; and Diana Suddreth, Utah State Office of Education.
3. The first two vignettes are reprinted with permission from the *Proceedings of the Seventh Annual Mathematics Teacher Education Partnership Conference* (Smith, Lawler, Strayer, & Augustyn, 2018).

REFERENCES

Aguirre, J., Mayfield-Ingram, K., & Martin, D. B. (2013). *The impact of identity in K–8 mathematics: Rethinking equity-based practices.* Reston, VA: National Council of Teachers of Mathematics.

Ali, N., Tretiakov, A., Whiddett, D., Hunter, I., & Ali, N. (2017). Knowledge management systems success in healthcare: Leadership matters. *International Journal of Medical Informatics, 97*, 331–340.

Association of Mathematics Teacher Educators. (2017). *Standards for preparing teachers of mathematics*. Raleigh, NC: Author. Retrieved from http://amte.net/standards

Birnbaum, R. (1991; 2nd ed. 1999). *How colleges work: The cybernetics of academic organization and leadership.* San Francisco: Jossey-Bass.

Bryk, A. S., Gomez, L., Grunow, Al., & LeMahieu, P. (2015). *Learning to improve: How America's schools can get better at getting better.* Boston: Harvard Education Publishing.

Bukowitz, W. R., & Williams, R. L. (2000). *The knowledge management fieldbook.* New York, NY: Prentice Hall.

Burnes, B. (2011). Introduction: Why does change fail, and what can we do about it? *Journal of Change Management, 11,* 445–450.

Carnegie Foundation for the Advancement of Teaching (n.d.). *The six core principles of improvement.* Retrieved from https://www.carnegiefoundation.org/our-ideas/six-core-principles-improvement

Conference Board of the Mathematical Sciences. (2012). *The mathematical education of teachers II.* Providence, RI and Washington, DC: American Mathematical Society and Mathematical Association of America.

Corbo, J. C., Reinholz, D. L., Dancy, M. H., Deetz, S., & Finkelstein, N. (2016). A framework for transforming departmental culture to support educational innovation. *Physical Review Physics Education Research, 12,* 01003. Retrieved from: http://arxiv.org/abs/1412.3034v3

Dalkir, K. (2017). *Knowledge management in theory and practice (3rd ed.).* Boston, MA: MIT Press.

Elrod, S., & Kezar, A. (2016). *Increasing student success in STEM: A guide to systemic institutional change.* Washington, DC: Association of American Colleges and Universities.

Fernandez, A., Crespo, S., & Civil, M. (Eds.). *Access & equity: Promoting high-quality mathematics grades 6–8.* Reston, VA: National Council of Teachers of Mathematics.

Franklin, C., Bargagliotti, A., Case, C., Kader, G., Schaeffer, R., & Spangler, D. (2015). *Statistical education of teachers.* Arlington, VA: American Statistical Association. Retrieved from http://www.amstat.org/

Fullan, M. G. (2016). *The new meaning of educational change* (5th ed.). New York, NY: Teachers College Press.

Gleason, J., Livers, S. D., & Zelkowski, J. (2015). *Mathematics classroom observation protocol for practices: Descriptors manual.* Retrieved from http://jgleason.people.ua.edu/mcop2.html

Hayward, C. N., Kogan, M., & Laursen, S. L. (2015). Facilitating instructor adoption of inquiry-based learning in college mathematics. *International Journal of Research in Undergraduate Mathematics Education, 1,* 1–24.

Hughes, M. (2011). Do 70 percent of all organizational change initiatives really fail? *Journal of Change Management, 11,* 451–464.

Kezar, A. (2014). *How colleges change: Understanding, leading, and enacting change.* New York, NY: Routledge.

McClellan, J. G. (2011). Reconsidering communication and the discursive politics of organizational change. *Journal of Change Management, 11,* 464–480.

Mejia, E. (2014). *Walking the talk, teaching the talk: Building a collective learning system at the Carnegie Foundation for the Advancement of Teaching.* Doctoral capstone, Harvard Graduate School of Education.

National Council of Teachers of Mathematics. (2014). *Principles to actions: Ensuring mathematical success for all.* Reston, VA: Author.

National Governors Association Center for Best Practices, Council of Chief State School Officers. (2010). *Common core state standards: Mathematics.* Washington, DC: Author.

Pugh, K., & Prusak, L. (2013). Designing effective knowledge networks. *MIT Sloan Management Review, 55*(1), 79–88.

Raelin, J. D., & Cataldo, C. G. (2011). Whither middle management? Empowering interface and the failure of organizational change. *Journal of Change Management, 11,* 481–507.

Sho-Fang, C., Ping-Jung, H., & Hui-Fang, C. (2016). Key success factors for clinical knowledge management systems: Comparing physician and hospital manager viewpoints. *Technology & Health Care, 24,* S297–S306.

Smith, W. M., Lawler, B. R., Strayer, J., & Augustyn, L. (Eds.). (2018). *Proceedings of the seventh annual Mathematics Teacher Education Partnership conference.* Washington, DC: Association of Public and Land-grant Universities.

Wang, Y., Meister, D. B., & Gray, P. H. (2013). Social influence and knowledge management systems use: Evidence from panel data. *MIS Quarterly, 37*(1), 299–313.

Wenger, E. (1999). *Communities of practice: Learning, meaning, and identity.* Cambridge, UK: Cambridge University Press.

CHAPTER 3

EQUITY AND JUSTICE IN THE PREPARATION OF SECONDARY MATHEMATICS TEACHERS[1]

Lorraine M. Males, Ruthmae Sears, and Brian R. Lawler

The Mathematics Teacher Education Partnership (MTE-Partnership) set forth to build a national consensus on how effective secondary mathematics teacher preparation programs might develop teacher candidates who promote mathematical excellence in their future students. To promote excellence in all students, equity and social justice must be underlying values for both secondary mathematics teacher candidates and mathematics teacher educators. To address the present inequities experienced in secondary mathematics classrooms at scale, reform movements need to "be enacted within a community of practice, using such approaches as a networked improvement community or design-based implementation research" (Confrey, 2017, p. 19). As the MTE-Partnership began to organize, it seemed well positioned to achieve this aim.

Underlying the MTE-Partnership's initial considerations of "effective secondary mathematics teacher preparation programs" and teacher candidates who "promote mathematical excellence" (MTE-Partnership, 2014, p. 1) were inferential statements related to equity and justice. Inferred statements regarding equity and justice could be found in several of the indicators of the MTE-Partnership's *Guiding Principles for Secondary Mathematics Teacher Preparation Programs* (2014), such as that teacher candidates view and convey that "mathematics is a living and evolving human endeavor" (p. 4), "maintain high expectations for all students" (p. 5), and "demonstrate a dedication to equitable pedagogy" (p. 5). Furthermore, the *Guiding Principles* recommend that a teacher preparation program "actively recruits teacher candidates representative of the broad diversity of students who they will teach and provides them with the support necessary for their success" (pp. 6–7). A final marker of the commitment to equity and justice was evident in the way many indicators in the *Guiding Principles* emphasize "all" students, specifically those in Principle 5: Candidates' Knowledge and Use of Educational Practices.

Given the underpinnings of equity in the *Guiding Principles*, in 2015, the fifth year of the MTE-Partnership, the Planning Committee determined that the elements of equity and social justice present in the *Guiding Principles* could not be sufficiently addressed solely by the five current Research Action Clusters (RACs). Thus, it was proposed that a working group be formed to address equity-related challenges in the preparation of secondary mathematics teachers. The larger MTE-Partnership community approved this effort at the 2016 annual conference, and each RAC nominated approximately two people to serve on the initial working group. The Equity and Social Justice Working Group (ESJWG) organized in early 2017, and since has worked to launch an improvement science research agenda to identify and understand equity-related concerns for secondary mathematics teacher preparation. Although the ESJWG buttresses the four current primary drivers of the MTE-Partnership (creating a vision, improving clinical preparation, increasing content knowledge, and improving recruitment and retention) and its members are committed to supporting the equity-related work of each of the RACs, its members also recommended that equity and social justice must stand apart as an additional, new secondary driver, pursuing a research agenda distinct from the other RACs. Figure 3.1 presents the current MTE-Partnership's driver diagram, showing equity and social justice as an additional primary driver that underlies the existing primary drivers as well as the ESJWG as a secondary driver.

Although one of the initial goals of the ESJWG was to define equity and social justice specifically for use in the MTE-Partnership, in its initial work, ESJWG used the following operational definitions of equity and social justice published by national mathematics education organizations to guide its efforts.

- Equity: The Association of Mathematics Teacher Educators (AMTE) defines equity as access to high-quality learning experiences; inclusion for all learners, mathematics educators, and mathematics teacher educators; and respectful and fair engagement with others (university colleagues, prospective and in-service teachers, future teacher educators, and P–12 students). This means actively working toward a more just and equitable mathemat-

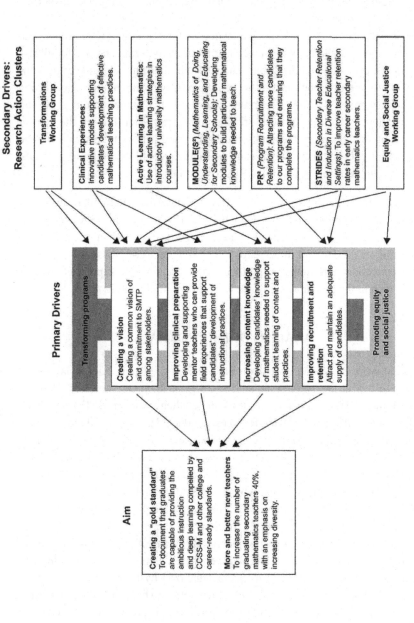

FIGURE 3.1. The current MTE-Partnership's driver diagram shows equity and social justice as an additional primary driver that underlies the existing primary drivers as well as the ESJWG as a secondary driver.

ics education free of systemic forms of inequality based on race, class, language, culture, gender, age, sexual orientation, religion, and dis/ability. (AMTE, 2015, p. 1)

- Social justice: A social justice stance requires a systemic approach that includes fair and equitable teaching practices, high expectations for all students, access to rich, rigorous, and relevant mathematics, and strong family/community relationships to promote positive mathematics learning and achievement. Equally important, a social justice stance interrogates and challenges the roles power, privilege, and oppression play in the current unjust system of mathematics education—and in society as a whole (The National Council of Supervisors of Mathematics [NCSM] and TODOS: Mathematics for ALL [TODOS], 2016, p. 1).

In this chapter, the initial work of ESJWG is described. Particularly, ESJWG discusses the following: an analysis of the problem, the resulting fishbone diagram, the relevant literature related to the analysis of the problem, its theories of change, and its current and future efforts. The summary describes how the mathematics education community might focus on improvement to support equity and social justice in the preparation of future secondary mathematics teachers.

ANALYSIS OF THE PROBLEM

The initial work of ESJWG was to make sense of the equity and social justice issues involved in the preparation of secondary mathematics teachers. Toward this goal, the ESJWG gathered concerns and challenges from MTE-Partnership members at the 2016 and 2017 conferences, conducted an additional member survey, and drew upon the expertise of the ESJWG members. The gathering and organizing of concerns was an inductive process in which members of the ESJWG individually identified concerns and challenges, as described above. These issues were next organized into themes, which included: definitions; policies; disconnections between school partners and higher education; expertise; resources; the question "what does 'it' look like?"; diversity of people; courageous conversations; deficit discourses; PSTs do not identify as agents of change toward a more just society; and mathematics is not a space where diverse ways of knowing and learning are valued. The challenges were organized by themes into a fishbone diagram (Bryk, Gomez, Grunow, & LeMahieu, 2015) to help ESJWG better name and understand the problems of equity and justice in the preparation of secondary mathematics teachers. The fishbone diagram in Figure 3.2 illustrates how ESJWG has mapped out the problem, that at present, the preparation of secondary mathematics teachers does not adequately attend to societal inequities and injustices.

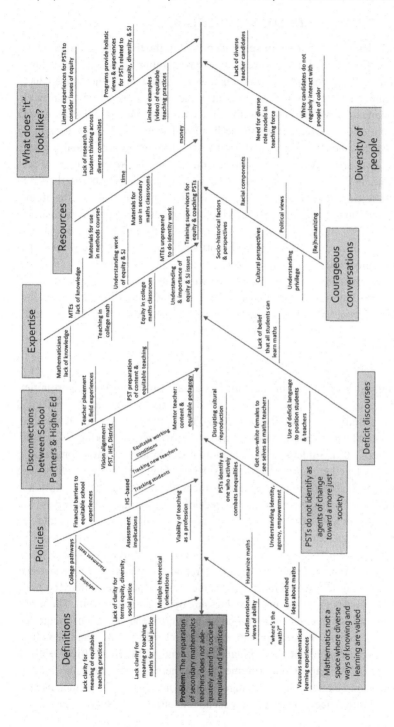

FIGURE 3.2.   The problem space organized into a fishbone diagram for ESJWG.

REVIEW OF THE LITERATURE

To better understand the problem space identified in the fishbone diagram, members of ESJWG engaged in a review of relevant literature. It was not an exhaustive review of the literature on equity and social justice in mathematics education, but rather it helped the group to understand and situate its potential work. In this review, the ESJWG first describes: the current social and political context of mathematics education, including the historical view of what it means to do mathematics as a school subject and what ways of knowing are valued (or not); the deficit discourses that have perpetuated our day-to-day work to prepare secondary mathematics teachers; and the notions of courageous conversations and teachers as agents of change. Second, the working group discusses educational systems, including the diversity that exists, the policies and practices that perpetuate inequality, and the disconnect between K–12 and higher education institutions. Finally, since the aim of the MTE-Partnership is to determine what preparation programs need to do in order to develop teacher candidates who promote mathematical excellence in their future students, the working group focuses on the issues faced by mathematics teacher educators including definitional issues, expertise, and resources. Thus, the group seeks to unpack the key problems identified within the ESJWG fishbone diagram (see Figure 3.2).

The Social and Political Context of Mathematics Education

Although many describe mathematics as a gateway to further opportunities, it also serves as a gatekeeper (Aguirre, Mayfield-Ingram, & Martin, 2013) and a marker of intelligence (Butterworth, 2018). It is quite common for inequalities in mathematical achievement scores to be conceptualized as rankings of competence (Flores, 2007). Students who have more opportunities to learn, compared to their less privileged counterparts, are considered to have higher aptitudes or are more capable. According to the U.S. Department of Education (2016), Black and Latinx students have unequal access to accelerated programs and courses. For instance, there are fewer Black and Latinx students enrolled in gifted and talented education programs and Advanced Placement (AP) courses, when compared to their White counterparts. Moreover, students of color and English language learners are retained more frequently than their White counterparts. These findings indicate that using test scores to discuss mathematical competence may be inappropriate because not all students have the same opportunities to learn. Therefore, focusing on the achievement gap can provide an erroneous image of inequities, locating the deficiency in students rather than the school structures and perpetuating a negative narrative about marginalized students (Gutiérrez, 2013). Without accounting for these gaps in opportunities, researchers and educators may employ a deficit model to explain below average performance using factors such as poverty, cultural differences, and parental educational attainment (Flores, 2007).

In addition, despite changing student demographics, high school mathematics has remained largely unchanged over the last few decades (National Council of Teachers of Mathematics [NCTM], 2018). The curriculum of high school mathematics, particularly the sequence of courses recommended for all students (e.g., two courses in algebra and one in geometry) has generally remained the same since it was recommended by the Committee of Ten over 120 years ago (Dossey, McCrone, & Halvorsen, 2016). Moreover, the ways in which mathematics educators expect students to learn mathematics and display mathematical competence are still rooted in methods that privilege certain ways of knowing over others (Nasir & de Royston, 2013). Overvaluing these privileged and particular ways of knowing have, in turn, perpetuated a notion that some students cannot think (mathematically) and are unable to engage in mathematics. Notions such as these are manifested into deficit discourses, both within classrooms and schools as well as in the ways we talk and write about mathematics education (Gutiérrez, 2013; Nasir & de Royston, 2013; Washington, Torres, Gholson, & Martin, 2012).

Privileged Ways of Knowing. In many mathematics classrooms, teachers and students have a narrow view of what counts as mathematics and assume that being smart means "knowing the procedures for getting accurate answers to computation problems quickly" (Featherstone et al., 2011, p. 15). In elementary schools where many students first experience formal school mathematics, knowing the facts is an important part of being smart in mathematics. However, according to Featherstone et al., this narrow conception of mathematical smartness excludes many children from feeling smart and looking smart to others and perpetuates an inaccurate vision of what it means to do mathematics. This conception focuses on individualism, speed, and other attributes that privilege certain ways of knowing. These attributes serve to organize mathematics classroom environments that do not provide opportunities for all students to learn.

For example, environments that focus on knowing facts and being able to answer computational problems accurately and quickly may impede the opportunities for learning for certain groups of students. When teachers provide opportunities for students to solve worthwhile mathematical tasks, and do this in an environment where they have opportunities to justify and communicate their mathematical thinking, more students are successful (Horn, 2005; Jackson & Delaney, 2017; Strutchens, 2000). According to Coleman, Bruce, White, Boykin, and Tyler (2017), African American students, in particular, value communalism, meaning they place importance "on their social bonds, interconnectedness, and interdependence with others as individuals consider duty to one's social group to be more important than individual rights and privileges" (p. 545). This value, consistent with Afro-cultural ethos, may position African American students to be in direct conflict with classroom norms and values that are rooted in traditional teacher-centered approaches where individuals generate and transmit knowledge.

A narrow list of attributes associated with doing mathematics, such as computing quickly and knowing facts, are often associated with standard algorithms and school-based approaches (Featherstone et al., 2011). Civil (2016) found that her prospective elementary teachers, who often claimed to not have successful K–12 school experiences, still privileged "school-looking methods over those that were grounded in everyday approaches" (p. 43). These school-based approaches in middle school and high school often privilege algebraic ways of knowing, whether they make sense to students or not. Algebra has an aura of prestige and is associated with high status and power (Aguirre et al., 2017; Healy & Hoyles, 2000). According to Civil (2016), this privileging is the result of the ways in which algebra is portrayed in school mathematics, in which every problem can be solved by a formula or fixed set of procedures, with no expectation for reasoning or sense making. These school-based methods can be at odds with approaches valued by immigrant students and their families (Civil, 2016), who often find their approaches more efficient than the ones being learned in U.S. schools (Civil & Planas, 2010). Teachers do not always value alternative approaches and hence place students in a difficult position navigating different practices at school and at home. A broader sense of what counts as mathematics is also diminished "in universities, which propagate the myth of one and only one kind of mathematics" (Fasheh, 2012, p. 94).

Deficit Discourses. The privileging of certain ways of knowing has resulted in deficit views of students, particularly ones that come from non-dominant populations. Rather than seeing students, parents, and communities as resources, our language positions "some people or communities as superior or inferior to others" (Aguirre et al., 2017, p. 136). Even understanding the notion of misconceptions, something that is pervasive in many teacher education programs (e.g., learning to anticipate misconceptions so that you can eliminate them), may perpetuate a deficit view of students or at least obscure the usefulness of using student conceptions productively to support students in developing expertise (Smith, diSessa, & Roschelle, 1993). Furthermore, labels such as *fast*, *high-level*, or *the cream*, are used to position some students as superior while other students are positioned as inferior using labels such as *slow*, *lazy*, *screw-ups*, or *bubble kids* (Horn, 2007; Suh, Theakston Musselman, Herbel-Eisenmann, & Steele, 2013).

Other labels that we use to describe students, such as English learners, also negatively position students. For example, in a conversation among members of the *Journal for Research in Mathematics Education*'s Equity Special Issue Editorial Panel (D'Ambrosio et al., 2013), Moskovitz noted "that the label *English learners* is not necessarily used to refer to students' language proficiency; many times it is being used as a tool to segregate student populations" (p. 24). Positioning some students as more and others as less capable leads to students having qualitatively different mathematical experiences, with certain students having limited access to important mathematics and robust approaches to instruction, such as inquiry-oriented methods (Horn, 2007). These perspectives along with institutional structures (discussed in the next section) further contribute to "cultural myths and stereotypes about who is 'smart' (or not) and what constitutes 'smartness'" (Aguirre et al., 2017, p. 136).

The results of these deficit perspectives in classrooms have also permeated our public discourse where we continue to focus on achievement and competence. Flores (2007) recommended reframing the discourse to consider opportunity gaps, which considers limited access to needed resources that may contribute to the success of the more privileged students. Similarly, Parks and King (2007) suggested that society departs from a conventional conception of competence and focuses on what students are doing as a means to accommodate diverse approaches and for students to demonstrate what they know. Therefore, as mathematics educators seek to address inequities within mathematics classroom settings, they need to consider social constructs within schools. Mathematics educators should also seek to ensure that the experiences afforded in the mathematics classroom be used to empower and liberate marginalized learners from the aforementioned examples of oppressive classroom norms and manifestations of a structural racism (Martin, 2009).

Courageous Conversations. Engaging in conversations about equity and social justice may be difficult. For instance, identity markers such as race/ethnicity, gender, and class and their connection with educational access and achievement have been the "elephant-in-the-room" topics in many U.S. classrooms (Mansfield & Jean-Marie, 2015). Castagno (2008) found that most White educators are reluctant to name things that are perceived as uncomfortable or threatening to the established social order. Thus, individuals often avoid addressing such issues, or discuss them superficially, resulting in little to no impact on teachers promoting equitable opportunities or advocating for social justice within mathematics.

Because matters of racism, sexism, and classism (and other -isms) are at the forefront of public discourse, it is crucial that those working in education assist students in processing, responding, and becoming active participants in the world around them (Boyd & Glazier, 2017). To attend to equity and social justice within a secondary mathematics environment and within teacher preparation programs, mathematics teacher educators are challenged to facilitate courageous conversations. Such conversations can be used to dismantle racism and equitably teach all students. For courageous conversations to be possible, all parties should agree to: (a) stay engaged; (b) speak their truth; (c) experience discomfort; and (d) expect and accept non-disclosure (Singleton & Linton, 2006). During courageous conversations, messages should be presented and discussed to facilitate heightened awareness of a problem situation, not to convert the receiver(s) to one's way of thinking (Amobi, 2007).

Agents of Change. Mathematics teachers can serve as gatekeepers by determining what mathematics is taught and ultimately learned, determining the benchmarks of competency for assessments, and determining who gets to progress to the next level. Even though mathematics teachers are positioned as gatekeepers, they are often reluctant to recognize the political impact of their teaching.

Reasons for this averseness may range from their belief that mathematics is a politically neutral subject (Felton-Koestler & Koestler, 2017), to their preference to avoid conflict with colleagues and stay within the "safe terrain of 'niceness'" (Dyches & Boyd, 2017), to their discomfort or reluctance with questioning their own status and privilege as someone who is good at mathematics (de Freitas, 2008). Moreover, mathematics teachers are unlikely to see themselves as responsible for taking on social justice issues considering it is not directly related to their duties of disseminating mathematical content (Langer-Osuna & Esmonde, 2017).

While these positions are also true of many veteran teachers, the issue is particularly salient for prospective teachers, because, as Price and Valli (2005) write, "as novices, they [preservice teachers] often have difficulty even thinking of themselves as teachers, much less as change agents" (p. 58). For instance, de Freitas (2008) found that prospective teachers do not consider that their roles and responsibilities encompass that of agents of change. Prospective teachers often outright resisted addressing issues of social justice in the mathematics classroom. For instance, teacher candidate responses included, "I'm just a math guy," or "I'm not one for social justice" (de Freitas, 2008, p. 49).

Auto-ethnographic assignments have been used to facilitate prospective teachers beginning to conceive of themselves as agents of change, to help them interrogate their own privilege associated with success in school mathematics (de Freitas, 2008). de Freitas found that the use of auto-ethnographic assignments "allowed for the kind of transformative learning that characterizes critical pedagogy" (p. 53). Alternatively, professional development and networking can also be used to empower prospective teachers to be agents of change. For example, Riley and Solic (2017) created an Urban Education Fellowship as a means of building a cadre of teachers committed to pursue fieldwork and careers in urban schools. Through attendance at conferences and teacher inquiry groups, the fellows gained perspectives on social justice and described a desire to learn more to address issues of social justice in the classroom and within their home communities. Auto-ethnographic assignments and fellowship programs are possible strategies that might be used to help prospective teachers "Understand Power and Privilege in the History of Mathematics Education," Indicator C.4.4, in the *Standards for Preparing Teachers of Mathematics* (AMTE, 2017, p. 23) and encourage them to position themselves as agents of change.

Educational Systems

The K–12 school environments help to empower students to enter the workforce and the world in which they live, and these environments are becoming increasingly diverse (Maxwell, 2014; Seah & Anderson, 2015). Thus, mathematics teacher preparation programs must consider means to better prepare future teachers to meet the needs of a diverse population and successfully exhibit competencies in both mathematical content and process standards (NCTM, 2014; National Governors Association Center for Best Practices, Council of Chief State School

Officers, 2010). In this section, the working group attends to diversity, policy, and the disconnect between K–12 schools and institutions of higher education, all of which impact the extent to which issues of equity and social justice are addressed in the preparation of secondary mathematics teachers.

Diversity. The diversity within a school setting can be perceived as polyhedral, with faces representing social, political, cultural, and linguistic perspectives (de Abreu, Gorgorió, & Björklund Boistrup, 2018). Thus, practitioners and researchers must attend to: students' identities, students' agency, how mathematics is socially represented, cultural representations in mathematics, the nature of discourse, and the mathematical practices emphasized at home and school. Similarly, Secada (1991b) suggested that when attending to diversity, careful consideration ought to be given to the learner, the curriculum utilized, the classroom practices observed, and the teachers' characteristics.

More specifically, Secada (1991a) advised that attention be given to the cultural mismatch between students and the school environment because the tacit power dynamics that exist within the school setting often result in the exclusion of some students. Moreover, teachers' preferences for particular student attributes can influence teachers' career decisions. For example, teachers' decisions to move from one school to another are strongly correlated to students' race and academic achievement (Simon & Johnson, 2015). Therefore, it is critical that individuals gain insight into teachers' beliefs and perspectives when examining diversity within a classroom setting.

Instructional strategies have been developed to embrace student diversity within the classroom setting, such as culturally relevant pedagogy (CRP; Ladson-Billings, 2014) or reconstructionist pedagogy (Leonard & Dantley, 2005). Culturally sustaining pedagogy (CSP; Paris, 2016) is a recent approach to fostering equitable classrooms, a refinement to CRP.

> CSP describes teaching and learning that seeks to perpetuate and foster linguistic, literate, and cultural pluralism as part of the democratic project of schooling and as a needed response to demographic and social change. CSP takes dynamic cultural and linguistic dexterity as a necessary good, and sees the outcome of learning as additive, rather than subtractive, as remaining whole rather than framed as broken, as critically enriching strengths rather than replacing deficits. CSP builds on decades of crucial asset-based pedagogical research that has countered pervasive deficit approaches, working against the backdrop of beliefs in White superiority and the systemic racism they engender, to prove that our practices and ways of being as students and communities of color are legitimate and should be included meaningfully in classroom learning. (p. 6)

Important elements of instructional strategies used to promote equity and social justice, such as those mentioned here, include encouraging students to participate in the classroom discourse (Parks & King, 2007), embracing the cultural and academic strengths children bring to the classroom, and furthering students' critical consciousness, specifically seeing the use of mathematics to act upon their world (Gutstein, 2006).

With respect to working in a diverse environment, Martin (2008) cautioned that a *color-blind* approach supports White privilege and cultural assimilation. The criteria for identifying effective mathematics teachers often overlook the teachers' social interactions, dispositions, and commitment to advocate for marginalized students (i.e. Latinx, African American, American Indian, and English learners) within the classroom setting (Gutiérrez, 2013). This limitation of constructs used to identify effective teachers can be problematic considering that students' mathematical abilities are socially constructed, and their identities are influenced via a dialectic between their lived experiences and social structures (Parks & King, 2007). Therefore, teachers are challenged to reflect on effective means of embracing diversity, rather than ignoring or suppressing students' diverse attributes—including mathematical ways of thinking.

Finally, with respect to diversity, school leadership—particularly school boards—rarely reflect the diversity of a district's student population, often resulting in the environment operating as a White institutional space (Martin, 2008). In this sort of space, too often race is used as a means to disaggregate data or as a variable to report mathematical achievement and is often the sole descriptor for diversity (Martin, 2009). Race must be conceptualized as not just a social construction, but as a sociopolitical construction, with historical underpinnings of racism and racialization.

Policy. The Supreme Court ruling in Brown vs. Board of Education of Topeka (*Brown v. Board of Education of Topeka*, 1954) opened the door for more diversity in student enrollments at schools nationwide by disallowing state-sponsored segregation, which initiated school busing programs. However, today our public schools, are on a path to racial re-segregation (Frankenberg & Orfield, 2012). Hence, establishing policies to address equity and social justice is critical to ensuring each and every student is afforded equitable opportunities to learn mathematics. Multiple national mathematics education organizations (AMTE, 2017; NCSM & TODOS, 2016), have placed "a renewed focus on access, equity, and empowerment" (Larson, 2016). AMTE specifically includes a set of five indicators that describe the knowledge, skills, and dispositions for teachers of mathematics with respect to the "Social Contexts of Mathematics Teaching and Learning (C.4)." These include:

- C.4.1 Provide Access and Advancement,
- C.4.2 Cultivate Positive Mathematical Identities,
- C.4.3 Draw on Students' Mathematical Strengths,
- C.4.4 Understand Power and Privilege in the History of Mathematics Education, and
- C.4.5 Enact Ethical Practice for Advocacy (p. 6).

Thus, principles of equity and social justice are present in both policy and vision documents that aim to outline the quality preparation of secondary mathematics teachers.

Moreover, policies relative to tracking can promote inequities in secondary mathematics. The practice of tracking students has been shown to not improve outcomes of low-achieving students and negatively impact students' mathematical identities (Boaler & Selling, 2017; Linchevski & Kutscher, 1998; Oakes, 1987, 1994). Additionally, the placement of students in various tracks can be predicted by race and teacher evaluations more than by a student's mathematical achievement (Faulkner, Stiff, Marshall, Nietfeld, & Crossland, 2014). The tracking of teachers further exacerbates student tracking. Teachers are often tracked based on seniority, with veteran teachers using their political or social power within the school to teach the "high" track, or those students with greater socioeconomic means who may be better positioned to purchase resources to enhance the instruction from these senior teachers. This tracking all too frequently leaves teachers with the least experience teaching those students that struggle the most and with the least resources (Dabach, 2015; Darling-Hammond, 2000; Grissom, Kalogrides, & Loeb, 2015; Isenberg et al., 2013; Kelly, 2004). Due to the negative impacts on student opportunities to learn mathematics as a result of tracking, NCTM has called for an end to both student and teacher tracking (NCTM, 2018).

Disconnection Among K–12 and Higher Education Institutions. Teacher preparation programs are tasked with preparing teachers to be effective in executing their instructional craft based on state and national standards in K–12 institutions. However, despite programs working with K–12 institutions, a common criticism of teacher preparation programs is the disconnect between what prospective teachers learn in their academic coursework and their experience during practicums and internships (Zeichner, 2010). Considering that prospective teachers often place a higher value on the clinical experience as this is perceived as the reality of school life, teacher preparation programs are challenged to bridge the experiences of institutions of higher education and K–12 environments (Allen & Wright, 2014; Campbell & Dunleavy, 2016; Little & Anderson, 2016). Given this context, teacher preparation programs must provide authentic examples of how to explicitly attend to equity and social justice within K–12 settings and consider means to readily infuse equity and social justice into their curriculum, assessment, and instruction, such that prospective teachers have models to support themselves in K–12 environments.

Mathematics Teacher Educators

Mathematics teacher educators are responsible for engaging in scholarly activities, while also preparing prospective teachers to develop their instructional craft. Admittedly, mathematics teacher educators' expertise and perspectives about equity and social justice can vary, which can influence how they ultimately attend to it within their practices (Dyches & Boyd, 2017; Parker, Morrell, Morrell, &

Chang, 2016; Slay, 2011; Vomvoridi-Ivanović & McLeman, 2015). For instance, a focus on social justice may not be a part of mathematics teacher educators' professional training (Dyches & Boyd, 2017). Thus, mathematics teacher educators need clear and concise definitions, as well as relevant resources, to attend to issues of equity and social justice. Moreover, they need a commitment to continue to develop expertise in the area of equity and social justice and to reflect on their beliefs and practices.

Definitions. The definition of equity can vary. Secada (1989) suggested that equity within an educational setting can provide insight into actions that can impact learning and can be used as a checklist for justice. Similarly, others have also sought to clarify how equity and social justice is conceptualized within a mathematics setting from a lens of critical pedagogy (Aguirre et al., 2013; Bartell, 2013; Civil, 2008; Gutiérrez, 2007; Gutstein, 2006). According to Pais (2012),

> Although notions of what it means to achieve equity diverge—and some authors prefer to use other terms such as "social justice" (e.g. Gutstein, 2003), "democratic access" (e.g. Skovsmose & Valero, 2008), or "inclusion/exclusion" (e.g. Knijnik, 1993)—it is usually acknowledged that research on equity requires social and political approaches that situate the problem in a broader context than the classroom or school (Anderson & Tate, 2008; Gates & Zevenbergen, 2009; Valero, 2004, 2007). (p. 50)

Pais emphasizes that although communities, including the mathematics education communities, have developed ways in which to talk about diversity, equity, and social justice, secondary mathematics teacher educators have yet to develop shared definitions from which to work. Furthermore, since equity and social justice can be conceptualized in different ways, and efforts to support equity and social justice during instruction can vary, there exists a need to provide prospective teachers with operational definitions and insights as to how to attend to diversity, equity, and social justice during enacted lessons.

Resources. Numerous publications about equity and social justice in mathematics education are available for mathematics teacher educators. However, mathematics teacher educators are often limited in their knowledge of or access to such resources or are adrift in navigating the overwhelming list of resources that might be considered. What is needed are well-structured databases of these resources and opportunities to learn with peers (Burton, 2003; Secada, Fenema, Adajian, & Byrd, 1995). Beyond just the need for well-structured opportunity to access resources, some subject-specific resources are lacking. There are limited vignettes and online video training that explicitly provide guidance as to how to enact equity within a secondary mathematics classroom, let alone how to engage prospective teachers with these issues in a secondary mathematics preparation program. Thus, there is a need for published materials, online webinars and videos, and other practitioner resources that explicitly describe how to attend to equity and social justice within a secondary mathematics environment.

Expertise. Mathematics teacher educators' expertise can vary based on their research agenda, doctoral preparation program, and professional learning opportunities (Reys & Kilpatrick, 2001). The variance in education can impact the nature of challenges faced and how they seek to resolve them. For instance, Vomvoridi-Ivanović and McLeman (2015) reported that mathematics teacher educators' loci of challenges with attending to equity were relative to: challenges with prospective teachers' willingness or ability to attend to equity or lack of a critical lens to discuss equity; challenges of facilitating instruction relative to equity; and challenges related to society, which may not allocate time to attend to equity.

Professional development can be used to enhance mathematics teacher educators' expertise for attending to equity. For instance, Parker, Morrell, Morrell, and Chang (2016) noted that after facilitating a multistage, equity-focused professional development training that allowed 28 mathematics, science, technology, and engineering instructors to implement ideas into their classrooms, their beliefs and understanding of issues of equity shifted. Thus, with intentional interventions mathematics teacher educators' abilities to attend to equity and social justice can be enhanced.

THEORY OF CHANGE

Review of the literature related to the themes that emerged in the initial analysis of the problem highlighted factors that impact how equity and social justice are attended to in the preparation of secondary mathematics teachers. Particularly, the review of the literature suggested that social and political influences, policies and institutional norms and practices, conceptualizations of equity and social justice, and the expertise of mathematics teacher educators can have implications on the extent equity and social justice are attended to within teacher preparation programs. Considering the challenges faced, careful planning is needed to ensure equity and social justice are appropriately addressed in K–12 mathematics settings and teacher preparation programs.

Upon careful examination of the problems of equity and social justice in the preparation of secondary mathematics teachers, through the construction of the fishbone diagram and review of the literature, the ESJWG generated the following aim statement (Bryk et al., 2015): "Preservice teachers' equity-driven sociopolitical dispositions, and knowledge and use of equitable teaching practices will improve over the course of their teacher education preparation program." This aim includes two specific goals for preparing future teachers. The first focuses on increasing prospective teachers' understanding of the social, historical, and institutional contexts of how mathematics affect teaching (AMTE, 2017) and their development of a disposition toward advocating for each and every student in both their classroom and school. The second goal attends to preparing prospective teachers to enact equitable instructional practices to improve the learning opportunities for each and every student.

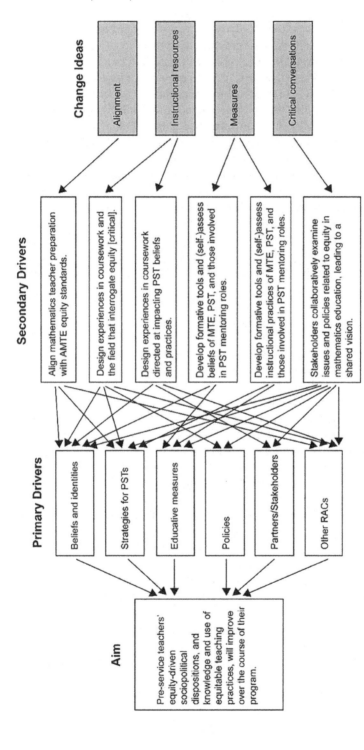

FIGURE 3.3. ESJWG Driver Diagram Naming the Aim and Theory for Change.

After articulating this aim, ESJWG created a driver diagram (see Figure 3.3) to visually represent its working theory of action toward achieving the aim. To construct the driver diagram, members of the ESJWG identified ways to address the concerns identified in the fishbone diagram. The themes identified in the fishbone diagram became the primary drivers, consolidating themes when appropriate, as these are the ESJWG's hypotheses about the main areas of influence necessary to advance the aim (Bryk et al., 2015). Challenges were then revisited and identified in the fishbone diagram to generate a list of secondary drivers, a small set of system components that are hypothesized to activate each primary driver. Finally, the working group generated change ideas or alterations to the system or process identified in the secondary drivers, which will serve as ideas to be tested through the Plan-Do-Study-Act (PDSA) Cycle in order to examine their potential to improve drivers (Bryk et al., 2015). This driver diagram established a common language and served to launch coordinated efforts among the ESJWG members. At present, the change ideas are general categories for the work the group feels likely to be most productive. The arrows on the diagram indicate relationships between the change ideas and the secondary drivers, which affect the primary drivers that have implications for the extent to which the aim is actualized.

DEFINING DIVERSITY, EQUITY, AND SOCIAL JUSTICE

The ESJWG driver diagram and literature review allowed the group to identify that a critical part of the work needed to involve the development of a set of shared definitions. Thus, before attempting to address the change ideas, the group's first task was to develop definitions for diversity, equity, and social justice. A subset of the members broke into three groups, each focusing on one of the definitions. These sub-groups began this work in their first in-person meeting, continued to meet virtually, and finalized their drafts in a second in-person meeting. Once each definition was drafted, the ESJWG solicited feedback on the definitions from each of the RACs. This feedback was used to further refine the definitions. Finally, members of the ESJWG solicited feedback from other knowledgeable colleagues including other mathematics educators and non-mathematics educators with expertise in the areas of diversity, equity, and social justice. This feedback is also being used to improve the definitions, which are in their final stages of revision. These definitions provide the grounding for the ESJWG to begin to initiate the change ideas identified in the driver diagram. Once finalized, these definitions will also serve as definitions for the work of the larger MTE-Partnership as it aims to address equity in the preparation of secondary mathematics teachers.

INITIATING CHANGE IDEAS

With these definitions in their final stages of revision, the ESJWG has shifted its focus to initiating the change ideas identified in the driver diagram. In this section, the working group describes its initial work on each change idea. It is important to note that this work has just begun, and the group has identified initial change ideas that may be more fruitfully addressed.

Addressing Alignment

To promote the alignment of mathematics teacher preparation programs with AMTE standards related to equity, one important first step was to determine how to engage other RACs with the ESJWG aim statement. The ESJWG facilitated a workshop at the 2018 MTE-Partnership annual meeting in which participants reflected on the extent they are explicitly addressing the tenets identified on the ESJWG fishbone diagram within their institutional programs and RAC. This activity highlighted that there were tenets that were seldom attended to, namely: policies, disconnect between higher education and K–12 school environments, and expertise of mathematics teacher educators. Thus, participants were subsequently asked to consider what could be done to ensure all tenets of the fishbone diagram were adequately addressed. This activity was a catalyst for RACs to consider how to refine their PDSA cycles to further support equity and social justice issues. For example, the Mathematics of Doing Understanding Learning and Educating for Secondary Schools (MODULE(S²)) RAC is building equitable instructional practices into its units, and the statistics team is including activities to promote critical statistical orientations and competence. In addition, the Active Learning Mathematics (ALM) RAC is designing research to examine the experiences of marginalized students in active learning classrooms at the collegiate level. In each of these RACs they are embracing equity and justice issues and are studying through the PDSA Cycle a question that impacts learning through an equity lens.

The ESJWG also solicited liaisons from each RAC at the 2018 MTE-Partnership meeting. The role of the liaison is to attend a small number of ESJWG meetings throughout the year in order to share their RAC's progress with respect to issues of equity and social justice, solicit feedback and assistance from the ESJWG, and further inform ESJWG of the complexities associated with equity and justice in the preparation of secondary mathematics teachers.

Addressing Instructional Resources

Considering that instructional resources significantly influence the content that will be taught and eventually learned (Stein, Remillard, & Smith, 2007), intentional efforts are needed to integrate equity and social justice definitions, tasks, and instructional guidelines into resources that are used in teacher preparation programs. While there may be a need for additional resources, our conversations within this working group and with other RACs indicated that there are a lot of existing resources, but that members of the partnership may not be aware of what already exists and/or know how to use these resources in their courses.

Therefore, as a first step in raising awareness and providing members of the MTE-Partnership with resources, at the 2018 annual meeting ESJWG introduced the five equity-based practices in mathematics classrooms (Aguirre, Mayfield-Ingram, & Martin, 2013) during a 90-minute working group session. During this workshop, MTE-Partnership members discussed how these mathematics teaching practices could be used to support equitable mathematics teaching and considered how these practices applied to the work of their RACs. Since this meeting, to provide the MTE-Partnership with additional resources, a subset of the ESJWG is working to develop an online repository for sharing activities that address issues of diversity, equity, and/or social justice. The repository will include links to materials including journal articles, activities, and online resources that members of the MTE-Partnership can use in their courses. In addition, to support the enactment of these resources, for each resource there will be a guide that will include the following information: (a) the objective(s), (b) a short description of how the resource was used (including with what population and in what course(s)), (c) the mathematical content addressed, (d) the *Common Core State Standards for Mathematics* (*CCSS-M*; National Governors Association Center for Best Practices & Council of Chief State School Officers, 2010) addressed, and (e) the 2017 *Standards for the Preparation of Teachers of Mathematics* (AMTE Standards) addressed.

A subset of the ESJWG is also working with the MODULE(S²) RAC (see Chapter 5) to provide feedback on the resources they develop for mathematics courses for secondary mathematics prospective teachers in order to address issues of diversity, equity, and social justice and support instructors in using the materials. These resources, which cover algebra, geometry, modeling, and statistics, are being designed and/or piloted and will be available for public use. Integrating issues of diversity, equity, and social justice into these resources will support mathematics instructors in having conversations with their prospective teachers so that future teachers begin to think about equity as it applies to mathematics content. Members of the ESJWG have agreed to pilot some of these resources in their courses to gain insight into how to support the writers of the modules in developing educative supports for instructors.

Addressing Measures

Instruments exist that attempt to capture equitable practices within the classroom setting (e.g., Louisiana State Personnel Development Grant, 2018) while other instruments, such as the MCOP² (Gleason, Livers, & Zelkowski, 2017), attempt to capture the degree of alignment of the mathematics classroom with mathematics practice standards (National Governors Association Center for Best Practices, Council of Chief State School Officers, 2010) and mathematical teaching practices (NCTM, 2014). These instruments can be useful, but for the purpose of our work, ESJWG needed to identify an instrument that readily aligns with its aim statement, primary and secondary drivers, and change ideas. Thus, a subset of the ESJWG is using the Equity Quantified in Participation (EQUIP) tool (Reinholz & Shah, 2018) in a series of PDSA cycles with prospective teachers in methods courses and with practicing teachers and administrators. Participants will work together to analyze mathematics teaching, allowing us to identify the extent to which this tool has the potential to measure the ESJWG's aim.

The EQUIP tool was designed to be a flexible and easy-to-use online application for collecting and generating equity analytics. Equity analytics are displays of quantitative data generated from collecting "relatively low-inference dimensions of classroom discourse," (Reinholz & Shah, 2018, p. 140) which can be cross-tabulated with a variety of user-defined markers such as, but not limited to, gender, race, socioeconomic status, and level of confidence. Research on the tool and its use indicates that it can be easily used by researchers and teachers to examine instruction with respect to equity. The ESJWG sees the potential in this tool for measuring its aim because it can be used to collect and generate data for teachers across time to monitor changes and longitudinal trends.

Addressing Critical Conversations

Lastly, a subset of the ESJWG is currently conducting PDSA cycles with respect to engaging prospective teachers in critical conversations in their methods courses. For example, at one institution, a mathematics teacher educator will ask prospective teachers to evaluate whether beliefs about students' learning are productive or unproductive (NCTM, 2014). At a second institution, a faculty member plans to assign a field-based portfolio (Crowley, 1993). A component of the portfolio is equity-focused where prospective teachers read and reflect on issues of equity in the classroom, then conduct a focused observation on student participation. The prospective teachers also will be asked to observe and reflect on the school and classroom culture, including funds of knowledge, tone, décor, what cultures and languages are represented, etc. The intent of the portfolio is to create an opportunity for prospective teachers to recognize issues of equity in the classroom, specifically related to participation and tracking. In addition to the portfolio, the mathematics teacher educator will engage prospective teachers in readings and discussions throughout the semester related to equity, tracking, and participation.

A third mathematics teacher educator in the ESJWG will examine the nature of equitable participation among prospective teachers after watching videos of secondary mathematics teaching. Particular attention will be placed on the nature of participation relative to gender, language, race, culture, and ethnicity. Thus, the mathematics teacher educator will document the amount of talk time between the prospective teachers and their peers, and between the teacher candidate and the instructor for the course. Additionally, data will be garnered via the type of questions asked and the foci of the questions. Each of these small studies helps

ESJWG to learn effective and impactful strategies to engage prospective teachers in challenging, critical conversations.

FOCUS ON IMPROVEMENT

The AMTE Standards articulate an aspirational vision for well-prepared beginning teachers' knowledge, skills, and dispositions and for what a strong teacher preparation program should entail. These standards rest on five assumptions, the first being that "Ensuring the success of every learner demands a deep, integrated focus on equity in every program that prepares teachers of mathematics" (AMTE, 2017, p. 1).

A deep, integrated focus on equity in every program requires a commitment from all stakeholders who prepare mathematics teachers, a commitment that requires stakeholders to understand what it means to focus on equity. The work of the ESJWG indicates that as a field we still have much to learn about equity, but that we also have numerous resources to support this work. For people interested in beginning to explore what it might mean to focus on equity, the group provides an annotated bibliography.

Reading List

For those interested in learning more about equity in mathematics education, the ESJWG provides a list of readings and short synopses below. There is a plethora of literature related to equity and social justice in mathematics education. This is not meant to be an exhaustive list, but merely a way to get started.

1. Aguirre, J., Herbel-Eisenmann, B., Celedón-Pattichis, S., Civil, M., Wilkerson, T., Stephan, M., Pape, S. & Clements, D. H. (2017). Equity within mathematics education research as a political act: Moving from choice to intentional collective professional responsibility. *Journal for Research in Mathematics Education, 48*, 124–147.
   - NCTM's Research Committee identified how the decisions of researchers in mathematics education to utilize an equity lens is not only a choice, but also a political act. The committee encourages mathematics education researchers, as human beings shaped by the political landscape of their different histories and experiences, to examine how these histories shape their choices to pose questions, understand, and, as a result, impact the field. The committee argues that equity should be an intentional, collective, and professional responsibility of the mathematics education research community.
2. Aguirre, J., Mayfield-Ingram, K., & Martin, D. B. (2013). *The impact of identity in K–8 mathematics: Rethinking equity-based practices*. Reston, VA: National Council of Teachers of Mathematics.

    – These authors examine how both teacher and student identity impacts, and is impacted in, the mathematics classroom. Rich possibilities for mathematical learning result when teachers bring forth these identities to offer high-quality, equity-based teaching, practices that are especially beneficial for students who have been marginalized by race, class, ethnicity, or gender. The authors identify five equity-based mathematics teaching practices: going deep with mathematics; leveraging multiple mathematical competencies; affirming mathematics learners' identities; challenging spaces of marginality; and drawing on multiple resources of knowledge. Classroom vignettes offer real-life examples to illuminate each teaching practice. Although the examples in this book come from K–8 classrooms, the ideas can be easily applied at the secondary level.

3.   Atweh, B., Graven, M., Secada, W., & Valero, P. (Eds.). (2011). *Mapping equity and quality in mathematics education.* New York, NY: Springer.

    – This book captures equity from an international perspective, with contributions from leading researchers in the field. It is grouped into four major themes: the theoretical landscape; mapping social construction and complexities; landmark of concerns; and no-highway, no destination. The book considers: equity as a political term, the various perspectives about equity, and qualitative studies that sought to unpack the nature of equity.

4.   Diversity in Mathematics Education Center for Teaching and Learning (DiME). (2007). Culture, race, power and mathematics education. In F. K. Lester (Ed.), *Second handbook of research on mathematics teaching and learning* (pp. 405–433). Charlotte, NC: Information Age.

    – This chapter presents a synthesis of the research examining the intersections of race, culture, and mathematics teaching and learning, emphasizing the complexity among students' cultural activity, racial experience, and mathematics education. Interspersed throughout the chapter are questions designed to "reveal subtle power relations, racialized experiences, and implicit ideologies and practices that constrain the development of a more equitable system of mathematics education" (p. 428). The chapter ends with implications for future research on culture, race, and power in mathematics education.

5.   Dyches, J. & Boyd, A. (2017). Foregrounding equity in teacher education: Toward a model of social justice pedagogical and content knowledge. *Journal of Teacher Education, 86,* 476–490.

    – Noting that Shulman's notion of Pedagogical Content Knowledge (PCK) significantly impacted the work of teacher preparation, these authors elaborate a model for Social Justice Pedagogical and Content Knowledge (SJPACK). The model articulates three domains of knowledge: Social Justice Knowledge, Social Justice Pedagogical

Knowledge, and Social Justice Content Knowledge. Implications for teacher preparation utilizing an SJPACK orientation are discussed.

6. Horn, I. S. (2007). Fast kids, slow kids, lazy kids: Framing the mismatch problem in mathematics teachers' conversations. *Journal of the Learning Sciences, 16*, 37–79.
   – Horn reports on the conversations among mathematics faculty at two high schools as they plan for efforts to implement research-based instructional strategies. Specifically, Horn examines the conceptual resources (i.e., beliefs, the high school mathematics teachers brought to their encounters with equity-geared reforms). Teachers were found to identify a mismatch of these equity reforms and their conceptions of students' abilities. Teachers at one school challenged these conceptions having success implementing the reforms, while teachers at the other school reworked the reforms to align with existing conceptions. In addition to this examination how teacher beliefs impact the implementation of quality instruction, the study demonstrates how teacher conceptions of students are negotiated and reified in the interactions with colleagues.

7. Langer-Osuna, J. & Esmonde, I. (2017). Insights and advances on research on identity in mathematics education. In Cai, J. (Ed.). *First Compendium for Research in Mathematics Education* (pp. 637–648). Reston, VA: National Council of Teachers of Mathematics.
   – This chapter reviews the different definitions of identity currently used in mathematics education research and explores the underlying theoretical frameworks. In particular, it addresses the lack of coherence for how the construct of identity is defined. The authors place emphasis on definitions and theories that attend to the social contexts of identity and view it as changeable over time. Four theoretical approaches to identity are discussed: discursive (post-structural), positional, narrative, and psychoanalytic, including definitions, theoretical underpinnings, and typical methodologies. Implications for mathematics teaching and learning spaces are presented for each approach.

8. Rodriguez, A. J., & Kitchen, R. S. (2005). *Preparing mathematics and science teachers for diverse classrooms: Promising strategies for transformative pedagogy*. Mahweh, NJ: Erlbaum.
   – Rodriguez and Kitchen have organized contributions from multiple authors to identify both a theoretical basis and practical strategies to support teachers learning to teach for understanding in cultural and gender-inclusive ways. The chapters offer rich narratives of mathematics and science teacher educators from across the U.S. describing their experiences with these aims, especially as they encounter resistance.

For current information on the ESJWG, please visit mtep.info/ESJWG.

Engaging Stakeholders

Another assumption underling the AMTE Standards is that "Multiple stakeholders must be responsible for and invested in preparing teachers of mathematics" (AMTE, 2017, p. 2). Consider reading this literature related to equity with others, which might involve reading this literature with mathematics educators; mathematicians; other teacher educators, particularly those in multicultural education; school administrators; or classroom teachers, particularly those that serve as cooperating or mentor teachers. To have a deeply integrated focus on equity in every program, all stakeholders need to be involved in developing a shared understanding and commitment to issues of equity in the preparation of secondary mathematics teachers.

CONCLUSIONS AND NEXT STEPS

As a working group, we have only just begun our work to investigate change ideas related to adequately attending to issues of equity and social justice in the preparation of secondary mathematics teachers. However, in this time, the ESJWG has learned a few important lessons with respect to this work and how the group might continue to build on and expand the work to engage with and support the entire MTE-Partnership.

Lessons Learned

First, the problem space for equity, diversity, and social justice work is extensive and complex—the many people, policies, and belief systems involved can make the system difficult to navigate. For stakeholders involved in the work of preparing secondary mathematics teachers, this space might be a new space, a space where they are not experts. Not only might this lack of knowledge make them uncomfortable, but also vulnerable as discussing issues of equity can often call into question the ideas and practices that have supported their own successes. Mathematics teacher educators, in particular, also feel tensions between studying issues of equity on their own and embedding issues of equity in all facets of research and practice. In addition, it can also be too easy to critique work, perspectives, or practices for not attending to issues of equity or not being equitable enough. To make progress in this complex space, it is essential that educators embrace an attitude of collective work and learning together.

Second, there are not common or shared meanings for the terms *diversity*, *equity*, and *social justice*, let alone a shared vision of what equitable mathematics teaching is. In fact, equity as a term is an empty signifier, in that it lacks an agreed upon meaning. Therefore, defining these terms for the MTE-Partnership is significant as the working group continues its work to improve the preparation of secondary mathematics teachers. Without shared definitions and vision, ESJWG runs the risk of not attending to these issues in meaningful ways.

It is clear that there is a nearly universal interest by members of the MTE-Partnership in attending to issues of equity, diversity, and social justice with regards to the preparation of secondary mathematics teachers. This interest is reflected in two ways: the desire to learn more about instructional and structural practices to address these concerns and the need for resources to do so.

NEXT STEPS

As a working group that is tasked to both support other RACs and its own activities, the ESJWG's primary challenge moving forward will be to develop a systematic way for engaging members from each of the RACs so that the working group can be mutually supportive of equity-related change activities. Particularly, the ESJWG will need to identify roles and responsibilities for members to ensure change ideas are adequately addressed as they are identified. Further, the knowledge and resources developed by ESJWG must be shared as teams continue efforts for program transformation. Thus, in subsequent PDSA cycles, the ESJWG will: survey its mathematics teacher education colleagues on the extent they align their course work to explicitly attend to the AMTE equity standards, design resources that could be used to facilitate experiences and reflections on issues of equity and social justice, measure the nature of how equity and social justice is attended to by prospective secondary mathematics teachers, and consider productive approaches on facilitating critical conversations.

The ESJWG goal of attending to equity and social justice in the preparation of future secondary mathematics teachers ultimately seeks to improve the K–16 school environment. Considering inequities that are evident in the preparation of mathematics teachers (Darling-Hammond, 2000), and the challenging aims to produce well-prepared beginning mathematics teachers (AMTE, 2017), the field of mathematics education can definitely benefit from these efforts. Notwithstanding that, prospective teachers often obtain clinical experiences under the guidance of individuals who might not have been adequately prepared, and prospective teachers are ultimately left to their own devices for refining their instructional craft with some choosing to leave the profession (Smith & Ingersoll, 2004). Thus, the ESJWG seeks to improve the nature of teacher preparation by providing resources, measurements, and means for explicitly attending to equity and social justice during enacted lessons.

ENDNOTE

1. All members of the Equity and Social Justice Working Group helped to conceptualize the paper. However, Lorraine M. Males, Ruthmae Sears, and Brian R. Lawler were lead authors, with contributions from Nancy Kress, Joshua R. Males, Julie McNamara, Farshid Safi, and Jamalee (Jami) Stone.

REFERENCES

Aguirre, J., Herbel-Eisenmann, B., Celedón-Pattichis, S., Civil, M., Wilkerson, T., Stephan, M., Pape, S. & Clements, D. H. (2017). Equity within mathematics education research as a political act: Moving from choice to intentional collective professional responsibility. *Journal for Research in Mathematics Education, 48*(2), 124–147.

Aguirre, J., Mayfield-Ingram, K., & Martin, D. B. (2013). *The impact of identity in K–8 mathematics: Rethinking equity-based practices.* Reston, VA: National Council of Teachers of Mathematics.

Allen, J. M., & Wright, S. E. (2014). Integrating theory and practice in the pre-service teacher education practicum. *Teachers and Teaching, 20*(2), 136–151.

Amobi, F. A. (2007). The message or the messenger?: Reflection on the volatility of evoking novice teachers' courageous conversations on race. *Multicultural Education, 14*(3), 2–7.

Association of Mathematics Teacher Educators (2015, September 15). *Position: Equity in Mathematics Teacher Education.* Retrieved from https://amte.net/sites/default/files/amte_equitypositionstatement_sept2015.pdf

Association of Mathematics Teacher Educators. (2017). *Standards for preparing teachers of mathematics.* Raleigh, NC: Author. Retrieved from http://amte.net/standards

Bartell, T. (2013). Learning to teach mathematics for social justice: Negotiating social justice and mathematical goals. *Journal for Research in Mathematics Education, 44*(1), 129–163.

Boaler, J., & Selling, S. K. (2017). Psychological imprisonment or intellectual freedom? A longitudinal study of contrasting school mathematics approaches and their impact on adults' lives. *Journal for Research in Mathematics Education, 48*(1), 78–105.

Boyd, A. S., & Glazier, J. A. (2017). The choreography of conversation: An exploration of collaboration and difficult discussions in cross disciplinary teacher discourse communities. *The High School Journal, 100*(2), 130–145.

Brown v. Board of Education of Topeka. 347 U.S. 483. Supreme Court of the United States (1954).

Bryk, A. S., Gomez, L., Grunow, Al., & LeMahieu, P. (2015). *Learning to improve: How America's schools can get better at getting better.* Boston, MA: Harvard Education Publishing.

Burton, L. (Ed.). (2003). *Which way social justice in mathematics education?* Westport, CT: Greenwood Publishing Group.

Butterworth, B. (2018). Mathematical expertise. In K. A. Ericcson, R. R. Hoffman, A. Kozbelt, & A. M. Williams (Eds.), *The Cambridge handbook of expertise and expert performance* (pp. 616–633). Cambridge, MA: Cambridge University Press.

Campbell, S. S., & Dunleavy, T. K. (2016). Connecting university course work and practitioner knowledge through mediated field experiences. *Teacher Education Quarterly, 43*(3), 49–70.

Castagno, A. E. (2008). "I don't want to hear that!": Legitimating whiteness through silence in schools. *Anthropology and Education Quarterly 39*(3), 314–333.

Civil, M. (2008). Language and mathematics: Immigrant parents' participation in school. In. O. Figueras, J. L. Cortina, S. Alatorre, T. Rojano, & A. Sepúlveda (Eds.), *Proceedings of the Joint Meeting of PME 32 and PME-NA XXX* (Vol. 2, pp. 329–336). México: Cinvestav-UMSNH.

Civil, M. (2016). Stem learning research through funds of knowledge. *Cultural Studies in Science Education, 11*, 41–59.

Civil, M., & Planas, N. (2010). Latino/a immigrant parents' voices in mathematics education. In E. Grigorenko & R. Takanishi (Eds.), *Immigration, diversity, and education* (pp. 130–150). New York: Routledge.

Coleman, S. T., Bruce, A. W., White, L. J., Boykin, A. W., & Tyler, K. (2017). Communal and individual learning contexts as they relate to mathematics achievement under simulated classroom conditions. *Journal of Black Psychology, 43*(6), 543–564.

Confrey, J. (2017). Research: To inform, deform, or reform? In. J. Cai (Ed.), *The compendium for research in mathematics education* (pp. 12–27). Reston, VA: National Council of Teachers of Mathematics.

Crowley, M. L. (1993). Student mathematics portfolio: More than a display case. *The Mathematics Teacher, 86*(7), 544–547.

Dabach, D. B. (2015). Teacher placement into immigrant English Learner classrooms: Limiting access in comprehensive high schools. *American Educational Research Journal, 52*(2), 243–274.

D'Abrosio, B., Frankenstein, M., Gutiérrez, R., Kastberg, S., Martin, D. B., Moschkovich, J., Taylor, E., & Barnes, D. (2013). Addressing racism. *Journal for Research in Mathematics Education, 44*(1), 23–36.

Darling-Hammond, L. (2000). Teacher quality and student achievement: A review of state policy evidence. *Education Policy Analysis Archives, 8*(1), 1–44.

de Abreu, G., Gorgorió, N., & Björklund Boistrup, L. (2018). Diversity in mathematics education. In T. Dreyfus, M. Artigue, D. Potari, S. Prediger, & K. Ruthven (Eds.), *Developing Research in Mathematics Education: Twenty Years of Communication, Cooperation and Collaboration in Europe* (pp. 211–222). New York: Routledge.

de Freitas, E. (2008). Troubling teacher identity: Preparing mathematics teachers to teach for diversity. *Teaching Education, 19*(1), 43–55.

Dossey, J. A., McCrone, S. S., & Halvorsen, K. T. (2016). *Mathematics education in the United States, 2016: A capsule summary fact book.* Reston, VA: National Council of Teachers of Mathematics.

Dyches, J., & Boyd, A. (2017). Foregrounding equity in teacher education: Toward a model of social justice pedagogical and content knowledge. *Journal of Teacher Education, 86*, 476–490.

Fasheh, M. J. (2012). The role of mathematics in the destruction of communities, and what we can do to reverse this process, including using mathematics. In O. Skovsmose & B. Greer (Eds.), *Opening the cage: Critique and politics of mathematics education* (pp. 93–106). Rotterdam, the Netherlands: Sense Publishers.

Faulkner, V. N., Stiff, L. V., Marshall, P. L., Nietfeld, J., & Crossland, C. L. (2014). Race and teacher evaluations as predictors of algebra placement. *Journal for Research in Mathematics Education, 45*(3), 288–311.

Featherstone, H., Crespo, S., Jilk, L. M., Oslund, J. A., Parks, A. N., & Wood, M. B. (2011). *Smarter together! Collaboration and equity in the elementary mathematics classroom.* Reston, VA: National Council of Teachers of Mathematics.

Felton-Koestler, M. D., & Koestler, C. (2017). Should mathematics teacher education be politically neutral? *Mathematics Teacher Educator, 6*, 67–72.

Flores, A. (2007). Examining disparities in mathematics education: Achievement gap or opportunity gap? *The High School Journal, 91*, 29–42.

Frankenberg, E., & Orfield, G. (2012). *The resegregation of suburban schools: A hidden crisis in American education.* Cambridge, MA: Harvard Education Press.

Gleason, J., Livers, S., & Zelkowski, J. (2017). Mathematics Classroom Observation Protocol for Practices (MCOP²): A validation study. *Investigations in Mathematics Learning, 9*(3), 111–129.

Grissom, J. A., Kalogrides, D., & Loeb, S. (2015). The micropolitics of educational inequality: The case of teacher-student assignments. *Peabody Journal of Education, 90*(5), 601–614.

Gutiérrez, R. (2007). (Re)defining equity: The importance of a critical perspective. In N. S. Nasir & P. Cobb (Eds.), *Improving access to mathematics: Diversity and equity in the classroom* (pp. 37–50). New York, NY: Teachers College Press.

Gutiérrez, R. (2013). The sociopolitical turn in mathematics education. *Journal for Research in Mathematics Education, 44*(1), 37–68.

Gutstein, E. (2006). *Reading and writing the world with mathematics: Toward a pedagogy for social justice.* New York: Routledge.

Healy, L., & Hoyles, C. (2000). A study of proof conceptions in algebra. *Journal for Research in Mathematics Education, 31*(4), 396–428.

Horn, I. S. (2005). Learning on the job: A situated account of teaching learning in two high school mathematics departments. *Cognition & Instruction, 23*(2), 207–236.

Horn, I. S. (2007). Fast kids, slow kids, lazy kids: Framing the mismatch problem in mathematics teachers' conversations. *Journal of the Learning Sciences, 16*(1), 37–79.

Isenberg, E., Max, J., Gleason, P., Potamites, L., Santillano, R., Hock, H., & Hansen, M. (2013). *Access to effective teaching for disadvantaged students: Executive summary* (NCEE 2013-4002). Washington, DC: National Center for Education Evaluation and Regional Assistance, Institute of Education Sciences, U.S. Department of Education.

Jackson, C., & Delaney, A. (2017). Mindsets and practices: Shifting to an equity-centered paradigm. In A. Fernandez, S. Crespo, & M. Civil (Eds.), *Access & equity: Promoting high-quality mathematics grades 6–8* (pp. 143–155). Reston, VA: National Council of Teachers of Mathematics.

Kelly, S. (2004). Are teachers tracked? On what basis and with what consequences. *Social Psychology of Education, 7*(1), 55–72.

Ladson-Billings, G. (2014). Culturally relevant pedagogy 2.0: a.k.a. the remix. *Harvard Educational Review, 84*(1), 74–84.

Langer-Osuna, J., & Esmonde, I. (2017). Insights and advances on research on identity in mathematics education. In Cai, J. (Ed.). *First Compendium for Research in Mathematics Education* (pp. 637–648). Reston, VA: National Council of Teachers of Mathematics.

Larson, M. (2016, September 15). *A renewed focus on equity.* [web log post]. Retrieved from https://www.nctm.org/News-and-Calendar/Messages-from-the-President/Archive/Matt-Larson/A-Renewed-Focus-on-Access,-Equity,-and-Empowerment/

Leonard, J., & Dantley, S. J. (2005). Breaking through the ice: Dealing with issues of diversity in mathematics and science education courses. In A. J. Rodriguez & R. S. Kitchen (Eds.), *Preparing mathematics and science teachers for diverse classrooms: Promising strategies for transformative pedagogy* (pp. 87–118). Mahweh, NJ: Erlbaum.

Linchevski, L., & Kutscher, B. (1998). Tell me with whom you're learning, and I'll tell you how much you've learned: Mixed-ability versus same-ability grouping in mathematics. *Journal for Research in Mathematics Education, 29*(5), 533–554.

Little, J., & Anderson, J. (2016). What factors support or inhibit secondary mathematics pre-service teachers' implementation of problem-solving tasks during professional experience? *Asia-Pacific Journal of Teacher Education, 44*(5), 504–521.

Louisiana State Personnel Development Grant (2018). *Equitable classroom practices observation checklist.* Retrieved from http://laspdg.org/files/Equitable%20Class-room%20Practices%20Observation%20Checklist.pdf

Mansfield, K. C. & Jean-Marie, G. (2015). Courageous conversations about race, class, and gender: Voices and lessons from the field. *International Journal of Qualitative Studies in Education, 28*(7), 819841.

Martin, D. B. (2008). E(race)ing race from a national conversation on mathematics teaching and learning: The National Mathematics Advisory Panel as white institutional space. *The Mathematics Enthusiast, 5*(2), 387–398.

Martin, D. B. (2009). Researching race in mathematics education. *Teachers College Record, 111*, 295–338.

Mathematics Teacher Education Partnership. (2014). *Guiding principles for secondary mathematics teacher preparation.* Washington, DC: Association of Public and Land-grant Universities. Retrieved from: http://mtep.info/guidingprinciples

Maxwell, L. A. (2014). U.S. school enrollment hits majority-minority milestone. *The Education Digest, 80*(4), 27–33.

Nasir, N. S., & de Royston, M. M. (2013). Power, identity, and mathematical practices outside and inside school. *Journal for Research in Mathematics Education, 44*(1), 264–287.

National Council of Supervisors of Mathematics & TODOS: Mathematics for ALL (2016). *Mathematics education through the lens of social justice: Acknowledgment, actions, and accountability.* Retrieved from https://www.todos-math.org/socialjustice

National Council of Teachers of Mathematics (2014). *Principles to actions: Ensuring mathematical success for all.* Reston, VA: Author.

National Council of Teachers of Mathematics (2018). *Catalyzing change in high school mathematics: Initiating critical conversations.* Reston, VA: Author.

National Governors Association Center for Best Practices, Council of Chief State School Officers. (2010). Common core state standards: Mathematics. Washington, DC: Author.

Oakes, J. (1987). Tracking in secondary schools: A contextual perspective. *Educational Psychologist, 22*(2), 129–153.

Oakes, J. (1994). More than misapplied technology: A normative and political response to Hallinan on tracking. *Sociology of Education, 67*(2), 84–89.

Pais, A. (2012). A critical approach to equity. In *Opening the cage* (pp. 49–91). Rotterdam, the Netherlands: Sense Publishers.

Paris, D. (2016). *On educating culturally sustaining teachers. Teaching Works.* Working papers. Ann Arbor, MI: University of Michigan. Retrieved from http://www.teachingworks.org/images/files/TeachingWorks_Paris.pdf

Parker, C., Morrell, C., Morrell, C., & Chang, L. (2016). Shifting understandings of community college faculty members: Results of an equity-focused professional development experience. *The Journal of Faculty Development, 30*(3), 41–48.

Parks, R., & King, C. S. (2007). Culture, race, power and mathematics education. In F. K. Lester (Ed.), *Second handbook of research on mathematics teaching and learning* (pp. 405–433). Reston, VA: National Council of Teachers of Mathematics.

Price, J. N., & Valli, L. (2005). Preservice teachers becoming agents of change: Pedagogical implications for action research. *Journal of Teacher Education, 56*(1), 57–72.

Reinholz, D. L., & Shah, N. (2018). Equity analytics: A methodological approach for quantifying participation patterns in mathematics classroom discourse. *Journal for Research in Mathematics Education, 49*(2), 140–177.

Reys, R. E., & Kilpatrick, J. (Eds.). (2001). *One field, many paths: US doctoral programs in mathematics education.* Washington, DC: American Mathematical Society/Mathematical Association of America.

Riley, K., & Solic, K. (2017). "Change happens beyond the comfort zone": Bringing undergraduate teacher candidates into activist teacher communities. *Journal of Teacher Education, 68*(2), 179–192.

Seah, W. T., & Andersson, A. (2015). Valuing diversity in mathematics pedagogy through the volitional nature and alignment of values. In A. Bishop, H. Tam, & T. N. Barkatsas (Eds.), *Diversity in Mathematics Education* (pp. 167–183). New York, NY: Springer.

Secada, W. G. (1989). *Equity in education.* Philadelphia: Falmer Press.

Secada, W. G. (1991a). Diversity, equity, and cognitivist research. In E. Fennema, T. P. Carpenter, & S. J. Lamon (Eds.), *Integrating research on teaching and learning mathematics* (pp. 17–53). Albany, NY: SUNY Press.

Secada, W. G. (1991b). Student diversity and mathematics education reform. In L. Idol & B. F. Jones (Eds.), *Educational values and cognitive instruction: Implications for reform* (pp. 297–332). New York: Routledge.

Secada, W. G., Fennema, E., Adajian, L. B., & Byrd, L. (Eds.). (1995). *New directions for equity in mathematics education.* Cambridge, MA: Cambridge University Press.

Simon, N. S., & Johnson, S. M. (2015). Teacher turnover in high-poverty schools: What we know and can do. *Teachers College Record, 117*(3), 1–36.

Singleton, G. E., & Linton, C. W. (Eds.) (2006). *Courageous conversations: A field guide for achieving equity in schools.* Thousand Oaks, CA: Corwin.

Slay, J. (2011). Being, becoming and belonging: Some thoughts on academic disciplinary effects. *Cultural Studies of Science Education, 6*(4), 841–844.

Smith, J. P., diSessa, A. A., & Roschelle, J. (1993). Misconceptions reconceived: A constructivist analysis of knowledge in transition. *The Journal of the Learning Sciences, 3*(2), 115–163.

Smith, T. M., & Ingersoll, R. M. (2004). What are the effects of induction and mentoring on beginning teacher turnover? *American Educational Research Journal, 41*(3), 681–714

Stein, M. K., Remillard, J., & Smith, M. S. (2007). How curriculum influences student learning. In F. K. Lester. Jr. (Ed.), *Second handbook of research on mathematics teaching and learning* (pp. 319-369). Reston, VA: National Council of Teachers of Mathematics.

Strutchens, M. E. (2000). Confronting beliefs and stereotypes that impede the mathematical empowerment of African American students. In M. E. Strutchens, M. L. Johnson, & W. F. Tate (Eds.), *In changing the faces of mathematics: Perspectives on African Americans* (pp. 7–14). Reston, VA: National Council of Teachers of Mathematics.

Suh, H., Theakston Musselman, A., Herbel-Eisenmann, B., & Steele, M. (2013). Teacher positioning and agency to act: Talking about "low-level" students. In M. V. Martinez & A. C. Superfine (Eds.), *Proceedings from the 35th annual meeting of the North American chapter of the international group for the psychology of mathematics education* (pp. 717–724). Chicago, IL: University of Illinois at Chicago.

U.S. Department of Education (2016, October 28). *2013–2014 Civil Rights data collection: A first look. Key data highlights on equity and opportunity gaps in our nation's public schools.* Retrieved from https://www2.ed.gov/about/offices/list/ocr/docs/2013-14-first-look.pdf

Vomvoridi-Ivanović, E., & McLeman, L. (2015). Mathematics teacher educators focusing on equity: Potential challenges and resolutions. *Teacher Education Quarterly, 42*(4), 83–100.

Washington, D., Torres, Z., Gholson, M., & Martin, D. B. (2012). Crisis as a discursive frame in mathematics education research and reform. In S. Mukhopadhyay & W. Roth (Eds.), *Alternative forms of knowing (in) mathematics* (pp. 53–69). Rotterdam, the Netherlands: Sense Publishers.

Zeichner, K. (2010). Rethinking the connections between campus courses and field experiences in college- and university-based teacher education. *Journal of Teacher Education, 61*(1–2), 89–99.

SECTION II

OPPORTUNITIES TO LEARN MATHEMATICS

This section includes a review of literature and presents current research endeavors focused on the mathematical preparation of teachers. Foundational research and current standards in mathematical preparation, including mathematical knowledge in and for teaching, are presented in Chapter 4. The next two chapters focus on work of two Research Action Clusters: Mathematics of Doing, Understanding, Learning, and Educating for Secondary Schools and Active Learning Mathematics, which describe work attending to mathematical knowledge in and for teaching in upper-level content courses and in the pre-calculus to calculus 2 sequence, respectively. Together, Chapters 5 and 6 attend to challenges encountered across mathematics content courses that are included in secondary mathematics teacher preparation programs.

CHAPTER 4

MATHEMATICAL PREPARATION OF SECONDARY MATHEMATICS TEACHER CANDIDATES[1]

Robert N. Ronau[2,3], David C. Webb, Susan A. Peters,
Margaret J. Mohr-Schroeder, and Eric Stade

INTRODUCTION

This chapter approaches improvement in the teaching and learning of secondary mathematics by examining the prescribed teacher knowledge of content and pedagogy in teacher preparation programs. Mathematical knowledge for teaching (MKT) will be discussed along with various historical documents and recommendations on secondary mathematics teacher preparation. In addition, this chapter will investigate the roles of mathematics departments and teacher education programs in the preparation of mathematics teachers through the examination of two surveys. This inquiry was guided by three goals: (a) to determine how teacher education programs should address secondary mathematics teacher knowledge, (b) to take stock of how programs currently address secondary mathematics teacher content knowledge, and (c) to suggest strategies for moving from the latter to the former. The discussion will be focused by the following questions: What knowledge is needed by secondary mathematics teachers? What are current recommendations for secondary mathematics teacher preparation? What

The Mathematics Teacher Education Partnership: The Power of a Networked Improvement Community to Transform Secondary Mathematics Teacher Preparation, pages 91–118.
Copyright © 2020 by Information Age Publishing
All rights of reproduction in any form reserved.

is the present state of secondary mathematics teacher preparation with regard to these recommendations? What actions might the Mathematics Teacher Education Partnership (MTE-Partnership) take to improve the mathematical preparation of secondary mathematics teachers to better align it with these recommendations?

REVIEW OF LITERATURE

The mathematical preparation of teachers has focused historically on differing interpretations of mathematics teacher content knowledge and how such knowledge is developed. This review of literature traces the development of this construct and the more contemporary articulation of mathematical knowledge for teaching. A summary of policy and research recommendations for mathematical preparation, followed by the role of teacher preparation programs in developing mathematical content knowledge, is then discussed.

What Knowledge is Needed by Secondary Mathematics Teachers?

Teacher knowledge is essential to student learning. As Darling-Hammond (2000) observes, "Student learning depends substantially on what teachers know and can do" (p. 10). But what does this mean, in the context of secondary mathematics teaching and learning? What does this dependence look like? What *should* secondary mathematics teachers know and be able to do with regard to mathematics?

The Development of Pedagogical Content Knowledge as a Construct. Teacher knowledge has long been a topic of interest to scholars, policymakers, and groups involved in teacher preparation. Early teacher education programs in the United States were based on the assumption that effective teaching requires knowledge of more than general content or pedagogy on the part of teachers (Donaghue, 2003). However, initial research on relationships between teachers' knowledge and student achievement in mathematics (e.g., Begle, 1979; Monk, 1994) produced contradictory findings regarding such relationships (e.g., Begle, 1972; Boardman, Davis, & Sanday, 1977; Eisenberg, 1977; Hanushek, 1972). Most of these studies, though, focused primarily on mathematical content knowledge (CK) *per se*, thus failing to consider the nature of knowledge used in teaching mathematics.

Later researchers took a more nuanced view of teacher knowledge and considered additional aspects of teacher knowledge such as curriculum knowledge and knowledge of students and their thinking (e.g., Shulman, 1987). Of particular note is Shulman's *pedagogical content knowledge* (PCK), which "embodies the aspects of content most germane to its teachability" (1986, p. 9). This notion of PCK came about from recognition that CK and general knowledge of teaching (i.e., pedagogical knowledge) alone are not sufficient for effective teaching; teachers need a combination of knowledge of content and pedagogy to teach mathematics to diverse learners (Shulman, 1987). In Shulman's 1987 formulation, PCK includes knowledge of powerful examples and demonstrations, useful representations of mathematical ideas, students' conceptions and preconceptions, and related teaching strategies.

The notion of PCK as a crucial feature of mathematics teacher knowledge resonated with many scholars and inspired other conceptualizations of teacher knowledge, such as technological pedagogical content knowledge (Mishra & Koehler, 2006; Niess, 2005); mathematical knowledge for teaching (Ball, Lubienski, & Mewborn, 2001; Hill, Schilling, & Ball, 2004); the Knowledge Quartet (Rowland, Huckstep, & Thwaites, 2005); and the mathematics competencies model of the Teacher Education and Development Study in Mathematics (Tatto et al., 2012). Describing the knowledge needed to teach mathematics has proven to be difficult, and the field lacks a single conception of what CK and PCK that teachers need.

Broadly, all frameworks for teacher knowledge indicate a general consensus among mathematics education researchers that mathematics teachers rely on multiple types and uses of specialized knowledge, including both CK and PCK, to do their work in the classroom effectively (Ball, Thames, & Phelps, 2008; Baumert et al., 2010; McCrory et al., 2012; Mishra & Koehler, 2006; Mohr-Schroeder et al., 2017). With respect to PCK, mathematics education researchers agree on the critical importance of teachers knowing how students think about mathematics as well as knowing how to represent and teach mathematics in ways that are meaningful for students. With regard to CK, although teachers need solid understandings of the mathematics they will teach (Ball et al., 2008), content knowledge alone may not be sufficient for successful mathematics teaching (Begle, 1972, 1979). Furthermore, research indicates teachers need to understand mathematics beyond that which they teach and must be able to engage in doing mathematics themselves (Hill, Ball, & Schilling, 2008).

Mathematical Knowledge for Teaching The mathematical knowledge for teaching (MKT) framework, articulated by Ball, Thames, and Phelps (2008), extends and refines the knowledge categories, CK and PCK introduced by Shulman in the 1980s. Ball and colleagues focused on two categories of content knowledge (Subject Matter Knowledge in Figure 4.1): Common Content Knowledge (CCK), mathematical knowledge and skill commonly used in area other than teaching, and Specialized Content Knowledge (SCK), knowledge commonly used in teaching. For example, calculating the mean test score of 60 from a set of four scores (0, 60, 80, 100) would involve common content knowledge (CCK), $(0 + 60 + 80 + 100)/4 = 60$; whereas determining a solution method that yields a mean of 80 that results from correctly calculating the mean after ignoring the data value of 0 score—a common student error—would involve specialized content knowledge (SCK), $(60 + 80 + 100)/3 \neq 60$.

According to Ball and colleagues (2008), PCK includes knowledge of students' thinking about mathematics; knowledge of powerful representations and pedagogy for teaching mathematics; and knowledge of curricula and instruc-

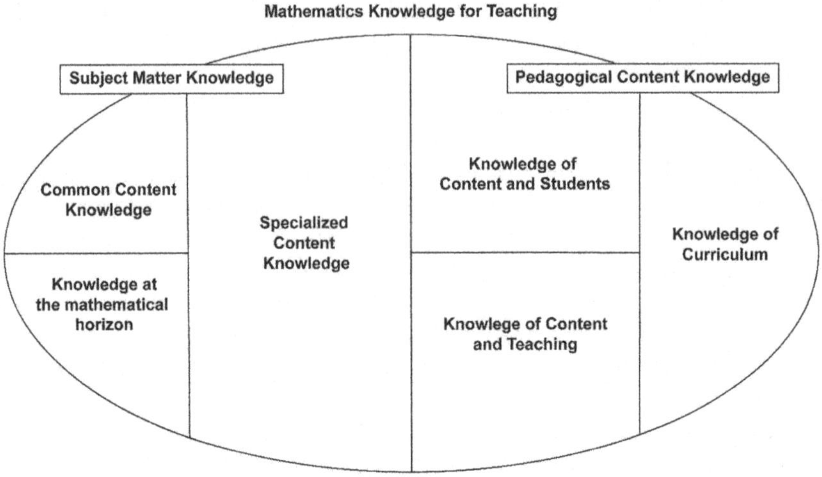

Mathematics Knowledge for Teaching

FIGURE 4.1. Mathematics Knowledge for Teaching Framework (Ball et al., 2008, p. 403). This model of Mathematics knowledge for teaching breaks apart Shulmans (1986) model of Content Knowledge (Labeled Subject Matter Knowledge) and Pedagogical Content Knowledge into specific categories. The model provides a more detailed map to more effectively study whether there are aspects of teachers' content knowledge that may predict students' achievement more than others.

tional materials for mathematics, including awareness of alignment within grades (for mathematics and other subjects) and across grades (for mathematics). For example, recognizing that students often fail to consider data values of 0 when analyzing data (Strauss & Bichler, 1988) would involve knowledge of students' thinking about mathematics. Designing instructional activities with appropriate representations to shift students' focus from individual data values, such as 0, to distributions as aggregate collections of data (e.g., Ben-Zvi & Arcavi, 2001) would involve knowledge of pedagogies for teaching exploratory data analysis. And further, sequencing instruction to scaffold students' development of various understandings of mean, from mean as a fair share to mean as a balance point, and selecting appropriate materials to facilitate the process, would involve knowledge of mathematics curricula and instructional materials.

Frameworks and Empirical Studies of the Knowledge Needed to Teach Mathematics. Along with the development of frameworks to posit components of teacher knowledge, researchers are gaining valuable insights into the knowledge needed to teach secondary mathematics effectively, though falling short of establishing the statistical significance of CK and PCK components. A study using the MKT framework found a significant correlation between measurements of elementary teachers' MKT and students' mathematical achievement gains (Hill,

Rowan, & Ball, 2005). However, correlations between student scores and measures of MKT subcategories or combinations of subcategories have been less evident (Hill, Ball, & Schilling, 2008; Schilling, Blunk, & Hill, 2007). Still, these subcategories illustrate the complexity of knowledge needed for teaching mathematics; moreover, further analysis has the potential to facilitate better isolation of their effects. Research related to MKT has largely focused on the knowledge of elementary teachers (e.g., Ball et al., 2001, 2008; Ma, 1999) or middle-level teachers (e.g., Hill, 2007; Izsák, Orrill, Cohen, & Brown, 2010; Shechtman, Roschelle, Haertel, & Knudsen, 2010; Tchoshanov, 2011); it has not been as robust in extending these investigations into the knowledge needed by secondary mathematics teachers to improve student achievement at the secondary level.

However, several research programs have been undertaken using similar, yet distinctly different frameworks, to examine the knowledge needed by secondary mathematics teachers to improve student achievement at the secondary level. For example, TEDS-M researchers studied the knowledge of teacher candidates preparing for elementary or lower secondary certification.[4] The TEDS-M framework for teacher knowledge focused on teachers' understanding of mathematics at the certification level and mathematical connections to higher levels (Döhrmann, Kaiser, & Blömeke, 2012). Specialized content knowledge was not included in the framework; however, cognitive domains of knowing, applying, and reasoning that respectively include mathematical work such as defining and enacting algorithms, problem solving, and proving were included. The TEDS-M conceptualization of PCK spans MKT categories related to students and teaching: (a) curricular and planning knowledge and (b) knowledge of enacting mathematics for teaching and learning (Döhrmann et al., 2012). Scores from prospective lower-secondary mathematics teachers in the U.S. also revealed considerable need for teacher preparation that better develops CK and PCK (Tatto et al., 2012).

Researchers from the Cognitive Activation in the Classroom (COACTIV) project investigated connections between German secondary mathematics teachers' CK and PCK, as well as associations of this knowledge with their students' learning gains (Baumert et al., 2010; Krauss, Baumert, & Blum, 2008; Krauss et al., 2008). In contrast to TEDS-M, COACTIV researchers conceptualized the CK needed to teach mathematics as including specialized content knowledge. The COACTIV group's conceptualization of PCK included three important types of knowledge: (a) knowledge about the mathematical learning potential of tasks; (b) knowledge of typical student misconceptions, struggles, and thinking; and (c) knowledge of powerful mathematics-specific representations and explanations. COACTIV researchers found empirical support for CK and PCK as distinct components of teacher knowledge (Krauss et al., 2008). Their findings also suggest that strong CK may facilitate the construction of PCK in ways that low levels of CK could not (Krauss et al., 2008). COACTIV studies also found that PCK has greater predictive power for student learning than CK (Baumert et al., 2010).

A framework for secondary mathematics teacher knowledge specific to algebra was developed by the Knowledge of Algebra for Teaching (KAT) project (McCrory et al., 2012). The KAT framework has two main categories: (a) CK and (b) its uses in teaching. The KAT conceptualization of knowledge for teaching includes the knowledge of school mathematics to be taught, knowledge of advanced mathematics relevant to that school mathematics, and knowledge of teaching school mathematics. This last subcategory is similar to specialized content knowledge (Hill et al., 2008). Uses of mathematical knowledge in teaching have three classifications: decompressing, trimming, and bridging (McCrory et al., 2012). *Decompressing*, or "unpacking" (Ball & Bass, 2000a), involves making tacit knowledge explicit. This use of mathematical knowledge might occur, for example, in connection with algorithms such as helping students understand "moving over" in an algorithm for multi-digit multiplication as an abbreviation for multiplying by 10. *Trimming* involves teaching concepts in ways suited to students' knowledge while maintaining mathematical integrity (Ball & Bass, 2000b) such as teaching with awareness that instantaneous and average rate of change are the same for linear functions, but not for all functions. *Bridging* connects content across topics and courses (e.g., connects base-10 expansions of natural numbers with polynomials in x in normal form). Although the KAT researchers clearly focused on CK, they also focused on PCK necessary for decompressing, trimming, and bridging in ways that enhance students' development of mathematical understanding. Measures of the KAT constructs have been developed, and analyses of KAT assessment data provide support for the multidimensional nature of knowledge of algebra for teaching without providing strong support for the specific structure of the KAT framework (Reckase, McCrory, Floden, Ferrini-Mundy, & Senk, 2015).

The Geometry Assessments for Secondary Teachers (GAST) project (Mohr-Schroeder et al., 2017) developed a framework for secondary mathematics teacher knowledge specific to geometry. The GAST framework combined aspects of the KAT framework (McCrory et al., 2012), the MKT framework (Ball et al., 2008; Hill, Rowan, & Ball, 2005), and cognitive complexity frameworks identified for the Diagnostic Teacher Assessments for Mathematics and Science (Saderholm, Ronau, Brown, & Collins, 2010). Additionally, the project team used depth of knowledge (Webb, 1997, 1999) and knowledge levels described in practice by the Trends in International Mathematics and Science Study (TIMSS) 2007 and 2011 assessment frameworks (Mullis et al., 2005; Mullis & Martin, 2009) to define geometry knowledge at three hierarchical cognitive complexity levels: *knowing* (e.g., recalling computing), *applying* (e.g., representing, solving), and *reasoning* (e.g., synthesizing, justifying). The GAST Knowledge for Teaching Geometry assessment framework and blueprint comprised: (a) knowledge of school geometry, (b) knowledge of advanced geometry, and (c) geometry pedagogical content knowledge. *Knowledge of school geometry* includes the geometry content typically found in secondary geometry curricula and taught in secondary schools. *Knowledge of advanced geometry* includes geometry content taught in postsecondary geometry courses and required of secondary mathematics teachers. *Geometry pedagogical content knowledge* includes the knowledge of pedagogies and practices needed to carry out the work of teaching high school geometry, including using appropriate technology in instruction, identifying valid student reasoning, and assisting students in identifying and overcoming reasoning errors. GAST researchers found that teachers' knowledge for teaching geometry had a positive, statistically significant impact on student achievement, which is consistent with COACTIV findings. GAST and COACTIV provide evidence that both strong content knowledge and pedagogical content knowledge seem to be necessary for teachers to impact student knowledge positively.

Finally, the Mathematical Understanding for Secondary Teaching framework consists of three perspectives: mathematical proficiency, mathematical activity, and the mathematical context of teaching (Kilpatrick et al., 2015). *Mathematical proficiency* includes teachers' deep understanding of and fluency with the mathematics that they teach and with elementary and college mathematics. *Mathematical activity* involves the doing of mathematics and includes three integrated strands (noticing, reasoning, and creating) that collectively involve individual and integrated activities such as observing mathematical structure, generalizing and justifying, and representing and defining. The third category, *the mathematical context of teaching*, involves the knowledge needed for teachers to facilitate students' development of mathematical proficiency. Mathematical proficiency and mathematical activity span other characterizations of CK, such as common and specialized content knowledge, and other characterizations of engaging in mathematics, such as knowing, applying, and reasoning. Similarly, the mathematical work of teaching spans PCK categories such as: accessing and understanding students' mathematical thinking; knowing and using curricula; and reflecting on mathematics teaching practice. Reflection on teaching assists researchers in characterizing mathematical understanding for secondary teaching as dynamic and subject to growth over the course of a teacher's career.

Firm agreement on the importance of both CK and PCK for secondary mathematics teachers is evident among these frameworks and their results. Collectively, these frameworks offer insights into important aspects of knowledge, including CK and PCK, that teacher preparation programs should strive to develop in prospective secondary mathematics teachers.

SECONDARY MATHEMATICS TEACHER PREPARATION: HISTORICAL DOCUMENTS, RECOMMENDATIONS, AND STANDARDS

Two main threads regarding secondary mathematics teacher preparation are discussed in this section. The first thread focuses on documents and recommendations for the teaching of secondary mathematics as articulated in content standards such as the *Common Core State Standards for Mathematics* (*CCSS-M*; National Governors Association Center for Best Practices & Council of Chief State School

Officers, 2010) and reports from the National Council of Teachers of Mathematics (NCTM). The second thread focuses on recommendations for mathematics majors and secondary mathematics teacher preparation programs. Together, these two threads offer differing but related perspectives regarding necessary foci for prospective secondary mathematics teachers.

Historically, prospective high school teachers studied the same content that mathematics majors studied (Ball & McDiarmid, 1989; Ferrini-Mundy & Findell, 2010). An unprecedented coordinated effort to address the mathematics preparation of K–12 students was launched with the NCTM report *An Agenda for Action: Recommendations for School Mathematics of the 80s* (1980). This report was followed by other recommendations from NCTM for changes to the school mathematics curriculum such as the *Curriculum and Evaluation Standards for School Mathematics* (NCTM, 1989) and for changes in teaching practices, as described in the *Professional Standards for Teaching Mathematics* (NCTM, 1991). At the turn of the 21st century, NCTM updated the curriculum standards in the *Principles and Standards for School Mathematics* (*PSSM*; NCTM, 2000). Soon after NCTM's publication of the *PSSM*, in 2007 the American Statistics Association released its *Guidelines for Assessment and Instruction in Statistics Education* (GAISE, Franklin et al., 2007). In 2010 the National Governors Association and the Council of Chief State School Officers joined forces and released the *CCSS-M*, which addressed K–12 mathematical content and practices.

The second set of policy documents takes into account the context of undergraduate mathematics and secondary teacher preparation programs to make recommendations regarding coursework and experiences for those preparing to teach mathematics. *The Mathematical Education of Teachers* (*MET*; Conference Board of the Mathematical Sciences [CBMS], 2001) strongly advocated for the co-participation of mathematics faculty and mathematics education faculty in mathematics education. With respect to teacher preparation in secondary mathematics, *MET* (CBMS, 2001) states:

> substantial mathematical understanding is needed even to teach whole number arithmetic well…. Middle grades curricula are even more demanding; for example, the structure of the rational numbers and the idea of proportionality require even more knowledge of teachers. High school mathematics is often considered more substantive than the mathematics of earlier grades, *but the challenges of developing a knowledge of it for teaching are often unacknowledged.* (p. xiii, emphasis added)

CBMS' *The Mathematical Education of Teachers II (MET II)*, released in 2012, reflects a more current understanding of these challenges and the need to address the mathematical content and practices requisite to building the knowledge that a secondary mathematics teacher needs to be effective in the era of the *CCSS-M*. In particular, it describes the mathematical knowledge that secondary teachers need beyond what is taught in the secondary mathematics curriculum. In 2015, the American Statistical Association released *Statistical Education of*

Teachers (*SET*; Franklin et al., 2015), which draws upon similar arguments to recognize the increasing role of probability and statistics in the 21st century and highlights subtle differences needed in preparation for teaching statistics based in differences between mathematics and statistics. The Association of Mathematics Teacher Educator's (2017) *Standards for the Preparation of Teachers of Mathematics* (AMTE Standards) builds from *MET, MET II*, and *SET* to further complete the picture of what robust secondary mathematics preparation programs should consider for preparing prospective teachers. These resources provide both a synthesis of research in mathematics and statistics education and recommendations from research mathematicians, statisticians and mathematics educators. We begin first with the recommendations outlined in *MET II*.

MET II: Recommendations for Secondary Teacher Preparation

The *MET II* report was intended as an update of the earlier MET report. It reiterates and elaborates themes articulated in *MET*:

- There is intellectual substance in school mathematics.
- Proficiency with school mathematics is necessary but not sufficient mathematical knowledge for a teacher.
- The mathematical knowledge needed for teaching differs from the mathematical knowledge needed for other mathematics-related professions.
- A teacher's knowledge for teaching mathematics can and should grow throughout a teacher's career.

Several recommendations of *MET II* encourage those involved in teacher preparation to rethink mathematics courses and, in particular, courses required for teachers, as too often the latter "emphasize preparation for graduate study or careers in business rather than advanced perspectives on the mathematics that is taught in high school" (CBMS, 2012, p. 5). The report provides suggestions for how its recommendations might be implemented in different types of institutional contexts.

MET II makes four overarching recommendations regarding what prospective mathematics teachers need to know. These recommendations include ensuring that prospective teachers complete mathematics courses that will develop their understanding of the mathematics they teach, and engaging prospective teachers in mathematical reasoning, explaining, sense making, and related habits of mind. *MET II* also recommends that teachers have opportunities to continue their development of mathematical knowledge and that preparation courses and professional development should model flexible, interactive styles of teaching that will enable teachers to develop these habits of mind in their students (CBMS, 2012).

With respect to the preparation of middle grades mathematics teachers, *MET II* states, "First and foremost, future teachers need courses that allow them to delve into the mathematics of the middle grades while engaging in mathematical practice as described by the CCSS" (CBMS, 2012, p. 46). More specifically, *MET II* recommends that prospective middle grades mathematics teachers take at least 24 credit hours of college-level mathematics. (The meaning of "college-level" is spelled out: "In no case should a course at or below the level of precalculus be considered a part of these 24 semester-hours" [p. 46].) *MET II* recommends that at least 15 of these 24 credit hours should comprise courses specifically designed for future middle grades teachers: Number and Operations (6 credit hours), Geometry and Measurement (3 credit hours), Algebra and Number Theory (3 credit hours), and Statistics and Probability (3 credit hours).

With respect to the preparation of high school mathematics teachers, *MET II* emphasizes that, particularly at the high-school level, even the most complete adherence to its college-level coursework recommendations does not obviate the need for further mathematics. As the authors write, "A reasonable goal for initial certification at this level is to create beginning teachers who are able to teach competently a portion of the high school curriculum and who are prepared to learn throughout their careers from their teaching and professional development experiences" (CBMS, 2012, p. 19).

K–12 Mathematics Standards

As outlined earlier, since the late 1980s a number of content standards have been produced that discuss the teaching of mathematics. These content standards include implications for the preparation of secondary mathematics teachers. This section will discuss only a few of these content standards documents and how they influenced understanding of the mathematics content that secondary mathematics should know to teach mathematics today.

The *Curriculum and Evaluation Standards for School Mathematics* (NCTM, 1989) initiated a national discussion on the recommended mathematics content for K–12 students. Drawing from contemporary theories of learning, NCTM standards and related publications that focused on teaching (NCTM, 1991) and assessment (NCTM, 1994) made explicit reference to shifts in the role of the teacher and students in classrooms that should be more student-centered. These shifts in classroom norms and roles, ideally, would position students as engaged learners in the construction and articulation of mathematical skills, concepts, and relationships. To achieve this vision for teaching mathematics, these proposals for curriculum, instruction, and assessment also recognized the need for mathematical preparation that would result in a more profound understanding of fundamental mathematics (Ma, 1999). For teachers to orchestrate instruction that takes into account students' reasoning about mathematical ideas, teachers would need to be more familiar with how students learned mathematics and how teachers could use tasks and instructional moves to develop the mathematical practices of students. When NCTM updated the *PSSM* in 2000, these documents greatly influenced the development of the *CCSS-M* (2010).

The *CCSS-M* have two major components:

1. The Standards for Mathematical Content, which describe what mathematics K–12 students should learn, understand, and know, at each grade level.
2. The Standards for Mathematical Practice, which describe how students are to engage with this mathematics in increasingly complex and sophisticated ways as they progress through the grades.

Influenced by *MET II* and by the *PSSM*, the *CCSS-M* organize middle grades mathematics[5] into domains and high school mathematics into conceptual categories as shown in Table 4.1. The *CCSS-M* detailed these content groups to demonstrate how specific topics in one grade level, within these content groups, would build capacity to address related topics in the next grade levels.

The Standards for Mathematical Practice were emphasized in the *CCSS-M* to the degree that this list of standards was placed on the opening page for every grade level. The eight Standards for Mathematical Practice emphasize the need to focus on problem solving, mathematical argumentation, and modeling with mathematics. These process standards also were recommended formally in NCTM's curriculum standards (1989, 2000). Some additions to this list of recommended practices include attending to precision, attending to and making use of structure, and recognizing and expressing regularity in repeated reasoning (National Governors Association Center for Best Practices & Council of Chief State School Officers, 2010).

In specifying the mathematical practices that students should acquire, the *CCSS-M* tacitly acknowledge what has been described above—that acquisition of mathematical content knowledge in the narrowest sense of the term may be insufficient for the use and application of mathematics. Together with the content standards, the standards for practice outline a school mathematics that may differ from what prospective teachers have learned. For example, rather than listing collections of special-purpose techniques, such as simplifying radicals or completing the square, the *CCSS-M* give greater attention to the principles that underlie them, allowing these and other specialized techniques to be seen as consequences of properties of operations (the *CCSS-M* term for the field axioms). In high school,

TABLE 4.1. CCSS-M Organization of Mathematics Content for Grades 6–12

Grade 6	Grade 7	Grade 8	High School
Ratios and Proportional Relationships			
The Number System			Number and Quantity
Expressions and Equations			Algebra
		Functions	
			Modeling
Geometry and Measurement			
Statistics and Probability			

proof need not be restricted to its traditional confines of axiomatic Euclidean geometry and trigonometric identities but can occur in other conceptual categories as a more mature form of the practice of constructing viable arguments in earlier grades.

Statistical Education of Teachers

SET was commissioned by the American Statistics Association and aligns with the recommendations from the *Guidelines for Assessment and Instruction in Statistics Education* (*GAISE*; Franklin et al., 2007) and *CCSS-M* as well as engages students in the practices and processes articulated in each. In concert with recommendations from NCTM and *MET II*, *SET* recommends the need for technology to develop concepts and analyze data such as using computer simulations to approach abstract concepts such as *p*-values and determine whether differences between experimental groups are significant. *SET* recommends that prospective teachers develop a deep conceptual understanding of statistics through activities that embed them in the statistical problem-solving process of formulating statistical questions, collecting data, analyzing data, and interpreting results in the context of data. *SET* recommendations also address the need for robust professional development for in-service teachers to develop the habits of mind of a statistical thinker and problem solver. Finally *SET* calls for institutions and statisticians to recognize the importance of emphasizing statistical reasoning at all levels of education and thus the critical importance of supporting it in teacher preparation programs.

The preceding recommendations in statistics education call for teachers' development of more than cursory knowledge of statistical content. As a result, they call for explicit attention to instruction that develops a deeper understanding of statistics among current and prospective teachers in ways that are consistent with the descriptions of MKT and the recommendations of *MET II* and that develop the unique aspects of knowledge related to statistics—statistical knowledge for teaching (Groth, 2007; 2013). Elsewhere in *SET* the *CCSS-M* mathematical practices are exemplified in the context of statistics education, both to demonstrate the consistencies between mathematics and statistics and to highlight the unique nature of the practices for doing statistics. Because statistics is a mathematical science that relies heavily on mathematics, the statistical preparation of teachers necessarily includes mathematical aspects; however, because statistics has distinct differences from mathematics, the statistical preparation of teachers also should reflect those differences.

Standards for the Preparation of Teachers of Mathematics

The AMTE Standards (2017) is one of the more recent attempts to define the knowledge that is needed by secondary mathematics teachers. At the opening of Chapter 2, the following guiding question is offered:

> Recognizing that learning to teach is an ongoing process over many years, what are reasonable expectations for the most important knowledge, skills, and dispositions that beginning teachers of mathematics must possess to be effective? (p. 7)

In this discussion of the AMTE Standards we focus primarily on an overview of the four standards that encompass these knowledge, skills, and dispositions. Each of these standards are further subdivided into anywhere from three to six indicators, that are further discussed in the contexts of early childhood, upper elementary, middle school, and high school instruction. To summarize the AMTE recommendations, these standards and their implications for secondary mathematics teacher preparation are discussed below.

The AMTE Standards focus primarily on defining well-prepared prospective teachers as having robust understandings of mathematics and statistics; demonstrating practices that promote Standards for Mathematical Practice and students' productive dispositions toward mathematics; and awareness of the historical and political realities of mathematics education so that they can serve as advocates for each and every student. These recommendations can be broadly characterized as focusing on MKT and the social context of schooling. One notable difference with the AMTE Standards' recommendations is the consistent mention of K–12 student dispositions. To address student dispositions, teachers need an emergent awareness of the role of student affect and attitudes toward mathematics. Well-prepared prospective teachers need to attend to student dispositions as they plan and enact instruction. Dispositions, an often unattended to psychological dimension of productive teaching practice, have been discussed (or more appropriately, lamented) by mathematicians and educators for over 100 years (cf. Perry, 1901; Polya, 1945). Teacher knowledge of the history of socio-historical aspects of mathematics education is necessary to understand how to interpret and enact content goals and Standards for Mathematical Practice with future generations of students in a way that values students' mathematical reasoning, insights, and contributions.

In Chapter 7 of the AMTE Standards (2017), there is additional elaboration of these standards (and others) for future secondary mathematics teachers. This chapter includes a strong recommendation for multiple mathematics courses designed to specifically develop teachers' deeper understanding of MKT:

> Simply providing a mathematics degree without attending to the specific needs of candidates preparing to teach high school mathematics will not suffice. Effective programs include the equivalent of three content courses specifically designed for teachers of high school mathematics, three mathematics-specific methods courses… and [relevant] clinical experiences. (p. 144)

Similar recommendations are made for significant support and experience with various instructional technologies, including the use of spreadsheets, dynamic geometry software, and tools that support statistical simulations. Overall, the AMTE Standards represent an amalgamation of recommendations emphasized in the various iterations of the content standards and secondary teacher preparation documents. To better understand how the mathematics education community has taken up these various recommendations, the next section summarizes results from two surveys of teacher education programs.

ALIGNMENT BETWEEN TEACHER EDUCATION PROGRAMS AND NATIONAL RECOMMENDATIONS: A REPORT OF TWO SURVEYS

The university context for the preparation of secondary mathematics teachers provides a unique opportunity to simultaneously address the above recommendations. As undergraduates seek to better understand their identities and future pursuits, a combination of clinical experiences and coursework in mathematics and education can be explored to better understand the expectations for decision making and problem solving that prospective teachers will later face. Even though the professional terrain of mathematics teaching can be complex, prospective teachers can better understand how to negotiate this complexity as they learn mathematics, observe others teach mathematics, and become better acquainted with various ways to coordinate, make sense of, and influence mathematical learning. To better understand the role of teacher preparation programs in accomplishing these goals, this section discusses findings from two surveys conducted by Newton, Maeda, Alexander, and Senk (2014) and the MTE-Partnership (2014b) to highlight the expected learning for mathematics content, mathematical practices, and professional knowledge.

Teacher Education MET II Coursework Alignment Survey 2012

In 2012, Newton, Maeda, Alexander, and Senk (2014) conducted a survey of teacher education programs to determine their alignment with four *MET II* recommendations for teacher candidates in mathematics teacher preparation programs: (a) mathematics courses that develop a solid understanding of school mathematics; (b) coursework that engages candidates in reasoning, explaining, and making sense of mathematics; (c) continual in-service professional growth in mathematics; and (d) mathematics courses that develop the habits of mind of a mathematical thinker and problem solver. The survey addressed middle- and high-school programs separately. In the discussion that follows the Teacher Education *MET II* Coursework Alignment Survey will be referred to as the Newton survey.

MTE-Partnership Teacher Education Coursework Alignment Survey 2014

A second survey was developed by the MTE-Partnership (2014b) to determine the nature of secondary mathematics teacher education content preparation in MTE-Partnership institutions. The MTE-Partnership survey questions were guided by the *CCSS-M* and *MET II* recommendations. The survey was sent to the mathematics departments of MTE-Partnership institutions, requesting one response per department or institution. Surveys were sent to 66 institutions, with 28 responses: 85% of respondents resided in mathematics departments, and 81% of respondents were tenure-track faculty. The survey was designed to answer three overarching questions:

1. What mathematical knowledge is deemed necessary for prospective secondary mathematics teachers?
2. What MKT is important?
3. How do the answers to 1 and 2 differ from our current aims of achieving articulated CK and PCK?

The first section of the MTE-Partnership survey addressed specific mathematics courses for prospective secondary mathematics teachers and their alignment with mathematics courses required for other mathematics-intensive majors. The following three questions were asked: Was this course required for all math majors? Do prospective secondary mathematics teachers take the course? Do students besides prospective secondary mathematics teachers take the course?

Similarities and Differences between the Two Surveys

The MTE-Partnership survey differed from the Newton Survey in that the MTE-Partnership survey used a single category of secondary rather than two categories for middle school and high school. The Newton survey addressed middle- and high-school program course requirements, whereas the MTE-Partnership survey compared the courses required for secondary mathematics teacher candidates with courses required for mathematics majors. The two surveys offer slightly different lenses into the courses required by mathematics teacher programs (see Table 4.2). In addition, the Newton survey focused exclusively on mathematics content courses and mathematics methods courses required in mathematics teacher education programs, whereas the MTE-Partnership survey delved more deeply into details about these courses. The MTE-Partnership survey addressed who taught these courses, what freedom instructors had to alter course curricula, and how mathematics courses in the teacher preparation program aligned with the Standards for Mathematical Practice. With respect to the latter, three questions guided the inquiry: Should the mathematical practice be emphasized in the program? Is the practice explicitly taught in the program (i.e., should instruction be designed to include this practice)? Is the practice explicitly assessed in the program (i.e., should assessment be designed to measure acquisition of this practice)? The MTE-Partnership survey also addressed the alignment of the teacher preparation program with prospective teachers' development of the four *MET II* lenses on mathematics: as scholars, educators, mathematicians, and teachers. For each, the survey asked: Should the lens be emphasized in the program? Is the lens explicitly taught in the program? Is the lens explicitly assessed in the program?

Survey Results

Required Mathematics Courses. The Newton survey suggests that courses required in secondary mathematics education programs meet the *MET II* guidelines for mathematics course hours. Although the MTE-Partnership survey revealed that middle-school programs typically required the mathematics courses recommended in *MET II* (e.g., Number Theory, History of Mathematics, or Functions and Modeling), it revealed little difference in mathematics preparation between mathematics-intensive majors and secondary mathematics teacher majors. On the other hand, some institutions reported courses with special sections for teachers such as Linear Algebra, Abstract Algebra, Discrete Mathematics, Probability and Statistics, and Reasoning and Proof. Only eight programs (10%) reported meeting the 9 credit hours of high school mathematics from an advanced perspective.

The percentage of institutions requiring Calculus, Linear Algebra, and Introduction to Proof/Abstract Mathematics in their high-school programs differed only by a few percentage points. A few key differences in course requirements were found in the areas of geometry, modeling, and statistics (Table 4.2). In addition, all 64 (100%) of the middle-school program respondents reported requiring at least 24 hours of mathematics courses and at least 9 hours of mathematics courses specifically designed for middle grades teachers; however, none of the institutions reported meeting the *MET II* recommendations of 15 credit hours of mathematics for middle grades teachers or providing at least two middle-grades methods courses. High-school program respondents ($n = 78$) reported meeting six coursework recommendations: a three-course calculus sequence (81%), an introductory statistics course (88%), and one introductory linear algebra course (97%). Of the 64 middle-school and 78 high-school programs surveyed, 62 were combined middle and secondary programs. Notably, this study revealed that programs combining middle- and high-school mathematics teacher education into a single program failed to meet *MET II's* recommendations for mathematics courses designed specifically for teachers. Eight programs (10%) reported that they met the 9 credit hours of high school mathematics from an advanced perspective.

Comparing Newton and MTE-P survey results revealed considerable agreement about required courses. Programs differed in the required courses for Abstract Algebra, Advanced Calculus, Calculus, Introduction to Proofs/Abstract Mathematics, and Linear Algebra by less than 10%. The largest differences in required courses were found for courses focused on functions, modeling, geometry, and history of mathematics.

Course Instructors, Curricular Freedom, and Inclusion of Pedagogical Issues. The MTE-Partnership survey overwhelmingly found that mathematics courses were taught by tenure-track faculty rather than non-tenure-track faculty (fixed-term and adjunct faculty), graduate students, or other instructors: Calculus (64%, $n = 16$ of 25), Introduction to Proofs (85%, $n = 17$ of 20), Discrete Mathematics (88%, $n = 21$ of 24), Geometry (88%, $n = 23$ of 26), and Functions and Modeling (82%, $n = 14$ of 17). Moreover, at most institutions, most of these

TABLE 4.2. Percent of Mathematics Courses Required by Institution Programs for Mathematics Majors and Mathematics Teacher Education Programs

| | MTE-Partnership DATA | | Newton DATA [3] | |
| | Secondary | | Teacher Prep | |
Required Courses	**Majors**	**Teacher Prep**	**Middle Grades[1]**	**High School**
Abstract Algebra	63%	65%		78%
Advanced Calculus/Analysis	63%	64%		29%
Calculus Series	100%	100%	98%	81%
Differential Equations	54%	42%		35%
Discrete Mathematics	33%	52%	90%	67%
Functions and Modeling	100%	38%	14%	15%
Geometry	26%	85%		90%
History of Mathematics	25%	50%	19%	18%
Intro to Proofs/Abstract Math	75%	78%		60%
Linear Algebra	100%	100%		97%
Number Theory	16%	36%	34%	31%
Probability	35%	56%	91%	88%
Statistics	56%	88%		
Additional course 1	63%	80%		46%[2]
Additional course 2	50%	100%		10%

[1] Most programs that prepared middle-school teachers required students to take Calculus, Statistics, and Discrete Mathematics. However, few required students to take History of Mathematics or Functions and Modeling.

[2] Mathematics Capstone Course.

[3] Blank table entries were not reported.

courses focused only on mathematics content and did not address issues of peda- gogy. Faculty reported varying degrees of freedom to modify curriculum across courses. Calculus and Linear Algebra were the most restrictive courses, with 12% and 16% of respondents indicating freedom to alter course curricula, respectively. The least restrictive courses with respect to curricular freedom were History of Mathematics, Geometry, Number Theory, and Functions and Modeling (required more often for secondary mathematics teachers than mathematics-intensive ma- jors).

Standards for Mathematical Practice. The next set of MTE-Partnership ques- tions focused on the level of agreement with respect to the *CCSS-M* Standards for Mathematics Practice. The survey used a Likert scale with categories of *strongly disagree, disagree, neutral, agree,* and *strongly agree* for three questions about whether the Standards for Mathematical Practice: (1) should be emphasized in the program; (2) were explicitly taught in the program; (3) were explicitly assessed in

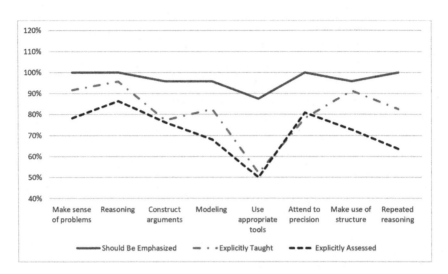

FIGURE 4.2. Mathematics Program Emphasis on Standards for Mathematical Practice. This figure shows results of a question about emphasis on Standards for Mathematics Practice from the MTE-Partnership survey of mathematicians and mathematics educators. Particularly interesting is the differences between what should be emphasized versus what is being taught and assessed.

the program. Although fewer respondents strongly agreed that the Standards for Mathematical Practice were explicitly taught and assessed, the combined strongly agree and agree categories showed little differences across these questions. For example, responses on the standard "Make Sense of Problems," all respondents agreed that this *should be taught* in the mathematics program (100%, $n = 25$ of 25), *explicitly taught* (92%, $n = 22$ of 24), and *explicitly assessed* (78%, $n = 18$ of 23). Greater differences among these categories were exhibited for the "Use Appropriate Tools" category, *should be taught* (88%, $n = 17$ of 24), *explicitly taught* (52%, $n = 12$ of 23), and *explicitly assessed* (50%, $n = 11$ of 22). Differences in the number of participants were due to neutral responses and non-responses (see Figure 4.2).

MET II **Ways of Knowing.** As discussed earlier in this chapter, *MET II* recommends that expert teachers know mathematics from four perspectives: (a) as scholars, (b) as educators, (c) as mathematicians, and (d) as teachers. The MTE-Partnership survey asked to what degree each lens should be emphasized and whether it was explicitly taught and explicitly assessed in the respondent's program. The responses showed some variation, but over 80% of the 24 respondents agreed that their programs should provide students with experiences to develop knowledge of mathematics as scholars, educators, mathematicians, and teachers. More disagreement occurred about whether these perspectives were taught. The percentage of responses indicating agreement (*strongly agree* and *agree*) regarding whether students were taught explicitly using each of these lenses were: scholars (87%, $n = 22$ of 23), educators (45%, $n = 10$ of 22), mathematicians (65%, $n = 15$ of 23), and teachers (39%, $n = 9$ of 23). Responses to "this lens is explicitly assessed in the program" were similar to those in the explicitly taught category. Overall support for statements that the *MET II* lenses were addressed in mathematics programs was mixed, although statements that programs should emphasize each perspective received considerable support.

Discussion. Course requirements for mathematics and teaching majors were similar across institutions. All of the respondents indicated that mathematics major programs and mathematics teacher education programs both require calculus (but not necessarily a three-course calculus sequence given the formulation of the survey question, as discussed in the Newton survey findings). All respondents reported requiring a linear algebra course in both the mathematics major and mathematics teacher education programs. After this set of courses, the alignment between programs decreased considerably. Both surveys showed high agreement with respect to the mathematics courses required by their teacher education programs. The Newton survey also indicated that almost all programs ($n = 74$, 95%) required at least one mathematics-specific methods course; the mean number of mathematics methods courses per program was 1.8.

Responses to the MTE-Partnership survey showed that a large percentage of the mathematics courses that are part of secondary mathematics teacher education programs were taught by tenure-track faculty. Responses indicated a feeling of more curriculum freedom for courses that primarily targeted teacher education majors than courses for mathematics majors.

The MTE-Partnership survey illustrated strong support for emphasizing the Standards for Mathematical Practice in their programs, but there was less agreement that they were explicitly taught and even greater disagreement that students' attainment of these standards was assessed. A similar pattern of responses occurred for the four ways of knowing mathematics described in *MET II*. Nearly all respondents agreed that these ways of knowing should be emphasized in their programs. They were much less inclined to agree that these lenses were explicitly taught and were balanced between agreeing and disagreeing about assessing them.

Overall, the MTE-Partnership survey shows that, although there are some similarities across institutions in teacher preparation programs, considerable differences are evident. Respondents included comments to explain some of their responses and offer greater insight into their programs. Some described master's-only secondary teacher education programs, which limits the impact of undergraduate mathematics courses. Other respondents mentioned the difficulty of predicting course assignments in very small mathematics departments in which "no one owns a course." Others indicated that most courses in teacher education programs were part of a larger mathematics program taken by mathematics majors that subsequently had little focus on pedagogy. Still others reported on teaching abstract algebra with manipulatives and using constructivist approaches to group theory.

THEORY OF IMPROVEMENT

The *CCSS-M* represents both an opportunity and a catalyst for significant change in the way that future mathematics teachers are prepared. *MET II, SET,* and the AMTE Standards provide a roadmap for accomplishing this change. However, the relationship between mathematics content knowledge and secondary mathematics teacher effectiveness may never be precisely understood. Such an understanding would require, among other things, a precise definition of secondary mathematics teacher effectiveness, which is undoubtedly a moving target. But *MET II* and other sources, including the research discussed in the first half of this article, describe many details of this relationship to drive toward a substantial and concerted course of action.

Because the *CCSS-M* or some variation of it has been widely adopted, students will be expected to meet these standards and mathematics teachers will be expected to support student learning of these standards. To support the achievement of these recommendations the MTE-Partnership seeks prompt solutions. Through this work, the MTE-Partnership hopes to understand what works and what does not, as well as what is required to work toward better solutions.

As noted in *MET II*, even the most robust university education experiences cannot provide prospective secondary mathematics teachers direct exposure to all topics that they might conceivably teach. Determination of those topics that require direct attention in university mathematics should be made judiciously to maximize prospective teachers' potential for teaching and learning additional topics at the in-service stage of their careers. Part of the work of the MTE-Partnership is to investigate how to help teachers make such judicious choices.

Research Action Clusters

The MTE-Partnership's *Guiding Principles for Secondary Mathematics Teacher Preparation Programs* (2014c) state in Guiding Principle 4: Candidates' Knowledge and Use of Mathematics, on the left-hand side of the driver diagram (see Figure 4.3), describes the MTE-Partnership's vision for content knowledge preparation of secondary mathematics teachers. To the immediate right of this principle, in the driver diagram, are the primary drivers for Principle 4. These drivers may be viewed as the impediments and obstacles to attaining the vision of the principle. Next, on the far right, are the secondary drivers, or actions and initiatives that might be applied toward mitigation of the primary drivers, and therefore, ultimately, toward achievement of the vision of Principle 4.

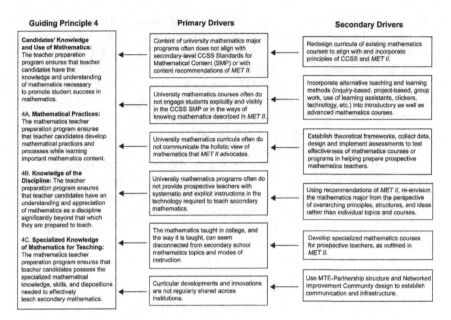

FIGURE 4.3. Driver Diagram for Mathematical Knowledge. This driver diagram addresses Guiding Principle 4 of the MTE-Partnership, Candidates' Knowledge and Use of Mathematics (MTE-Partnership, 2014c). The center column of primary drivers shares key issues that need to be addressed to meet this principle, and the right column of secondary drivers lists proposed strategies to guide MTE-Partnership actions.

To address the secondary drivers, the MTE-Partnership initially launched three Research Action Clusters (RACs). Each RAC was to carry out an intervention or set of interventions associated with a number of these secondary drivers.

One of these three RACs, Assessing Prospective Teachers' Knowledge of Connections Between Secondary and College-level Mathematics, was later renamed to be Knowledge for Teaching Mathematics Tasks. Its focus was on developing formative assessment tasks to measure the mathematics knowledge for teaching of prospective and in-service teachers. Unfortunately, this RAC never was able to articulate a clear plan of action and was disbanded in spring 2015.

As the results of the MTE-Partnership survey in the previous section indicate, college mathematics offerings: (a) do not always succeed in promoting the *CCSS-M* Standards for Mathematical Practice and (b) are frequently insufficient to meet the recommendations that *MET II* makes for the mathematical preparation of future secondary mathematics teachers. Of particular concern is the small proportion of institutions that report meeting *MET II* recommendations for courses designed specifically for prospective middle- and high-school teachers. The Building Communities and Courses RAC was designed to "build communities among mathematicians, mathematics educators, and K–12 collaborators that can work together to establish common content courses for mathematics teachers that are relevant to their professional needs" (MTE-Partnership, 2014a). Because knowledge of the *CCSS-M, MET II, SET*, and AMTE Standards is far from widespread among mathematics faculty who determine and teach mathematics courses for teachers, working groups of stakeholders were organized to determine and implement strategies for increasing faculty awareness and appreciation of the content and implications of these documents. From those original working groups, the general charge of this RAC was to develop courses and related curricular materials. Their initial aim was to develop at least three modules that could be used in a variety of contexts to improve prospective students' "deep understanding of secondary mathematics concepts from advanced perspectives while engaging in the *CCSS-M* Practice Standards" (MTE-Partnership, 2014a). This RAC met its initial aim and was reconstituted as the Mathematics of Doing, Understanding, Learning and Educating for Secondary Schools (MODULE(S^2)) in the summer of 2014 to pursue an extended aim. Through the MTE-Partnership's networked improvement infrastructure the materials developed by the MODULE(S^2) RAC were disseminated, and RAC partners continue to explore how they can be further developed and integrated into existing mathematics programs with minimal disruption, addressing the secondary drivers in the driver diagram related to redesigning curricula and developing specialized mathematics courses (see Figure 4.3).

The Active Learning Mathematics (ALM) RAC was organized by research mathematicians and mathematics educators to promote the mathematical content and process recommendations of the *CCSS-M, MET II, SET*, and AMTE Standards through active, student-centered approaches to teaching and learning in lower-division mathematics classes, especially those taken by STEM majors— i.e., precalculus through calculus 2 (referred to as the P2C2 sequence). The ALM RAC adopted the perspective that college-level mathematics should serve as a gateway, rather than a gatekeeper, to further study in STEM fields. By expanding the use of active learning approaches (Webb, 2016), the ALM RAC endeavored to increase opportunities for mainstream STEM students to develop the skills, dispositions, and habits of mathematical thinkers and problem solvers while learning the mathematical content material requisite for their fields of study. The primary target of the ALM RAC has been to incorporate alternative teaching and learning methods, which is one of the secondary drivers in the driver diagram (see Figure 4.3). The other chapters of Section II: Opportunities to Learn Mathematics in this book address the continuing development of the ALM RAC (Chapter 5) and the MODULE(S^2) RAC (Chapter 6).

FINAL COMMENTS

The goals for this chapter were to determine how teacher education programs address secondary teacher knowledge, to take stock of how programs currently meet this target, and to suggest strategies for moving from the latter to the former. These goals were addressed through the lens of various policy documents, whose recommendations represent, in part, a response to the content and practice standards delineated in the *CCSS-M*.

The grade-level content described in the *CCSS-M* are more intellectually challenging at all levels than earlier sets of content standards for school mathematics. Further, the *CCSS-M* Standards for Mathematical Practice require students to gain deeper insight into the nature of mathematics and to engage more substantially in mathematical activity. In articulating the mathematical preparation of teachers that will enable their students to rise to the challenges of the *CCSS-M*, the teacher preparation recommendations described in *MET II, SET,* and the AMTE Standards set a correspondingly high bar for the levels of mathematical and pedagogical understanding, insight, and engagement that must be developed in prospective secondary mathematics teachers.

In particular, prospective teachers need to know more mathematics than that which will be expected of their students. To have a clear vision of how to approach the teaching of secondary mathematics, these teachers need to know the relevant mathematical content from an advanced standpoint. That is, they require a broad, deep understanding of the mathematical structure underlying this content, and an appreciation of both the mathematics that foreshadows this content and the upcoming, advanced topics that connect to this content.

Additionally, prospective teachers must engage in authentic mathematical activity and develop mathematical habits of mind, so that they are equipped to encourage the same in their students. Further, the mathematical learning and experiences that these prospective teachers receive must cultivate their potential not only as doers and knowers, but also—and just as importantly—as teachers of mathematics. As recommended by *MET II*, their courses need to develop the mathematics understanding required for their teaching, experience in reasoning, explaining, and making sense of mathematics. Also, their knowledge development does not stop with their preparation courses; teachers need opportunities to develop and enhance their mathematical knowledge through their career in ways that model flexible and interactive ways of teaching (CBMS, 2012). Moreover, in most mathematics teacher preparation programs, this approach will require changing most of the undergraduate mathematics classes for majors (President's Council of Advisors on Science and Technology, 2012).

What does this mean for the university-level mathematics courses that prospective secondary mathematics teachers must take? What varieties, levels, and mixtures of mathematical knowledge (e.g. CCK and SCK), pedagogical knowledge (KCS and KCT), and curricular knowledge should these courses cultivate to produce qualified and successful secondary mathematics teachers? What is meant by qualified and successful, and how should such properties be measured? To what extent do existing courses, curricula, and programs for prospective teachers already promote these ways of learning and knowing mathematics, and of executing the teaching of mathematics? In cases where courses do not promote such knowledge and understanding, or do so inadequately, how ready are the institutions for meaningful change? Who will enact that change, according to what principles, and what will that change look like?

As indicated above, some progress has been made toward answering many of these questions. But the work in improving secondary mathematics teacher education has only begun. The MTE-Partnership, guided by the improvement science principles of the Networked Improvement Community model, represents a significant, concerted, and auspicious effort toward a robust, significant, and scalable—yet flexible—set of solutions for university-based mathematics teacher programs nationwide.

ENDNOTES

1. Work on this chapter was supported in part by a grant from the National Science Foundation (DUE 1624610, 1624628, 1624639, and 1624643) and The Helmsley Charitable Trust. All findings and opinions are those of the authors, and not necessarily those of the funding agency.
2. Other contributors to this chapter include: William S. Bush.
3. Robert N. Ronau was an employee at the National Science Foundation while this material was developed. Any opinions, findings, and conclusions or recommendations expressed in this material are his and do not necessarily reflect the views of the National Science Foundation.
4. National education systems vary. In the U.S., "lower secondary" is middle grades. Teachers of these grades may be prepared as middle grades specialists or in programs that include elementary teachers or high school teachers.
5. Definitions of middle level vary by state. In most states, middle level includes Grades 6–8, but in some states, middle level can span as much as Grades 4–9.

REFERENCES

Association of Mathematics Teacher Educators. (2017). *Standards for preparing teachers of mathematics.* Raleigh, NC: Author.

Ball, D. L., & Bass, H. (2000a). Making believe: The collective construction of public mathematical knowledge in the elementary classroom. In D. C. Phillips (Ed.), *Ninety-ninth yearbook of the National Society for the Study of Education: Part I* (pp. 193–224). Chicago, IL: University of Chicago Press.

Ball, D. L., & Bass, H. (2000b). Interweaving content and pedagogy in teaching and learning to teach: Knowing and using mathematics. In J. Boaler (Ed.), *Multiple perspectives on mathematics teaching and learning* (pp. 83–104). Westport, CT: Ablex.

Ball, D. L., Lubienski, S., & Mewborn, D. S. (2001). Research on teaching mathematics: The unsolved problem of teachers' mathematical knowledge. In V. Richardson

(Ed.), *Handbook of research on teaching* (4th ed., pp. 433–456). Washington, DC: American Educational Research Association.

Ball, D., & McDiarmid, G. (1989). *The subject matter preparation of teachers* (Issue paper, 89-4). East Lansing, MI: National Center for Research on Teacher Education.

Ball, D. L., Thames, M. H., & Phelps, G. (2008). Content knowledge for teaching: What makes it special? *Journal of Teacher Education, 59*, 389–407.

Baumert, J., Kunter, M., Blum, W., Brunner, M., Voss, T., Jordan, A., Klusmann, U., Krauss, S., Neubrand, M., & Tsai, Y. (2010). Teachers' mathematical knowledge, cognitive activation in the classroom, and student progress. *American Educational Research Journal, 47*, 133–180.

Begle, E. G. (1972). *Teacher knowledge and student achievement in algebra* (School Mathematics Study Group Report No. 9). Palo Alto, CA: Stanford University.

Begle, E. G. (1979). *Critical variables in mathematics education: Findings from a survey of the empirical literature*. Washington, DC: Mathematical Association of America and National Council of Teachers of Mathematics.

Ben-Zvi, D., & Arcavi, A. (2001). Junior high school students' construction of global views of data and data representations. *Educational Studies in Mathematics, 45*, 35–65.

Boardman, A. E., Davis, O. A., & Sanday, P. R. (1977). A simultaneous equations model of the educational process. *Journal of Public Economics, 7*, 23–49.

Conference Board of the Mathematical Sciences. (2001). *The mathematical education of teachers*. Providence, RI, and Washington, DC: American Mathematical Society and Mathematical Association of America.

Conference Board of the Mathematical Sciences. (2012). *The mathematical education of teachers II*. Providence, RI, and Washington, DC: American Mathematical Society and Mathematical Association of America.

Darling-Hammond, L. (2000). Teacher quality and student achievement: A review of state policy evidence. *Education Policy Analysis Archives, 8*(1), 1–42.

Döhrmann, M., Kaiser, G., & Blömeke, S. (2012). The conceptualization of mathematics competencies in the international teacher education study TEDS-M. *ZDM Mathematics Education, 44*, 325–340.

Donoghue, E. F. (2003). The emergence of a profession: Mathematics education in the United States, 1890–1920. In G. M. A. Stanic & J. Kilpatrick (Eds.), *A history of school mathematics* (vol. 1, pp. 159–193). Reston, VA: National Council of Teachers of Mathematics.

Eisenberg, T. A. (1977). Begle revisited: Teacher knowledge and student achievement in algebra. *Journal for Research in Mathematics Education, 8*, 216–222.

Ferrini-Mundy, J., & Findell, B. (2010). The mathematical education of prospective teachers of secondary school mathematics: Old assumptions, new challenges. In Committee on the Undergraduate Program in Mathematics (Ed.), *CUPM discussion papers about mathematics and the mathematical sciences in 2010*, (pp. 31-41). Washington, DC: Mathematical Association of America.

Franklin, C., Bargagliotti, A. E., Case, C. A., Kader, G. D., Schaeffer, R. L., & Spangler, D. A. (2015). *The statistical education of teachers*. Alexandria, VA: American Statistical Association.

Franklin, C., Kader, G., Mewborn, D., Moreno, J., Peck, R., Perry, M., & Schaeffer, R. (2007). *Guidelines for assessment and instruction in statistics education (GAISE) report: A pre-K–12 curriculum framework.* Alexandria, VA: American Statistical Association.

Groth, R. E. (2007). Toward a conceptualization of statistical knowledge for teaching. *Journal for Research in Mathematics Education, 38,* 427–437.

Groth, R. E. (2013). Characterizing key developmental understandings and pedagogically powerful ideas within a statistical knowledge for teaching framework. *Mathematical Thinking and Learning, 15,* 121–145.

Hanushek, E. A. (1972). *Education and race: An analysis of the educational production process.* Lexington, MA: D. C. Heath.

Hill, H. C. (2007). Mathematical knowledge of middle school teachers: Implications for the No Child Left Behind Policy initiative. *Educational Evaluation and Policy Analysis, 29,* 95–114.

Hill, H. C., Ball, D. L., & Schilling, S. G. (2008). Unpacking pedagogical content knowledge: Conceptualizing and measuring teachers' topic-specific knowledge of students. *Journal for Research in Mathematics Education, 39*(4), 372–400.

Hill, H. C., Schilling, S. G., & Ball, D. L. (2004). Developing measures of teachers' mathematics knowledge for teaching. *Elementary School Journal, 105,* 11–30.

Hill, H. C., Rowan, B., & Ball, D. L. (2005). Effects of teachers' mathematical knowledge for teaching on student achievement. *American Educational Research Journal, 42,* 371–406.

Izsák, A., Orrill, C. H., Cohen, A. S., & Brown, R. E. (2010). Measuring middle grades teachers' understanding of rational numbers with the mixture Rasch model. *The Elementary School Journal, 110*(3), 279–300.

Kilpatrick, J., Blume, G., Heid, M. K.,, Wilson, J., Wilson, P., & Zbiek, R. M. (2015). Mathematical understanding for secondary teaching: A framework. In M. K. Heid & P. S. Wilson (with G. W. Blume) (Eds.), *Mathematical understanding for secondary teaching: A framework and classroom-based situations* (pp. 9–30). Charlotte, NC: Information Age Publishing & National Council of Teachers of Mathematics.

Krauss, S., Baumert, J., & Blum, W. (2008). Secondary mathematics teachers' pedagogical content knowledge and content knowledge: Validation of the COACTIV constructs. *ZDM Mathematics Education, 40,* 873–892.

Krauss, S., Brunner, M., Kunter, M., Baumert, J., Blum, W., Neubrand, M., & Jordan, A. (2008). Pedagogical content knowledge and content knowledge of secondary mathematics teachers. *Journal of Educational Psychology, 100,* 716–725.

Ma, L. (1999). *Knowing and teaching elementary mathematics: Teachers' understanding of fundamental mathematics in China and the United States.* Mahwah, NJ: Lawrence Erlbaum Associates.

Mathematics Teacher Education Partnership. (2014a). *The building communities and courses research action cluster.* Unpublished paper presented at the 2014 MTE-Partnership Conference, Milwaukee, WI.

Mathematics Teacher Education Partnership. (2014b). *Mathematics content survey for MTEP member teacher preparation programs.* Working Group 3: Teacher Knowledge.

Mathematics Teacher Education Partnership. (2014c). *Guiding principles for secondary mathematics teacher preparation.* Washington, DC: Association of Public and Land-grant Universities. Retrieved from mtep.info/guidingprinciples

McCrory, R., Floden, R., Ferrini-Mundy, J., Reckase, M. D., & Senk, S. L. (2012). Knowledge of algebra for teaching: Framework of knowledge and practice. *Journal for Research in Mathematics Education, 5,* 584–615.

Mishra, P., & Koehler, M. J. (2006). Technological pedagogical content knowledge: A new framework for teacher knowledge. *Teachers College Record, 108*(6), 1017–1054.

Mohr-Schroeder, M. J., Ronau, R., Peters, S., Lee, C. W., & Bush, W. (2017). Predicting student achievement using measures of teachers' knowledge for teaching geometry. *Journal for Research in Mathematics Education, 48*(5), 520–566.

Monk, D. H. (1994). Subject area preparation of secondary mathematics and science teachers and student achievement. *Economics of Education Review, 13*(2), 125–145.

Mullis, I. V. S., & Martin, M. O. (2009). *TIMSS 2011 item writing guidelines.* (Available from the TIMSS & PIRLS International Study Center, Boston College, Boston, MA).

Mullis, I. V. S., Martin, M. O., Ruddock, G. J., O'Sullivan, C. Y., Arora, A., & Erberber, E. (2005). *TIMSS 2007 assessment frameworks.* Chestnut Hill, MA: TIMSS & PIRLS International Study Center, Boston College.

National Council of Teachers of Mathematics (NCTM). (1980). *An agenda for action: Recommendations for school mathematics of the 1980s.* Reston, VA: Author.

National Council of Teachers of Mathematics (NCTM). (1989). *Curriculum and Evaluation Standards for School Mathematics.* Reston, VA: Author.

National Council of Teachers of Mathematics (NCTM). (1991). *Professional standards for teaching mathematics.* Reston, VA: Author.

National Council of Teachers of Mathematics (NCTM). (2000). *Principles and standards for school mathematics.* Reston, VA: Author.

National Governors Association Center for Best Practices, Council of Chief State School Officers. (2010). *Common core state standards: Mathematics.* Washington, DC: Author.

Newton, J., Maeda, Y., Alexander, V., & Senk, S. L. (2014). How well are secondary mathematics teacher education programs aligned with the recommendations made in MET II? *Notices of the American Mathematical Society, 61*(3), 292–295.

Niess, M. L. (2005). Preparing teachers to teach science and mathematics with technology: Developing a technology pedagogical content knowledge. *Teaching and Teacher Education, 21,* 509–523.

Perry, J. (1901/1970). Discussion on the teaching of mathematics. In J. K. Bidwell & R. G. Clason (Eds.), *Readings in the history of mathematics education* (pp. 220–245). Washington, DC: NCTM.

Polya, G. (1945/2004). *How to solve it: A new aspect of mathematical method.* Princeton, NJ: Princeton University Press.

President's Council of Advisors on Science and Technology. (2012). *Engage to excel: Producing one million additional college graduates with degrees in science, technology, engineering, and mathematics.* Washington, DC: Executive Office of the President.

Reckase, M. D., McCrory, R., Floden, R. E., Ferrini-Mudy, J., & Senk, S. L. (2015). A multidimensional assessment of teachers' knowledge of algebra for teaching: Developing an instrument and supporting valid inferences. *Educational Assessment, 20,* 249–267.

Rowland, T., Huckstep, P., & Thwaites, A. (2005). Elementary teachers' mathematics subject knowledge: The knowledge quartet and the case of Naomi. *Journal of Mathematics Teacher Education, 8,* 255–281.

Saderholm, J. C., Ronau, R. N., Brown, E. T., & Collins G. (2010). Validation of the Diagnostic Teacher Assessment of Mathematics and Science (DTAMS) instrument. *School Science and Mathematics Journal, 110*(4), 180–192.

Schilling, S. G., Blunk, M., & Hill, H. C. (2007). Test validation and the MKT measures: Generalizations and conclusions. *Measurement: Interdisciplinary Research & Perspectives, 5,* 118–128.

Shechtman, N., Roschelle, J., Haertel, G., & Knudsen, J. (2010). Investigating links from teacher knowledge, to classroom practice, to student learning in the instructional system of the middle-school mathematics classroom. *Cognition and Instruction, 28*(3), 317–359.

Shulman, L. S. (1986). Those who understand: Knowledge growth in teaching. *Educational Researcher, 15*(2), 4–14.

Shulman, L. S. (1987). Knowledge and teaching: Foundations of the new reform. *Harvard Educational Review, 57,* 1–22.

Strauss, S., & Bichler, E. (1988). The development of children's concepts of the arithmetic mean. *Journal for Research in Mathematics Education, 19,* 64–80.

Tatto, M. T., Schwille, J., Senk, S. L., Ingvarson, L., Rowley, G., Peck, R., & Reckase, M. (2012). *Policy, practice, and readiness to teach primary and secondary mathematics in 17 countries: Findings from the IEA Teacher Education and Development Study in Mathematics (TEDS-M).* Amsterdam, The Netherlands: International Association for the Evaluation of Educational Achievement.

Tchoshanov, M. A. (2011). Relationship between teacher knowledge of concepts and connections, teaching practice, and student achievement in middle grades mathematics. *Educational Studies in Mathematics, 76,* 141–164.

Webb, D. (2016). Applying principles for active learning to promote student engagement in undergraduate calculus. *Proceedings of the 13th International Congress of Mathematics Education (ICME).* Hamburg, Germany: ICME.

Webb, N. (1997). *Research monograph number 6: "Criteria for alignment of expectations and assessments on mathematics and science education."* Washington, DC: CCSSO.

Webb, N. (August 1999). *Research monograph no. 18: "Alignment of science and mathematics standards and assessments in four states."* Washington, DC: CCSSO.

CHAPTER 5

DEVELOPING MATHEMATICAL KNOWLEDGE IN AND FOR TEACHING IN CONTENT COURSES[1]

Alyson E. Lischka, Yvonne Lai,
Jeremy F. Strayer, and Cynthia O. Anhalt

The mathematical preparation of secondary mathematics teachers is a complex endeavor best carried out through collaboration among a variety of stakeholders. This complexity arises from the reality that mathematics teacher educators must "cultivate their [prospective mathematics teachers'] potential not only as doers and knowers, but also—and just as importantly—as teachers, of mathematics" (Ronau, Webb, Peters, Mohr-Schroeder, & Stade, Chapter 4 of this book). Moreover, mathematics teacher preparation occurs in an environment where "[d]ominant cultural beliefs about the teaching and learning of mathematics continue to be obstacles to consistent implementation of effective teaching and learning in mathematics classrooms" (National Council of Teachers of Mathematics [NCTM], 2014, p. 9). In this environment, many prospective secondary teachers find ideas in required upper-level mathematics courses disconnected from both the mathematics they will teach and the teaching practices they are encouraged to enact (Goulding, Hatch, & Rodd, 2003; Moreira & David, 2008; Ticknor, 2012;

The Mathematics Teacher Education Partnership: The Power of a Networked Improvement Community to Transform Secondary Mathematics Teacher Preparation, pages 119–141.

Wasserman, Villanueva, Mejia-Ramos, & Weber, 2015; Zazkis & Leikin, 2010). Upper-level undergraduate mathematics courses can do more to provide opportunities for prospective teachers to connect advanced mathematical understandings to school mathematics and to draw on those understandings when they engage in the work of secondary mathematics teaching.

Working within this complex problem environment, the Mathematics of Doing, Understanding, Learning, and Educating for Secondary Schools (MODULE(S^2)) Research Action Cluster (RAC) uses a Networked Improvement Community (NIC) model (Bryk, Gomez, Grunow, & LeMahieu, 2015) to refine content and reshape instruction in upper-level mathematics courses commonly required of prospective secondary mathematics teachers. In the MODULE(S^2) RAC, mathematics faculty, mathematics education faculty, and K–12 personnel are collaborating to create and disseminate curriculum materials that develop mathematical knowledge for teaching (MKT; Ball, Thames, & Phelps, 2008; Rowland, 2013) in the context of upper-level mathematics courses.

In this chapter, we argue that providing opportunities for prospective teachers to build MKT requires educative curricular materials (Davis & Krajcik, 2005) that simultaneously support the learning of university mathematics instructors as well as the prospective teachers themselves. In addition, we describe an emergent, research-based design for developing and implementing these curriculum materials and propose priorities for those seeking to capitalize on the NIC model in future work. The efforts described in this chapter contribute to the overall vision for the transformation of secondary mathematics teacher preparation programs presented in this book by addressing the Mathematics Teacher Education Partnership's *Guiding Principles* (MTE-Partnership, 2014) that focus on teacher candidate knowledge, skills, and dispositions.

THE GOALS OF THE RAC AND THEORY OF CHANGE

During the initial work of the MTE-Partnership, members formed RACs to address each the *Guiding Principles* (MTE-Partnership, 2014). The initial organization of the partnership led to the creation of the Building Courses and Communities RAC, which was later renamed as MODULE(S^2). From the onset, this RAC attended to Guiding Principle 4: Candidates' Knowledge and Use of Mathematics (MTE-Partnership, 2014). In particular, the RAC aimed to design materials that focused on building deep understanding of secondary mathematics concepts from an advanced perspective by engaging prospective teachers in the *Common Core State Standards for Mathematics'* (*CCSS-M*; National Governors Association Center for Best Practices & Council of Chief State School Officers, 2010) Standards for Mathematical Practice. The design of the materials is modular, built as stand-alone units so that they can be flexibly used within existing courses, and includes connections to the practices of teaching using simulations or approximations of teaching practice (Grossman et al., 2009).

Early on, the RAC focused its work in two areas. First, the group surveyed secondary teacher preparation programs to determine the content areas of greatest need for such materials. The results of this survey demonstrated a need for materials for geometry, algebra, statistics, and mathematical modeling. Second, the group focused on building communities in which collaboration between mathematics educators and mathematicians would provide fertile ground for implementing materials. Thus, writing teams developed, which included mathematics education faculty, mathematicians, and K–12 teachers for each content area, building on the work of all stakeholders in this endeavor. Measures of effectiveness of this work were centered on the number of implementations and the amount of communication occurring between collaborators. As personnel in the RAC shifted and attempts to seek funding for the work were drafted, the aim was refined from a focus on products to a focus on the relationships between products and outcomes. The work of the RAC developed to focus not only on developing materials for upper-level content courses that prospective secondary mathematics teachers take, but also understanding how prospective teachers develop MKT and how the conditions of content course instruction contribute to that development.

The MODULE(S^2) RAC's revised aim is to improve prospective secondary teachers' opportunities to develop MKT in ways that are connected to content taught in upper-level undergraduate mathematics courses. These may include mathematics courses designed specifically for teachers, such as capstone courses, as well as mathematics courses that all mathematics majors take, such as statistics or abstract algebra. This aim addresses the Association of Mathematics Teacher Educators' (2017) *Standards for the Preparation of Teachers of Mathematics* (AMTE Standards) on candidate knowledge, skills, and dispositions (C.1, C.2, C.3, and C.4) along with program standards for providing opportunities to learn mathematics (P.2).

One promising strategy for enhancing the effectiveness of the aforementioned mathematics courses is to use tasks that elicit mathematical knowledge in explicitly pedagogical contexts (Lai, 2018; Lai & Howell, 2016; Stylianides & Stylianides, 2010; Wasserman, Fukawa-Connelly, Villanueva, Mejia-Ramos, & Weber, 2017). Following this strategy, the RAC works to position more mathematics instructors to teach with such tasks, and to do so in ways that align with the mathematical aims of their courses. We refer to these tasks as *approximations of mathematical teaching practice*, in reference to the term approximations of practice (Grossman et al., 2009) and emphasizing a focus on developing MKT. For example, in a Geometry Module task, prospective teachers consider incorrect student work involving the reflection of a line segment over a given line of reflection. The task requires prospective teachers to consider the ways in which they might respond to the learners' thinking in order to move students' mathematical thinking forward, thereby providing an opportunity to activate and develop prospective teachers' MKT through interaction with this approximation of mathematical teaching practice.

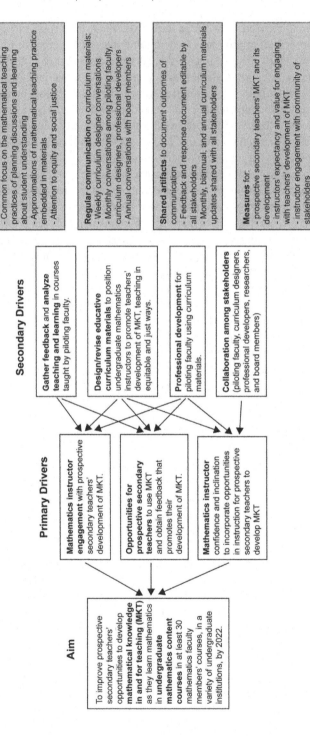

FIGURE 5.1. MODULE(S²) Driver Diagram. The MODULE(S²) Research Action Cluster driver diagram summarizes how the work of the RAC is structured to support the overall aim.

In order to accomplish the change needed to achieve the RAC aim, the work of MODULE(S²) attends to the primary drivers shown in the driver diagram (see Figure 5.1). We test this theory through the Plan-Do-Study-Act Cycle (PDSA; Bryk et al., 2015) to improve curriculum and instruction within the NIC. We write curriculum materials and design professional development with the primary drivers in mind. The curriculum materials include opportunities for prospective teachers to use and develop their MKT through approximations of teaching practice, which attends to the second primary driver and is informed by feedback from piloting faculty, a secondary driver, and collaboration among stakeholders, another secondary driver. Instructors participate in professional development activities and teach with the curriculum materials as other secondary drivers (see Figure 5.1). We also study enactment of instruction using the materials in collaboration with board members and piloting faculty by gathering feedback and analyzing teaching and learning in mathematics courses and professional development activities, two of the secondary drivers. Based on this process, we act to revise curriculum materials and professional development activities to strengthen alignment with drivers and aims. Throughout this process, we use measures, including those shown in Figure 5.1, to determine whether our primary drivers support the overarching MODULE(S²) aim, and we refine drivers as needed.

This theory of improvement has the potential to reconstruct the ways in which prospective secondary mathematics teachers experience learning mathematics, especially with regard to the connections they construct between advanced mathematics and their personal teaching practices. The work of the MODULE(S²) RAC is but one piece that fits within the larger systemic transformation of secondary mathematics teacher preparation envisioned by the MTE-Partnership. Nevertheless, the possibility of shaping prospective teachers' content knowledge and the pedagogical practices of both instructors and prospective teachers is heartening and necessary.

PRODUCTS AND EMERGING RESEARCH OF THE RAC

The need for mathematics instructors and prospective secondary teachers to attend to and activate MKT in university mathematics courses drives this work (see Figure 5.1). Additionally, instructors need support in doing this MKT-focused work. In this section, we describe how we address these drivers in greater detail as we outline a rationale for developing educative curriculum materials, the principles underlying materials development, and emergent research efforts.

Rationale for Educative Curriculum Materials

The notion that a set of curriculum materials can be *educative*, that they might simultaneously support the learning of teachers and students, has been suggested by Bruner (1960) and Ball and Cohen (1996) and taken up by K–12 mathematics curriculum writers in the reform era of the 1990s and 2000s (Davis & Krajcik,

2005). Similarly, we hold that educative curriculum materials can contribute to reforming undergraduate mathematics teacher education so that teachers of mathematics courses (mathematics faculty members) can intentionally teach for developing the MKT of their students (prospective secondary mathematics teachers).

An emergent community of mathematicians welcomes resources featuring approximations of mathematical teaching practice in mathematics courses (Lai, 2018; Lai & Howell, 2016). However, despite this welcoming posture, members of this community may also lack confidence in their abilities to teach with approximations of mathematical teaching practice and may therefore opt not to teach with them, even if they were made available (Lai, 2018). Exacerbating the situation is that it is unlikely that most mathematics faculty members would already know the MKT intended to be learned from an approximation of secondary mathematics teaching practice, let alone the knowledge needed for developing a novice's MKT. As a careful reading of Chapter 4 shows, multiple frameworks describing the complexity of the interrelated notions of content knowledge, pedagogical content knowledge, and MKT communicate the intricacies that may seem truly daunting for faculty who may be teaching for the development of MKT for the first time. Moreover, mathematicians typically do not have secondary teaching experience, and there are substantial differences among mathematicians' professional knowledge and MKT both at secondary and tertiary levels (e.g., Speer, King, & Howell, 2015).

However, there is some evidence that mathematicians with mentored experiences teaching with approximations of mathematical practice may feel more confident in using these tasks in subsequent courses (Lai, 2018). We see the educative components of our curriculum materials, coupled with professional development activities, as providing mathematics faculty with experiences that will build their confidence in teaching with approximations of mathematical teaching practice. Combining mathematics faculty members' desire for using approximations of mathematical teaching practice with an increased confidence in carrying them out creates a powerful vehicle for transforming undergraduate mathematics courses.

Guiding Principles for MODULE(S²) Educative Curriculum Development

MODULE(S²) is developing and refining educative curriculum materials in four areas of mathematics: algebra, geometry, statistics, and mathematical modeling. Table 5.1 describes overarching topics that the 12 modules address, three for each area of mathematics. Across all areas, we are guided by several common curriculum writing principles that align with our primary drivers. When reading the description below, notice that the fourth principle focuses the materials development efforts on providing opportunities for prospective secondary teachers to use MKT and obtain feedback that will promote their MKT development—our second primary driver. The remaining principles relate to two of MODULE(S²)' primary drivers and are focused on ensuring that the materials support mathematics

TABLE 5.1. Curriculum Modules for Upper Division Mathematics Courses

	Module 1	Module 2	Module 3
Algebra	Functions and Relations	Exponential Growth and Exponentiation	Fields
Geometry	Axiomatic Systems	Transformational Geometry	Similarity and Measurement
Mathematical Modeling	The Mathematical Modeling Process	Building Competencies in Mathematical Modeling	Advanced Perspectives in Mathematical Modeling
Statistics	Study Design and Exploratory Data Analysis	Inference	Association

instructors as they both engage in providing opportunities for MKT development and develop greater confidence and inclination to provide such opportunities.

Our first curriculum writing principle is that modules are designed as stand-alone replacement units, with all three modules in one area designed to serve as an entire course curriculum. This flexible design allows for the adoption of one module, or multiple modules depending on the needs of the preparation program. Moreover, studies in K–12 education have found that, when combined with professional development, replacement units can support teachers' learning (Cohen & Hill, 2001; Wilson, 2003). We posit that replacement units, used in mathematics courses for prospective secondary teachers, can support undergraduate faculty development of MKT for teaching at the secondary level as well as ways of supporting prospective secondary teachers' development of MKT.

Second, we rely on national policy recommendations for mathematicians and mathematics education faculty to guide the content and pedagogy represented in the curriculum materials. Policy documents include:

- *Standards for the Preparation of Teachers of Mathematics* (AMTE, 2017);
- *Common Core State Standards for Mathematics* (National Governors Association Center for Best Practices & Council of Chief State School Officers, 2010)
- *Mathematical Education of Teachers II* (*MET II;* Conference Board of the Mathematical Sciences [CBMS], 2012);
- *Instructional Practices Guide* (Mathematical Association of America, 2018); and
- *Statistical Education of Teachers* (*SET;* Franklin, Bargagliotti, Case, Kader, Scheaffer, & Spangler, 2015)

We used these documents to guide the pedagogical vision of the materials so that instructional objectives were most likely to benefit mathematics teachers. For a full discussion of these policy recommendations and how they shape the U.S. mathematics education landscape see Chapter 4 of this book.

Third, we leverage several principles to promote equity in the design of the materials. Specifically, the materials are written to support instructors in maintaining cognitive demand while supporting learners with varying skills to be able to contribute to solutions, thus helping all learners gain confidence in their mathematics abilities. The materials also support instructors in centering learners' authentic experiences and knowledge as legitimate intellectual spaces for investigation of mathematical ideas. These principles align with equity-based instructional practices established in the field such as going deep with mathematics, affirming mathematics learners' identities, and drawing on multiple resources of knowledge (Aguirre, Mayfield-Ingram, & Martin, 2013). For instance, the Algebra Module 1 instructor materials encourage the instructor to teach a course where prospective teachers "recognize what is worthwhile about others' reasoning" and "view their own mistakes courageously and with an open mind" so that "they accept that making errors and learning from them is part of the mathematical process" (Lai, 2019). These materials articulate and explain teaching moves to support this vision of learning, such as "praise thoughtful errors" and "relinquish your authority to the students and the mathematics" (Lai, 2019). The Mathematical Modeling Module 2 materials focus on naming competencies needed in modeling so as to give both instructors and prospective teachers language for going deep with the mathematics and affirming the identity of teachers as learners.

Fourth, we utilize approximations of mathematical teaching practice to serve several functions. They support prospective teachers in learning to teach concepts that are found in key strands of the *CCSS-M*; they feature high-leverage teaching practices (e.g., Ball, Sleep, Boerst, & Bass, 2009); and they support content that can be taught in either a capstone course or mathematics major course. For instance, Figure 5.2 shows an example of an approximation of mathematical teaching practice centered on using a linear function to represent a bivariate relationship, a central topic in school algebra (National Governors Association Center for Best Practices & Council of Chief State School Officers, 2010).

We have found that engaging teachers in this approximation of practice supports their understanding of covariation and correspondence views of algebraic relationships, a distinction which is arguably part of MKT for the secondary level (Thompson & Thompson, 1994). This approximation of practice also provides a frame for reflecting on how functions have been treated in secondary mathematics courses and undergraduate courses such as calculus, real analysis, or topology. The concepts of covariation and correspondence views of functions are thus fitting topics for a capstone course and serve to connect secondary and undergraduate mathematics (CBMS, 2001).

Finally, following Davis and Kracjik (2005), our materials strive to be transparent about pedagogical judgments, help instructors notice learners' development over time, and use class activities to promote productive change in learners' understandings. As these authors noted, sustained and meaningful shifts in teaching can arise from supporting instructional reasoning. To accomplish this trans-

---

Ms. Allen is working with her students to review for an end of year Algebra I exam. She included the following problem on a worksheet of practice problems.

While visiting New York City, John kept track of the amount of money he spent on transportation by recording the distance he traveled by taxi and the cost of the ride.

Distance, d, in miles	Cost, C, in dollars
3	8.25
5	12.75
11	26.25

Show that the data can be represented by the linear function $C = 2.25d + 1.5$.

Two of Ms. Allen's students used different methods to solve the problem.

Jing	Matt

$$\frac{12.75 - 8.25}{5 - 3} = \frac{4.50}{2} = 2.25$$

$$
\begin{aligned}
C - 26.25 &= 2.25(d - 11) \\
C - 26.25 &= 2.25d - 24.75 \\
C &= 2.25d + 1.5
\end{aligned}
$$

$$
\begin{aligned}
2.25(3) + 1.5 &= 8.25 \\
2.25(5) + 1.5 &= 12.75 \\
2.25(11) + 1.5 &= 26.25
\end{aligned}
$$

Which student(s) demonstrated a method that is a mathematically valid solution to the review problem?

FIGURE 5.2. Sample Approximation of Teaching Practice. Example of an approximation of mathematical teaching practice, from the Algebra Module 1 curriculum materials (Lai, 2019). This approximation of mathematical teaching practice is based on the Allen mini-case, part of the Content Knowledge for Teaching Mini-cases project of the Educational Testing Service (Howell et al., 2017).

parency, MODULE(S^2) materials include notes to the instructor that highlight specific and general opportunities to develop content connections to secondary mathematics and its teaching, as well as pedagogical reasoning that can apply beyond the specific situation in which it first arises. For instance, Algebra Modules 1 and 2 provide discussions of covariation and correspondence views of specific functions, and their affordances and limitations. The modules then explain how these discussions generalize to other functions, ways in which these points of view can be seen in mathematics from secondary to graduate contexts, and why the materials emphasize the distinction between these points of view. In doing so, we intend for the instructor both to develop understanding of these views and to be alert to opportunities for noticing these points of view when they arise elsewhere in the curriculum materials and in tasks that instructors may construct or adapt for the prospective teachers in other courses.

In Algebra Module 2 the instructor notes suggest how to lead a discussion while students complete a contextualized table task, a type of task that is common to secondary curriculum materials. We decompose the discussion into launch-explore-summarize components (e.g., Hirsch, Coxford, Fey, & Schoen, 1995), with the launch component based on Jackson, Garrison, Wilson, Gibbons, and Shahan's (2013) findings about how to set up word problems, and the explore and summarize components based on questioning strategies for middle school students working on pattern problems (Rivera & Becker, 2009). Then, we address the discussion structure in general, why this structure can be helpful across middle grades and high school grades, and why the questioning strategies are mathematically productive. From these discussions, we offer opportunities for the instructor to develop capacity to model for prospective teachers one way to facilitate discussions; to draw on this decomposition of leading discussions in future lessons both within and outside of the module; and to build capacity for responding to prospective teachers' approximations of mathematical teaching practice in which they plan a discussion.

Emerging Research

While developing educative curriculum materials, the MODULE(S²) RAC simultaneously conducts research concerning implementing instruction with the materials. The RAC carries out this research using methodology aligned with the NIC-based theory of improvement for the ongoing work: research is hypothesized based on existing research, tested in rapid-cycle implementation with measurements of impact during testing, and followed by revision and re-implementation. The emerging research focuses on two areas: efforts to understand how MKT develops so as to improve the effectiveness of the curriculum materials in positioning instructors to cultivate prospective teachers' MKT, and efforts to understand how faculty implement instruction using the curriculum materials. Each of these is described in the following section.

Emerging Research on the Development of MKT. Two theoretical tools necessary to enable the design of educative curriculum are knowledge of how learners' conceptions develop, and frameworks for what to look for in learners' actions as evidence of development. Translating this to the teaching of MKT via approximations of mathematical teaching practice, two resources are key for educative curriculum design: first, theories of how MKT develops, and second, strategies for how to promote instructors' capacity to observe and interpret change in prospective teachers' thinking in terms of MKT development.

With regard to the first resource, efforts to understand how prospective teachers develop MKT are in nascent stages (Hoover, Mosvold, Ball & Lai, 2016). Among existing research, Silverman and Thompson's (2008) framework for studying the development of MKT, which is grounded in research about both mathematics education and the learning sciences, provided the foundation for our work. Within this foundation, we integrate the *knowledge quartet* work from Rowland (2013), which theorizes where in the work of teaching teachers draw on their mathematical knowledge to make instructional decisions.

We draw on these existing theories of how MKT develops to establish the second resource, a novel integration of existing observational frameworks for examining MKT and its development. In what follows, we discuss each framework, why each existing framework alone does not suffice for analyzing and observing teachers' development of MKT, and how we interpret the relationships among each framework to construct our emerging framework, shown in Table 5.2.

Silverman and Thompson (2008) theorize that MKT development is the result of robust personal understandings of mathematics combined with the capacity to reflect on students' thinking. This reflection results in both multiple models of how students may understand an idea and an understanding of multiple instructional sequences for supporting different mathematical understandings. The salient feature of this reflection is that it involves *decentering*—the ability to think outside one's own initial cognition.

Although Silverman and Thompson (2008) describe components of the development of MKT, they do not elaborate where these components might be observed in teaching practice or in an approximation of teaching practice. Hence observing the components proposed by Silverman and Thompson is difficult using only the descriptions in their framework. To determine where MKT is used in teaching, we turn to Rowland and colleagues' (2016) work on the knowledge quartet, which identifies teaching situations that activate MKT and then sorts the teaching situations into dimensions of MKT. The knowledge quartet features four dimensions: (a) foundation, which includes knowledge of mathematics and its nature; (b) transformation, the presentation of ideas to learners in the form of analogies, illustrations, examples, explanations, and demonstrations; (c) connection, the sequencing of material for instruction and an awareness of the relative cognitive demands of different topics and tasks; and (d) contingency, the ability to respond to unanticipated events in the work of teaching, including student thinking (the latter three are contexts in which foundation knowledge is brought to bear).

When teachers have robust understandings and can engage in the decentering needed to develop MKT, they can design instructional activities to be more responsive to student thinking as well as analyze students' knowledge more acutely. We interpret foundation to include a teacher's personal understandings and transformation, connection, and contingency as actions informed by decentering.

Although Silverman and Thompson's (2008) work provides a theory for how MKT develops, and the knowledge quartet (Rowland et al., 2016) provides a description of where different types of knowledge are applied to aspects of teaching, neither elaborate how one instance of an application of MKT to teaching practice may be more sophisticated than another instance. For this, Ader and Carlson's (2018) work provides a mechanism for distinguishing levels of the sophistication of activation of MKT by describing patterns in observable behaviors during teaching practice that indicate the extent to which the teacher has decentered. In their framework, there are four levels, in which a teacher is:

TABLE 5.2. Framework for Observing and Analyzing Development of MKT in Approximations of Mathematical Teaching Practice

Component of Developing MKT[a]	Dimension of MKT Activated, with Example Teaching Situations that Activate Dimension[b]	Developmental Level, in Terms of Observable Actions[c]
Personal Understandings of Mathematics	Foundation (Descriptions of mathematical ideas; explanations of why procedures work)	Note: Levels here depend on the mathematics featured in the approximation of mathematical teaching practice. This is just one possible example of how levels may appear. Level 0: Specific reference to mathematics is not present OR Performs procedures incorrectly and describes underlying concepts incorrectly (lacking in CCK) OR does not describe relevant procedures completely Level 1: Performs relevant procedures correctly to get an answer OR Describes relevant procedures with inaccuracies Level 2: Describes relevant procedures accurately, either for themselves or for learners Level 3: Connects procedures to underlying concepts, with mostly mathematically precise and appropriate language Level 4: Connects structure of procedure to underlying concepts, with mostly mathematically precise and appropriate language
Decentering applied to Activities and Analyzing Potential for Student Understandings of Mathematics	Transformation (Proposed ideas to explain to students; explanations directed to students)	Gives explanations, representations, and examples, that: Level 0: Do not provide any explanations, representations, or examples to students Level 1: Describe only procedures or echoes key phrases Level 2: Describe own way of thinking of the mathematics Level 3: Attempt to change students' current thinking Level 4: Build on respects students' understanding toward the intended KDU

Connection (Proposed questions or activities for students; explicit statements of connections among questions and activities)	Prompts students to say or do things in ways that:	
	Level 0:	Do not ask students to say or do anything
	Level 1:	Focus only on procedures or echoing key phrases
	Level 2:	May reveal student thinking, but then gives explanations while not asking students to provide reasoning
	Level 3:	Attempt to change students' current thinking
	Level 4:	Build on respect students' understanding toward the intended KDU
Contingency (References to given student work; interpretations of or commentary on given student work; responses to given student work)	Uses student thinking in ways that:	
	Level 0:	Do not act in any visible way upon the thinking
	Level 1:	Evaluate the mathematical validity of the thinking but do not use the thinking in teaching
	Level 2:	Reference the thinking to guide students toward teacher's way of thinking
	Level 3:	Follow up on students' responses to perturb students to change their thinking
	Level 4:	Frame questions or explanations in terms of students' thinking to help move students' understanding toward the intended KDU.

Notes: [a]Silverman and Thompson, 2008; [b]Rowland, 2013; [c]extends Ader and Carlson 2018.

- Level 1: Interested in students' answers but not their thinking;
- Level 2: Interested in students' thinking, so as to guide students toward the way the teacher is thinking;
- Level 3: Seeks to make sense of students' thinking and attempts to move the thinking; and
- Level 4: Seeks to make sense of students' thinking and builds on that thinking during instruction.

To extend these levels to each of the knowledge quartet dimensions, we used Ader and Carlson's levels in an initial pass on coding prospective teachers' responses to approximations of mathematical teaching practice in MODULE(S²) materials. Then, we adapted our descriptions of levels to address the teaching actions that were visible in our data in a way that was consistent with observation protocols that have been validated as measuring quality of teaching (see Table 5.2; Junker et al., 2006; Learning Mathematics for Teaching, 2011).

Research Supporting the Enactment of Educative Curricula That Develops MKT. In the previous section, we made the case that a framework for observing the development of MKT, premised on theories of how MKT develops, is useful for curriculum designers and instructors because such a framework and associated theory can guide planning as well as in-the-moment decisions of teachers. We also described our example of such a framework. In this section, we illustrate how this sort of framework is also useful for achieving the research and practice goals of the MODULE(S²) RAC. These goals include examining how written and enacted curricula (Stein, Remillard, & Smith, 2007) shape the development of prospective teachers' MKT, and positioning instructors to enact curriculum that promotes the development of prospective teachers' MKT.

We use our emerging framework (see Table 5.2) to investigate one of the guiding hypotheses behind our curriculum design work and then use the results of the investigation to inform the curriculum materials' revision, thereby creating a cycle of improvement informed by data and improvement science (Bryk et al., 2015). Our hypothesis states that how an instructor teaches a mathematics course will influence the types of responses prospective teachers provide to assignments for that course. More specifically, the more an instructor enacts instruction that builds on and enhances prospective teachers' thinking, the more likely it is that the prospective teachers will be able to build on and enhance sample student thinking in their approximations of mathematical teaching practice assignments. Results of this analysis inform the curriculum materials design, providing ways to improve the instructor notes aimed at helping instructors promote prospective teachers' thinking in each lesson, as well as discuss how specific instances are an example of techniques that can be applied across lessons. For instance, in the Algebra Module 1, when discussing graphical representations of an inverse relation, the materials suggest for the instructor to encourage prospective teachers' explicit reasoning with the definition of a graph of a relation and the definition of an inverse relation. As the materials discuss, the technique of attending to and eliciting reasoning based on a definition is useful both for fostering understanding and for mathematics teaching more broadly. In future years, we will continue to use the results of our framework related analysis to expand this educative aspect of the materials.

Our framework for observing the development of MKT is an analytic tool for investigating this hypothesis; it allows us to parse data about curriculum use and prospective teachers' MKT in systematic and strategic ways. Our framework, as an analytic tool, will allow us to see whether different aspects of MKT in prospective teachers' approximation of practice responses are consistent within and across sites or over time, or how they may vary across sites and over time. It will also provide a frame for analyzing the instructor's use of curriculum materials so as to examine potential influence of each site's instruction on prospective teachers' responses at that site. Such an analysis provides potential explanations for how enacted curriculum influences prospective teachers' MKT and its development.

Our framework for observing the development of MKT is also a conceptual tool for examining the relationship between written and enacted curricula. Using the framework, we can organize conversations with piloting faculty focused on how and whether the instructors' interpretations of written curriculum shaped teaching actions described in the framework. For example, one issue that can be explored is whether instructors notice where the materials are intended to help prospective teachers develop their own personal knowledge as opposed to engaging prospective teachers in envisioning different ways that students develop mathematical knowledge. A systematic investigation into written curriculum, enacted curriculum, and prospective teachers' knowledge can point us to how and where to revise the curriculum materials.

Summary. Through the creation of a framework for observing and analyzing development of MKT in approximations of mathematical teaching practice, the RAC is making progress toward both research goals: understanding the development of MKT and developing tools to aid in the analysis of the implementation of the modules. Our framework provides concrete descriptions of teaching actions that activate MKT and suggests how these actions represent development into more sophisticated MKT. As Moss (2011) argued, "conceptions of quality" are essential for enacting learning opportunities and assessments (p. 2879). In this sense, the framework is a curricular design resource. We have used it to improve the design of approximations of mathematical teaching practice to ensure that opportunities to elicit MKT across multiple dimensions exist in the assignments in effective ways. The framework is also an analytic tool in that it may allow us to describe relationships between instructors' teaching and prospective teachers' responses to approximations of practice. Finally, the framework is a conceptual tool for facilitating conversations with piloting faculty about their interpretations of the curriculum materials. Our work on this framework provides a foundation for future research and practice.

FUTURE DIRECTIONS FOR THE RAC

The work of the MODULE(S²) RAC described in this chapter is the first step toward improving the mathematical experiences and opportunities offered in content courses for prospective secondary mathematics teachers and toward meeting the RAC aim: to improve opportunities for prospective teachers to develop MKT as they learn mathematics (see Figure 5.1). The descriptions provided in this chapter represent both the state of the work as of the writing of this chapter and the vision for the short-term growth of the work. A greater vision for impacting content and instruction in upper-level mathematics content courses includes additional efforts in both research and curriculum development.

Original choices for the focus of the content of the materials were determined through surveys of MTE-Partnership institutions along with reviewing national policy recommendations as described in Chapter 4 of this book. The survey results pointed to a need for curriculum materials in the four content areas that the MODULE(S²) materials address. However, expansion to other areas of mathematics might also be considered in the future. The mathematical preparation of secondary teachers does not only take place in the areas of algebra, geometry, mathematical modeling, and statistics. In addition, as calls for a shifted focus in high school mathematics curricula gain momentum and support (NCTM, 2018), it may be necessary to include the new foci more centrally in the mathematical preparation of secondary teachers.

As the curriculum materials development continues, so does what can be learned about the development of MKT among prospective teachers. The proposed framework for observing and analyzing the development of MKT described in this chapter is a first step toward understanding how this construct is at play in courses that prepare secondary teachers. Through RAC data collected on the development of prospective teachers' MKT, this construct can be further explored. Future work might explore the connection between hypothetical learning trajectories for various mathematical concepts and the development of MKT. Additionally, research could explore the variation in development of MKT across different concepts. Each line of questioning serves to inform improvements in the educative curriculum that is developed for these courses.

Besides possible expansions to other content areas and providing a deeper understanding of the development of MKT, the work of understanding curriculum implementation among university mathematics and mathematics education faculty is complex, ongoing, and critical to the primary driver of mathematics instructor confidence and inclination to incorporate new opportunities (see Figure 5.1). A primary focus of the work of this RAC will continue to explore the ways in which the implementation of educative curriculum can be supported across various settings and faculty experiences. Gathering measures of instructors' expectancy and value for engaging in the development of prospective teachers' MKT along with data gathered through professional learning conversations can inform the progress of this work and open new avenues for improvement. Opportunities for joining in all areas of this work and locating access to the available curriculum materials can be found at the MTE-Partnership website (http://mtep.info/MODULES2).

A FOCUS ON IMPROVEMENT

The initiatives described in this chapter, taken together, represent an ambitious goal of transforming the content and instruction in upper-level mathematics content courses for prospective secondary teachers. A program interested in vast reconstruction might choose to implement all four sets of complete courses as the mathematics curriculum for their prospective secondary teachers. However, this sweeping change is likely not possible in most settings. In addition to revamping program requirements, the willingness and time commitment required of necessary university mathematics and mathematics education faculty may not be present. Therefore, a focus on improvement aimed at transforming the way in which upper-level mathematics content courses are taught—and that may utilize the work of the MODULE(S^2) RAC—can occur in a variety of forms. For programs considering transformation of upper-level content courses, the following three recommendations are offered and elaborated upon in the following paragraphs:

1. Gather data to guide your decisions on which courses should be revised and consider which of the MODULE(S^2) materials can assist in this revision.
2. Allow for flexible redesign of courses that permits a range of changes from the partial redesign of a course (one unit of a course) to a complete course overhaul.
3. Offer support to implementing faculty, either through supports offered by the MODULE(S^2) RAC or through collaboration between mathematicians and mathematics educators at your institution.

In each type of implementation, the decisions about which changes to implement should be based on the needs of the program in improving the experiences of the prospective teachers. For example, at Middle Tennessee State University, prospective teachers were struggling to find success on state-required content tests for certification. The decision to include the MODULE(S^2) materials was made to improve the content knowledge learning and opportunities for building connections to secondary mathematics for these prospective teachers. Program faculty continue to track the success of prospective teachers on required certification exams and draw connections between content course design and assessment data.

When considering redesign of courses, allow for flexibility in the depth and breadth of revision. In some cases, a complete overhaul of curriculum is necessary, and the faculty have the knowledge and willingness to do so. In other cases, making small changes is called for as faculty learn to implement new instructional techniques and focus on new constructs. The flexible design of the MODULE(S^2) curriculum materials as stand-alone units allows for implementation in a variety of forms. At Middle Tennessee State University, the mathematics education faculty wanted a greater focus on the development of MKT in the mathematics content courses in their program. Therefore, materials from all content areas are used across the program. All three geometry modules form the entire curriculum for the junior level college geometry course that is required of prospective teachers and open to all mathematics majors. In addition, individual modules from each of the other areas are implemented in a capstone mathematics content course for prospective secondary teachers. As of the writing of this chapter, a new statistics course is under development for prospective secondary teachers. The planned curriculum for this course is the complete set of statistics modules. At other institutions, modules from different content areas have been combined to form the content for a capstone course for secondary teachers. In others, one module from geometry (Transformational Geometry) has been used along with other curriculum in a geometry course.

Another consideration for choosing to transform upper-level mathematics content courses is the support of faculty as they enact the transformation of courses. Transitioning curriculum to encourage the development of MKT among prospective teachers can be overwhelming for faculty and requires support and collaboration. Again, the flexible design of the MODULE(S^2) materials allows for different ways of approaching this issue. First, the work of the RAC includes implementation support through online discussion groups and instructor guides. Each of these supports can ease the transition to a new curriculum for faculty. Whether implementing the MODULE(S^2) materials or working to transform courses through other means, discussion groups and collaboration with instructors are essential. Second, consider ways that a flexible redesign can support faculty in transforming courses. An instructor might start by implementing one new module within a course as they learn more about developing MKT and connecting to secondary content, or developing whatever new focus is necessary. In addition, some institutions are finding ways to support co-teaching of courses in which there is a redesign. Courses co-taught by mathematics and mathematics education faculty allow for the educative curriculum to deepen the understanding of both faculty members and enrich the experiences of the prospective teachers in the courses.

In addition to the above recommendations, we offer the following selection of readings that may support program faculty in understanding the value and impact of implementation of MODULE(S^2) materials as one means of transformation of upper-level content courses. The following research literature can further aid faculty in stating an argument for revision of upper-level content courses for prospective teachers.

The potential of the work of MODULE(S^2) for transforming the experiences of prospective secondary teachers in mathematics content courses through an iterative and principled development process that intentionally incorporates perspectives across the system of mathematics education is promising. Addressing an issue as complex as the mathematical preparation for secondary teaching requires both a set of ideas and a community with which to engage those ideas. This work has the capability to grow a community invested in building capacity for developing prospective teachers' MKT in ways that connect to and may transform existing course structures.

READING LIST

The following annotated bibliography provides readings to support beginning implementation of MODULE(S^2) materials.

1. Anhalt, C. O., Cortez, R., & Bennett, A. B. (2018). The emergence of mathematical modeling competencies: An investigation of prospective secondary mathematics teachers. *Mathematical Thinking and Learning, 20*, 202–221.
   - This study of prospective teachers without prior mathematical modeling experience sheds light on how their newly developed conceptual understanding of modeling manifested itself in their work on the final task of a modeling module. The module provided opportunity for prospective teachers to experience the Common Core Mathematical Practice Model with Mathematics and begin to develop competency in modeling. The results suggest that infusing modules in existing courses can be an effective way to elevate prospective teachers from unfamiliarity with modeling to noticeable levels of proficiency in various modeling sub-competencies.
2. Casey, S., Albert, J., & Ross, A. (2019). Developing knowledge for teaching graphing of bivariate categorical data. *Journal of Statistics Education, 27*(1).
   - In this pilot efficacy study, Casey, Albert, and Ross found that following use of the materials, teachers were more likely to correctly use relative frequencies in their analysis and expanded their knowledge of graphs to include segmented bar graphs. They also improved in their analysis of a student's graph and proposed responses to students.
3. Lai, Y. & Howell, H. (2016). Conventional courses are not enough for future high school teachers. *Post for the AMS Blog on Teaching and Learning Mathematics.* Retrieved from http://blogs.ams.org/matheducation/2016/10/03/conventional-courses-are-not-enough-for-future-high-school-teachers/
   - This blog post walks through an example of an assessment item for MKT at the secondary level to illustrate what MKT might look like at the secondary level, an underdeveloped area of research. This same assessment item forms the basis for the approximation of mathematical teaching practice featured in this chapter. The post then presents three arguments for why MKT can be productively thought of as a form of applied mathematics, and why mathematicians teaching

mathematics to prospective teachers might have a stake in thinking about it this way.

4. Lai, Y. (2018). Accounting for mathematicians' priorities in mathematics courses for secondary teachers. *Journal of Mathematical Behavior, 53,* 164–178.

– One account for inattention to MKT in mathematics courses for teachers is that mathematicians value developing teachers' MKT less than they value developing teachers' pure mathematical knowledge. This study provides another account: mathematicians may value developing teachers' MKT and using approximations of mathematical teaching practice as much or more than developing teachers' pure mathematical knowledge and using purely mathematical tasks— however, they lack confidence and resources for teaching to develop teachers' MKT as compared to teaching to develop teachers' more purely mathematical knowledge. The study implies that improving mathematics courses for teachers may involve building mathematicians' confidence and skill in teaching with approximations of mathematical teaching practice.

5. Silverman, J. & Thompson, P. (2008). Toward a framework for the development of mathematical knowledge for teaching. *Journal of Mathematics Teacher Education*, 11, 499–511.

– This article lays out the notion that MKT might be developed by (a) first coming to deep personal understanding of a mathematical idea, (b) reflecting on the ways someone else might come to understand this idea, and (c) reflecting on different sets of tasks that could be used to help someone come to this understanding. This working theory informs the research and development of the MODULE(S^2) RAC. As Silverman and Thompson suggested, these three practices may be akin to habits of mind that support the development of MKT.

For current information on the MODULE(S^2) RAC, please visit mtep.info/MODULES2.

ENDNOTES

1. Work on this chapter was supported in part by a grant from the National Science Foundation IUSE (Improving Undergraduate STEM Education) multi-institutional collaborative grant #1726707 (APLU), #1726098 (University of Arizona), #1726252 (Eastern Michigan University), #1726723 (Middle Tennessee State University), #1726744 (University of Nebraska–Lincoln), and #1726804 (Utah State University). All findings and opinions are those of the authors, and not necessarily those of the funding agency.

2.   Other contributors to this chapter include: Emina Alibegovic (Rowland Hall School), Jason Aubrey (University of Arizona), Stephanie Casey (Eastern Michigan University), Ricardo Cortez (Tulane University), Brynja Kohler (Utah State University), and Andrew Ross (Eastern Michigan University).

REFERENCES

Ader, S. B., & Carlson, M. P. (2018). *Observable manifestations of a teacher's actions to understand and act on student thinking.* Paper presented at the annual meeting of the Special Interest Group of the Mathematical Association of America on Research in Undergraduate Mathematics Education, San Diego, CA.

Aguirre, J., Mayfield-Ingram, K., & Martin, D. (2013). *The impact of identity in K–8 mathematics learning and teaching: Rethinking equity-based practices.* Reston, VA: National Council of Teachers of Mathematics.

Association of Mathematics Teacher Educators. (2017). *Standards for preparing teachers of mathematics.* Retrieved from http://amte.net/standards

Ball, D. L., & Cohen, D. K. (1996). Reform by the book: What is—or might be—the role of curriculum materials in teacher learning and instructional reform? *Educational researcher, 25*(9), 6–14.

Ball, D. L., Sleep, L., Boerst, T. A., & Bass, H. (2009). Combining the development of practice and the practice of development in teacher education. *The Elementary School Journal, 109,* 458–474.

Ball, D., Thames, M. H., & Phelps, G. (2008). Content knowledge for teaching: What makes it special? *Journal of Teacher Education, 59,* 389–407.

Bruner, J. (1960). *The process of education.* Cambridge, MA: Harvard University Press.

Bryk, A., Gomez, L. M., Grunow, A., & LeMahieu, P. (2015). *Learning to improve: How America's schools can get better at getting better.* Cambridge, MA: Harvard Education Press.

Cohen, D., & Hill, H. C. (2001). *Learning policy: When state education reform works.* New Haven, CT: Yale University Press.

Conference Board of the Mathematical Sciences. (2001). *The mathematical education of teachers.* Providence, RI: American Mathematical Society.

Conference Board of the Mathematical Sciences. (2012). *The mathematical education of teachers II.* Providence RI and Washington DC: American Mathematical Society and Mathematical Association of America.

Davis, E. A., & Krajcik, J. (2005). Designing educative curriculum materials to promote teacher learning. *Educational Researcher, 34*(3), 3–14.

Franklin, C., Bargagliotti, A. E., Case, C. A., Kader, G. D., Scheaffer, R. L, & Spangler, D. A. (2015). *The statistical education of teachers.* Alexandria, VA: American Statistical Association. Retrieved from www.amstat.org/education/SET

Grossman, P., Compton, C., Igra, D., Ronfeldt, M., Shahan, E., & Williamson, P. W. (2009). Teaching practice: A cross-professional perspective. *Teachers College Record, 111,* 2055–2100.

Goulding, M., Hatch, G., & Rodd, M. (2003). Undergraduate mathematics experience: Its significance in secondary mathematics teacher preparation. *Journal of Mathematics Teacher Education, 6,* 361–393.

Hirsch, C. R., Coxford, A. F., Fey, J. T., & Schoen, H. L. (1995). Teaching sensible mathematics in sense-making ways with the CPMP. *The Mathematics Teacher, 88*, 694–700.

Hoover, M., Mosvold, R., Ball, D. L., & Lai, Y. (2016). Making progress on mathematical knowledge for teaching. *The Mathematics Enthusiast, 13*(1–2), 3–34.

Howell, H., Nabors-Olah, L., Lai, Y., DeLucia, M., & Kim, E. M. (2017). *The Allen minicase: An instructional case to support learning of MKT.* Princeton, NJ: Educational Testing Service.

Jackson, K., Garrison, A., Wilson, J., Gibbons, L., & Shahan, E. (2013). Exploring relationships between setting up complex tasks and opportunities to learn in concluding whole-class discussions in middle-grades mathematics instruction. *Journal for Research in Mathematics Education, 44*, 646–682.

Junker, B., Weisberg, Y., Matsumura, L. C., Crosson, A., Wolf, M. K., Levison, A., & Resnick, L. (2006). *Overview of the Instructional Quality Assessment.* CSE Technical Report 671. Los Angeles, CA: Center for the Study of Evaluation National Center for Research on Evaluation, Standards, and Student Testing (CRESST).

Lai, Y. (2018). Accounting for mathematicians' priorities in mathematics courses for secondary teachers using expectancy-value theory. *Journal of Mathematical Behavior, 53*, 164–178.

Lai, Y. (2019). *MODULE(S²): Algebra for secondary teaching, module 1- functions and relations.* Washington, DC: MODULE(S²).

Lai, Y., & Howell, H. (2016). Conventional courses are not enough for future high school teachers. *Post for the blog of the American Mathematical Society.* Retrieved from https://blogs.ams.org/matheducation/2016/10/03/conventional-courses-are-not-enough-for-future-high-school-teachers/

Learning Mathematics for Teaching. (2011). Measuring the mathematical quality of instruction. *Journal of Mathematics Teacher Education, 14*, 25–47.

Mathematical Association of America (2018). *Instructional practices guide.* Retrieved from https://www.maa.org/programs-and-communities/curriculum%20resources/instructional-practices-guide

Mathematics Teacher Education Partnership. (2014). *Guiding principles for secondary mathematics teacher preparation.* Washington, DC: Association of Public and Land-grant Universities. Retrieved from mtep.info/guidingprinciples

Moreira, P. C., & David, M. M. (2008). Academic mathematics and mathematical knowledge needed in school teaching practice: Some conflicting elements. *Journal of Mathematics Teacher Education, 11*, 23–40.

Moss, P. A. (2011). Analyzing the teaching of professional practice. *Teachers College Record, 113*, 2878–2896.

National Council of Teachers of Mathematics. (2014). *Principles to action: Ensuring mathematical success for all.* Reston, VA: Author.

National Council of Teachers of Mathematics. (2018). *Catalyzing change in high school mathematics: Initiating critical conversations.* Reston, VA: Author.

National Governors Association Center for Best Practices, Council of Chief State School Officers. (2010). *Common core state standards: Mathematics.* Washington, DC: Author.

Rivera, F. D., & Becker, J. R. (2009). Algebraic reasoning through patterns. *Mathematics Teaching in the Middle School, 15*, 212–221.

Ronau, R. N., Webb, D. C., Peters, S. A., Mohr-Schroeder, M. J. & Stade, E. (this volume). Mathematical preparation. In W. G. Martin et al. (Eds.) *The mathematics teacher education partnership: The power of a networked improvement community to transform secondary mathematics teacher preparation.*

Rowland, T. (2013). The knowledge quartet: The genesis and application of a framework for analysing mathematics teaching and deepening teachers' mathematics knowledge. *Sisyphus-Journal of Education, 1*(3), 15–43.

Rowland, T., Thwaites, A., & Jared, L. (2016). *Analysing secondary mathematics teaching with the knowledge quartet.* Paper presented at the 13th International Congress on Mathematical Education, Hamburg, Germany.

Silverman, J., & Thompson, P. W. (2008). Toward a framework for the development of mathematical knowledge for teaching. *Journal of Mathematics Teacher Education, 11*, 499–511.

Speer, N. M., King, K. D., & Howell, H. (2015). Definitions of mathematical knowledge for teaching: Using these constructs in research on secondary and college mathematics teachers. *Journal of Mathematics Teacher Education, 18*, 105–122.

Stein, M. K., Remillard, J., & Smith, M. S. (2007). How curriculum influences student learning. In F. K. Lester, (Ed.), *Second handbook of research on mathematics teaching and learning*, (pp. 319–370). Charleston, NC: Information Age Publishing.

Stylianides, G. J., & Stylianides, A. J. (2010). Mathematics for teaching: A form of applied mathematics. *Teaching and Teacher Education, 26*(2), 161–172.

Thompson, P. W., & Thompson, A. G. (1994). Talking about rates conceptually, Part I: A teacher's struggle. *Journal for Research in Mathematics Education, 25*, 279–303.

Ticknor, C. S. (2012). Situated learning in an abstract algebra classroom. *Educational Studies in Mathematics, 81*, 307–323.

Wasserman, N. H., Fukawa-Connelly, T., Villanueva, M., Mejia-Ramos, J. P., & Weber, K. (2017). Making real analysis relevant to secondary teachers: Building up from and stepping down to practice. *PRIMUS, 27*, 559–578.

Wasserman, N., Villanueva, M., Mejia-Ramos, J.-P., & Weber, K. (2015). Secondary mathematics teachers' perceptions of real analysis in relation to their teaching practice. In *Annual Conference of Research in Undergraduate Mathematics Education (RUME)*. Pittsburgh, PA.

Wilson, S. (2003). *California dreaming: Reforming mathematics education.* New Haven, CT: Yale University Press.

Zazkis, R., & Leikin, R. (2010). Advanced mathematical knowledge in teaching practice: Perceptions of secondary mathematics teachers. *Mathematical Thinking and Learning, 12*, 263–281.

CHAPTER 6

IMPROVING FRESHMAN-LEVEL MATHEMATICS COURSES VIA ACTIVE LEARNING MATHEMATICS STRATEGIES[1]

Wendy M. Smith, Kadian M. Callahan,
Tabitha Mingus, and Angie Hodge

Student success in undergraduate mathematics has significant implications for whether students choose to continue in STEM (science, technology, engineering, or mathematics) majors and future related careers (e.g., National Research Council [NRC], 2012). Even for those students who do not choose to major in STEM, success in entry-level undergraduate mathematics courses, such as calculus, can make or break students' decisions to persist in postsecondary education (Bressoud, Carlson, Mesa, & Rasmussen, 2013). The *Characteristics of Successful Programs in College Calculus* project (Bressoud et al., 2013) showed the percentage of students with grades of D, F or Withdraw (DFW) in Calculus I ranged from an average of 25% at Ph.D.-granting universities to an average of 37% at regional comprehensive universities. Unfortunately, uninspiring and unengaging instructional practices in these entry-level courses compound the loss of STEM majors (Freeman et al., 2014), and disproportionately impact populations that are already underrepresented in STEM (Olson & Riordan, 2012). These concerns,

The Mathematics Teacher Education Partnership: The Power of a Networked Improvement Community to Transform Secondary Mathematics Teacher Preparation, pages 143–175.
Copyright © 2020 by Information Age Publishing

together with a commitment to developing students' mathematical content knowledge, led the Mathematics Teacher Education Partnership (MTE-Partnership) to create an Active Learning Mathematics Research Action Cluster (ALM RAC).

Although the ALM RAC is not as obvious a component of teacher preparation as some of the other MTE-Partnership RACs, the ALM RAC's focus on integrating active learning strategies into the calculus sequence aligns with recommendations by the Conference Board of the Mathematical Sciences (Conference Board of the Mathematical Sciences [CBMS], 2012) and the Association of Mathematics Teacher Educators (Association of Mathematics Teacher Educators, [AMTE], 2017) for the preparation of secondary mathematics teachers. These recommendations assert that prospective teachers need to fully experience mathematics themselves by "struggling with hard problems, discovering their own solutions, reasoning mathematically, modeling with mathematics, and developing mathematical habits of mind" (CBMS, 2012, p. 54). These experiences are central for developing students' own mathematics content knowledge and expose prospective teachers to the type of pedagogy that they are called to use with their future students (see National Council of Teachers of Mathematics, NCTM, 2014). Thus, integrating active learning into collegiate mathematics courses helps prospective teachers experience first-hand how mathematics teaching and learning can look when not purely lecture-based.

This chapter provides some background on the rationale and research related to active learning in mathematics, and then situates the work of the ALM RAC. In addition to discussing the ALM RAC in general, the chapter includes four vignettes of change efforts related to active learning. Finally, the chapter concludes with a focus on improvement, describing how other programs seeking to infuse more active learning in their mathematics courses may initiate such efforts.

Defining Active Learning

The work of the ALM RAC is guided by research on active learning and sociocultural theories of learning and builds upon current research on the teaching and learning of mathematics from K–12 education and at the undergraduate level. Active learning engages learners in using higher-order thinking processes like analyzing, synthesizing, and creating (Krathwohl, 2002) and involves talking, listening, writing, reading, and reflecting (Bonwell & Eison, 1991; Hobson, 1996; McGuire, 2015; Meyers & Jones, 1993). These learning processes optimize the creation of new mental structures or the incorporation of new information into existing mental structures. In active learning classrooms, learners are intimately involved in their knowledge development through the critical analysis of course material, discussion of ideas and problem-solving strategies with peers, and justification of their reasoning to others (Cuoco, Goldenberg, & Mark, 1996; Davidson, 1971; Legrand, 2001; Millett, 2001; Wahlberg, 1997; Weissglass, 1993; Yackel & Cobb, 1996). Thus, active learning approaches align well with pedagogical implications of sociocultural theories of learning, which posit that cognitive development occurs within both social and cultural contexts (Vygotsky, 1978, 1986). The social components of active learning pedagogical practices address learners' psychological needs for cognitive development: classroom social norms shape learners' beliefs about their own role, others' roles, and the general nature of mathematical activity; sociomathematical norms shape learners' mathematical beliefs and values; and the establishment of classroom mathematical practices shape learners' mathematical interpretations and reasoning. From this perspective, mathematics learning is conceptualized as being influenced by the microculture of the classroom and by external factors, such as learners' participation in their local, home, and global communities (Cobb & Hodge, 2002; Pourdavood, Carignan, Martin, & Sanders, 2005).

There is not a universal definition for *actively learning mathematics* in the research literature; the term seems to have come into regular use following the first recommended national K–12 standards for mathematics (NRC, 1989). One definition includes "actively engage students in their learning [of] mathematics by dynamically involving them... Activities developed and implemented place the responsibility of learning mathematics... on the student" (Jardine, 1997, p. 115). Often researchers and practitioners use a broad definition of active learning, such as, "Active learning involves students engaging with course content beyond lecture" (Kerrigan, 2018, p. 35), or, "Active learning is generally defined as any instructional method that engages students in the learning process... active learning requires students to do meaningful activities and think about what they are doing" (Prince, 2004, p. 223).

The ALM RAC initially developed a broad definition of active learning mathematics as well: teaching and learning aligned with the Standards for Mathematical Practice in the *Common Core State Standards for Mathematics* (*CCSS-M*; National Governors Association Center for Best Practices & Council of Chief State School Officers, 2010). The central problem focusing the ALM RAC was originally stated as: Promote *active learning* in freshmen-level undergraduate mathematics courses, as emphasized in the *CCSS-M*. That definition has evolved over time, drawing on Freeman et al. (2014) and an emerging focus on equity.

The ALM RAC scope of work has broadened to consider how to shift the paradigm of undergraduate mathematics teaching and learning toward actively engaging learners through sociocultural practices. The ALM RAC envisions active learning classrooms that intentionally strive for equitable outcomes, by fostering greater academic success and persistence in STEM for a broad range of learners with diverse backgrounds and experiences. Specifically, the ALM RAC now defines active learning as instruction that adheres to the principles: (a) students learn meaningful mathematics by engaging in challenging, cognitively demanding problems; (b) students deepen and clarify their thinking through routinely discussing their own reasoning and considering the reasoning of others (peer-to-peer interactions); (c) instructors deepen their own understanding of mathematics content and pedagogical content knowledge when they elicit and make use of student thinking to advance the mathematical agenda; and (d) instructors foster a sense of belonging when they explicitly attend to issues of diversity, equity, and inclusion. Together, these active learning principles enhance teaching and learning experiences in undergraduate mathematics classrooms. Each of these principles are discussed in more detail in the sections that follow.

Challenging and Cognitively Demanding Problems

The design and implementation of problems and tasks are essential for meaningful active learning in mathematics because they shape the quality of students' thinking (Hiebert et al., 1997). Cognitively demanding problems move students beyond the memorization and algorithmic thinking that are characteristic of the lowest levels of Bloom's Taxonomy (Krathwohl, 2002), and instead challenge students to make connections among different concepts, representations, and processes (Smith & Stein, 1998). Pressing students to grapple with mathematics in this way encourages them to reason about the underlying structures of mathematics and engages them in metacognition. Engaging in mathematical reasoning— "the development, justification, and use of mathematical generalizations" (Russell, 1999, p. 1)—is an important process for fostering meaningful mathematics learning for K–12 students and their teachers (CBMS, 2012; NCTM, 2000, 2014), and includes more than formal proofs (Cai & Cirillo, 2014; Stylianides, 2008, 2009). Metacognition—the knowledge and regulation of cognition (Schraw & Dennison, 1994)—fosters students' ability to improve on their own learning and has been linked to academic success in undergraduate STEM courses (e.g., Zhao, Wardeska, McGuire, & Cook, 2014). Nevertheless, scholars acknowledge that the quality of the task itself is not enough to guarantee that students will reason mathematically and develop metacognitive skills that the task is intended to bolster. Discussing students' ideas resulting from engaging with the task is critical for developing meaningful mathematical thinking and understanding as it creates opportunities for students to clarify their understandings, learn from others' thinking, and consider how and why the mathematics works (NCTM, 2000; Smith & Stein, 2011, 2018; Stein, Grover, & Henningson, 1996).

Peer-to-Peer Interactions

Peers play an important role in the undergraduate experience. According to Pascarella and Terenzini (1991), peer interactions influence cognitive development, identity development, self-confidence, self-efficacy, and social and academic integration into the university environment. Academically centered peer-to-peer interactions also support students' retention and persistence in undergraduate degree programs (Astin, 1999; Bank, Slavings, & Biddle, 1990; Howell, 2006; Liu & Liu, 2000; Loo & Rolison, 1986). In the classroom context, exchanging ideas with peers creates an opportunity for learners to reconsider their own thinking and understanding of mathematics relative to other ways of thinking and knowing

(Callahan, 2016a; Cobb, Boufi, McClain, & Whitenack, 1997; Ellis, 2011; Pierson, 2008). Reasoning about different representations with peers can strengthen prospective secondary mathematics teachers' content knowledge by requiring them to make connections between the representations and can strengthen their knowledge of how to teach that content by providing them greater access to others' thinking as they interpret different representations (Callahan, 2016b; Callahan & Hillen, 2012; Herbel-Eisenmann & Phillips, 2005; Izsák & Sherin, 2003).

Eliciting and Using Learners' Thinking

The ways that instructors elicit and use learners' mathematical ideas plays a critical role in effective mathematics instruction (e.g., Empson & Jacobs, 2008; Leatham, Peterson, Stockero, & Van Zoest, 2015; Pierson, 2008; Smith & Stein, 2011; White, 2003). For example, the use of purposeful questions shapes "the cognitive opportunities afforded to students" (Boaler & Brodie, 2004, p. 780) and can be used to "assess and advance students' reasoning and sense making about important mathematical ideas and relationships" (National Council of Teachers of Mathematics, 2014, p. 35). Effectively using mathematical ideas to shape learning involves creating opportunities for learners to express their mathematical thinking (Boaler & Brodie, 2004), noticing opportunities to further mathematical understanding (Leatham et al., 2015), and responding to learners' ideas in mathematically productive ways (Smith & Stein, 2011, 2018). In addition to supporting learning, scholars describe several benefits of eliciting learners' mathematical thinking that enhances instructors' knowledge and practice. Scholars have found that examining learners' thinking supports instructors' content knowledge and pedagogical content knowledge by changing the way they think about the mathematics and learners' reasoning about the mathematics, and by expanding their understanding of ways to represent and teach that content (Chamberlin, 2005; Driscoll & Moyer, 2001; Herbel-Eisenmann & Phillips, 2005; Kazemi & Franke, 2004; Laursen, Hassi, Kogan, Hunter, & Weston, 2011).

Diversity, Equity and Inclusion

The presence of active learning strategies does not automatically lead to equitable and inclusive student experiences. Equity-based practices in mathematics classrooms include: going deep with mathematics, leveraging multiple mathematical competencies, affirming mathematics learners' identities, challenging spaces of marginality, and drawing on multiple resources of knowledge (Aguirre, Mayfield-Ingram, & Martin, 2013). For instructors to engage in equity-based practices, they need to both value equity and hold strength-based beliefs about students. If instructors hold deficit beliefs about students, making broad assumptions about what students cannot do (e.g., my students cannot handle group work), this often results in instructors reducing the mathematical rigor of lessons (e.g., Battey, Neal, Leyva, & Adams-Wiggins, 2016; Ladson-Billings, 1997; Lubienski, 2002).

Conversely, enacting equitable and inclusive practices that draw upon the diversity of ideas and student experiences fosters a sense of belonging for students that is positively related to effort, participation, and emotional engagement (Wilson et al., 2015). Thus, for active learning strategies to be effective for all students, considerations of equity and inclusion must be part of instructors' mindsets, with an intentional focus on ensuring equitable outcomes. For more information on equity, see Chapter 3 in this book.

EXPANDING COMMITMENTS TO
ACTIVE LEARNING IN MATHEMATICS

Exploring ideas of cognitively demanding problems, establishing peer-to-peer communication, instructor use of student thinking, all keeping a focus on equitable student outcomes, collectively provides a foundation for active learning. However, many mathematics instructors have only experienced learning mathematics through a lecture environment, in which students struggled on their own through procedural homework problems and to make their own connections. It can be very difficult for someone who has only experienced mathematical lectures to envision how mathematics teaching and learning can occur through non-lecture activities, such as student-collaborative problem solving accompanied by small-group and whole-class discussions. Thus, work to adopt active learning strategies also needs to consider what is known about effectively changing teaching practices.

Making profound changes to instructional practices is difficult. In particular, engaging students actively in learning mathematics is not necessarily intuitive and may be counter to instructors' own experiences as learners of mathematics. Instructors need professional development and ongoing support in order to learn how to intentionally focus on equitable student outcomes, use student thinking for making mathematical decisions about lessons, enact peer-to-peer discussions, and maintain mathematical rigor. The need for such professional development and ongoing support is confounded by the small number of people trained in such activities, as well as the dearth of research literature about how to support instructors in diverse contexts to enact positive changes to their instructional practices.

As noted above, when implementing active learning practices, not only are instructors working to understand new ways of shaping learners' thinking, learners are also developing new ways of engaging with the material and with one another. To effectively integrate this type of change, it is recommended that institutions "use both a 'top-down' and a 'bottom-up' approach, take into consideration the relevant factors that affect faculty work, and strategically use multiple change levers" (National Research Council, NRC, 2012, p. 183). Change efforts must involve commitment from all levels of the institution and incorporate consistent messages about and in support of the change. This commitment should include re-examining reward structures, such as tenure and promotion guidelines and processes for determining salary increases; implementing flexible structures that allow time and space to experiment with new ideas; and incorporating meaningful measures of learning that can counter resistance from some learners as they adjust to unfamiliar instructional practices (Dee, Henkin, & Hearne, 2011; Jacobsen, 1997; NRC, 2012; Weiman, 2017).

Research indicates that communities of practice are especially important for fostering instructional change and for sustaining those changes over time. Communities of practice provide support and feedback for instructors while they navigate the realities of making instructional changes (Henderson, Dancy, & Niewiadomska-Bugaj, 2012), and they have the potential to change instructors' conceptions of teaching and learning and the value they place on research-based instructional practices (Dee, Henkin, & Hearne, 2011; Laursen et al., 2011; NRC, 2012). Communities of practice also serve as a way to expand the impact of instructional change efforts to influence departmental cultures. As Weiman's (2017) work on the adoption of research-based instructional practices shows,

> The best indicator of the overall impact on teaching and departmental culture is the fraction of the faculty that have made large changes in their teaching. This indicates both a willingness to consider thinking about teaching in a different way as well as learning how to actually teach differently. (p. 98)

To extend the reach of change initiatives, it is critical to include innovators and early adopters who can speak from their own experience and serve as mentors for other instructors—they hold the greatest potential for influencing other instructors to adopt new instructional practices (Jacobsen, 1997).

An effective community of practice has a vision of desirable outcomes, hypothesized levers for change, and specific change strategies hypothesized to influence the change levers. A focus on shifting instructional practices in mathematics courses needs to include curricular innovation (e.g., Mathematical Association of America, MAA, Instructional Practices Guide, 2018). Instructor and student attitudes and beliefs are also crucial to effectively implement active learning strategies (e.g., Lipnevich, MacCann, Burrus, & Roberts, 2011; Ma & Kishor, 1997; Richardson, 1996; Sherman & Christian, 1999; Taraban, Box, Myers, Pollard, Bowen, 2007). Finally, attention to sustainability of changes needs to be present from early in the efforts. For example, having or increasing the amount of coordination across multi-section mathematics courses can improve the potential for changes to be sustained over time (e.g., Bressoud, Mesa, & Rasmussen, 2015; Kezar, 2014).

INITIATING CHANGE EFFORTS

The work of the ALM RAC is built on the foundation of broader MTE-Partnership work, particularly work related to mathematics (see Chapter 4 of this book). Initial ALM RAC activities included defining actively learning mathematics, exploring the problem space (e.g., what are the barriers to ALM implementation), agreeing on a common aim, developing a driver diagram to illustrate change strategies, and determining how to measure the intended outcome of improved student success.

Along with the definition of active learning in mathematics provided earlier, ALM RAC members adapted design principles to guide the implementation of active learning (see Table 6.1): mathematics content, instructional activities, classroom discourse, instructional environment, instructional decisions, and formative assessments. These design principles emphasize that active learning pedagogies need to be supported by rich mathematical tasks that collectively present a coherent view of mathematics and lead students to deeper conceptual understanding. Instructional tasks need to be designed to go beyond procedures, so students engage in mathematical reasoning. Establishing effective norms for when and how to communicate is essential for fostering active learning so that students have time to process their ideas and to talk through their emergent mathematical reasoning with others during class. The incorporation of formative assessments is essential for providing meaningful feedback to both instructors and students. Information gathered from formative assessments equip instructors to adapt lessons as needed to advance student understanding based on students' current needs, and provides students with an understanding of what is important for them to learn and serves as a preview for summative assessments.

From a foundation of these design principles, the overarching goal of the ALM RAC's work is to:

> …improve student success with undergraduate mathematics, starting with the Precalculus to Calculus 2 (P2C2) sequence. This goal of student success is accomplished through effective teaching practices, which are supported by learning environments that are more conducive to student interaction, reasoning, and problem solving and the use of instructional resources to support ALM. Faculty buy-in and institutional leadership supports training for Graduate Teaching Assistants and other P2C2 instructors. Also, for many campuses, undergraduate learning assistants are used to support student work with group activities and enhance student engagement in mathematical activity. (Smith & Webb, 2017, p. 49)

Aligned with the design principles and overarching goal, the original ALM RAC aim was: Through the use of active learning design principles, increase the percent of students succeeding in targeted freshman-level math courses (Precalculus, Calculus 1 and/or Calculus 2) by Y% by December 2014. In this initial aim, the improvement target "Y" varied by institution, depending on current levels of student success. For instance, institutions with lower than 50% passing rates had a higher goal for percentage of improvement than institutions with over 75% passing rates. Over time, this goal has remained the same, with the target date shifting to the future, and institutions increasing their improvement target.

Since aims need to be measurable, early ALM RAC discussions also centered on defining student success and determining appropriate common measures of that success. While universities often consider student success to mean passing courses with a grade of C or better, other measures of success include positive attitudes toward mathematics, choosing and persisting in a STEM major, and tak-

TABLE 6.1. Design Principles for Active Learning Mathematics Implementation

Principle	Explanation
Mathematics content	While skills and procedural knowledge are a necessary goal of undergraduate calculus, the overarching principle toward the design of courses should be toward key ideas and coherence
Instructional activities	Questions posed should promote active construction of meaning, sense making, and relational reasoning
Norms for classroom discourse	Students should be encouraged to share reasoning-in-process, including partially developed conjectures, explanations, and representations of solution strategies (Jansen, Cooper, Vascellaro, & Wandless, 2016)
Instructional environment	Classroom norms and organization of the lesson should "support interaction in small groups, whole-class discussion, and individual seatwork in accordance with the needs of the learner and the learning task" (Roj-Lindberg, 2001, p. 8)
Instructional decisions	Choices made in lesson design and adaptation should favor the perspective of the learners; instructors consider student background knowledge and anticipate common difficulties with the content
Formative assessments	Reflect emphasis on understanding meaning and concepts, not just procedures

ing additional mathematics courses after freshman-level courses. The ALM RAC discussed ways to measure student attitudes, course-taking patterns, and grades, along with instructor attitudes and beliefs and classroom practices.

To achieve this aim, the ALM RAC members were each committed to implementing and sustaining active learning practices, including curricular changes, expanding instructors' knowledge, supporting positive student dispositions, and fostering ongoing commitments by instructors, departments, and institutions. Thus, building on the design principles, there are five primary drivers in the work of the ALM RAC:

1. Creating equitable curriculum and assessment materials that support active learning and the coordination of multiple sections ("horizontal") and across courses ("vertical"). When curricula focus only on procedures, there needs to be attention to adding rich mathematical tasks that elicit and support student reasoning. Assessments need to align with such rich mathematical tasks, to gauge student conceptual understanding in addition to procedural fluency.

2. Building the capacities of mathematics instructors to effectively use active learning techniques in ways that are responsive to learners' needs. Telling instructors to spend less class time lecturing is not sufficient to induce pedagogical change; instructors need significant support and opportunities to practice new pedagogies.

3. Supporting positive learner dispositions that foster a sense of belonging, a growth mindset, and intrinsic motivation to continue to pursue a career in STEM. Retaining students in STEM majors is difficult if students fear

or dislike mathematics; instructors likely need support and professional development to learn strategies to better develop positive mindsets with students. Encouraging the sharing of incomplete reasoning and praising process rather than focusing solely on correct answers (although correct mathematics is still ultimately important) can help develop a positive classroom climate.

4. Expanding commitments to strengthen teaching and learning in undergraduate mathematics courses in ways that promote equitable student outcomes to mathematics departments at other institutions. Active learning strategies alone are not sufficient to guarantee equitable outcomes. In setting norms for classroom discourse, instructors need to attend to power dynamics and ensure students each have a voice in the classroom.

5. Coordinating content and instruction in multi-section courses (horizontal coordination) and in sequences of courses (vertical coordination) to promote a common, equitable, and positive experience among students as they progress through their program. For changes to be sustained across multiple (and changing) instructors, multiple dimensions of each course need to be coordinated (e.g., common exams and grading); adopting similar strategies across the P2C2 sequence helps students experience mathematics more coherently. Although a department does not necessarily need to adopt a new textbook in order to start using active learning strategies, some tasks are more group-worthy than others; group-worthy tasks are more likely to prompt discussion and elicit students to communicate mathematical reasoning.

In 2018, the MTE-Partnership overall adopted a dual focus on program transformation coupled with equity and social justice, and the ALM RAC worked to align with this vision. The ALM RAC driver diagram was revised to infuse a focus on equity into all aspects of ALM RAC work (see Figure 6.1).

In addition to the primary drivers, described previously, the ALM RAC also moved toward operationalizing the primary drivers by defining secondary drivers and a theory of change. The design principles articulated previously, while providing a foundation for the primary drivers, also define efforts each campus would undertake, so appear in the driver diagram as secondary drivers. Secondary drivers also include resources for supporting active learning pedagogies: instructor professional development, increased course coordination, physical environment, and student supports outside of class. The ALM RAC recognized that change efforts take time, and the pedagogical changes may seem overwhelming. Thus, secondary drivers also include the idea to start small (and show success), and to develop a coalition of stakeholders who will work toward a common goal.

In conjunction with its driver diagram, the ALM RAC articulated a theory of change (see Figure 6.2). The theory of change suggests each institution seek to make changes within its local contexts that can lead to improved teaching,

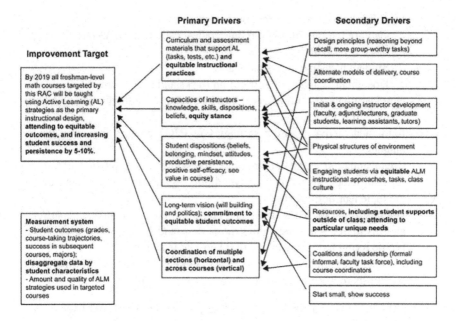

FIGURE 6.1. Active Learning Mathematics Driver Diagram. The ALM RAC Driver Diagram, Updated June 2018 to Include a Focus on Equity.

which in turn leads to improved student success. With the ALM RAC working as a Networked Improvement Community (NIC), the group can collectively identify pathways for change. The plan was that common data collected would allow ALM RAC members to connect effective teaching and student success back to the embedded contextual factors and change efforts.

Aligned with the overarching theory of change, and the primary and secondary drivers, the ALM RAC has prioritized several specific change strategies:

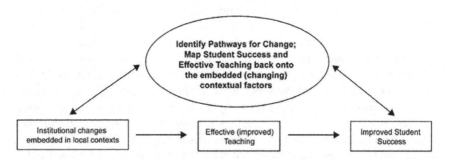

FIGURE 6.2. ALM TAC Theory of Change. The ALM RAC theory of change supports efforts to move toward a common goal.

- Establish baselines of student outcomes:
  - Capture the rate of non-passing grades, called the DFW rate for grades of D, F, or Withdraw;
  - Student measures of attitude; and
  - Document analyses of curriculum (tasks and assessments).
- Hire learning assistants to assist lead instructors with facilitating collaborative group work.
- Implement active learning strategies in P2C2 courses in a coordinated manner.
- Create and offer initial and ongoing professional development and training for instructors, including faculty, adjuncts, graduate students, and undergraduate learning assistants.
- Document the current climate and culture of the mathematics department and institution—understanding the system and contexts is a crucial step before trying to improve the culture around teaching.
- Enhance communication between and among ALM RAC members as a way to share learnings and accelerate change efforts.

The ALM RAC members have extensively discussed viable strategies to accomplish each of these priorities. For instance, part of ongoing training of instructors would involve observing the instructor, offering feedback, and then observing again (e.g., small Plan-Do-Study-Act (PDSA) Cycles[2] with each instructor). As another example, ALM RAC members each planned to return to their departments and have conversations with other instructors to gauge their interest in and familiarity with active learning strategies. Such conversations can lead to collective visioning of effective instruction, along with *will-building* for the implementation of active learning strategies; will-building is the process of getting stakeholders to collectively agree to a common vision, and thus to support the associated change efforts.

Since 2013, the ALM RAC members have worked collaboratively to improve instruction in introductory calculus courses, using change strategies and the driver diagram to make progress toward a common aim. While the contexts across the 20 ALM RAC member campuses are quite different, requiring somewhat different approaches to implementing ALM, the NIC model has been very beneficial, providing a structure to develop a collective wisdom. The ALM RAC has shared strategies for accessing and analyzing student course-related data, exchanged and co-developed instructional resources, used common measures to document shifts in student dispositions, and have regularly discussed local models used to support learning environments. Several campuses followed the University of Colorado's lead and hired learning assistants. Most campuses began by focusing on a single course, and over time have expanded their efforts to include additional P2C2 courses, prerequisite courses to Precalculus, and Calculus 3. Discussions across campuses have helped to identify key features of effective approaches and have confirmed the critical role of supporting instructors to learn how to successfully adopt active learning strategies. On some campuses, efforts are now sustainable, whereas on other campuses the efforts are just beginning, or expanding. Ongoing work includes additional coordinated data collection about student and instructor experiences and attitudes toward active learning.

ALM RAC EXTERNAL FUNDING

Twice, the ALM RAC received external funding to support change efforts. First, the ALM RAC received funding in 2013, as part of a larger grant from the Helmsley Charitable Trust to the MTE-Partnership (see Chapter 1 of this book). Based on the ALM RAC membership in 2013, the funding was divided among the five active member universities. The goal for the Helmsley funding to the ALM RAC was to transform instruction (aligned with ALM RAC strategies) in targeted freshman-level courses. Each campus had its own budget; most institutions funded researcher time to collect and analyze common data, hired learning assistants in targeted courses, and supported travel to collaborate across the teams and learn from one another.

The Helmsley grant, along with funding from the Academy of Inquiry-Based Learning to some individual ALM RAC members, served as a springboard for larger projects within the ALM RAC. For example, the University of Nebraska at Omaha partnered with the University of Colorado Boulder in the creation of TAC-Tivities (tactile learning activities that were originally created for the calculus sequence; see next section). Many universities within the ALM RAC transformed all or parts of their calculus sequence curriculum as a result of this work. A greater focus on student engagement and active learning with less focus on instructor lecture was at the heart of these transformations.

The second funding opportunity for the ALM RAC sought to broaden views of change efforts and was built on the experiences from the Helmsley funding. Some departments were able to use the relatively small Helmsley awards and translate those into large and sustained changes, while other campuses were unable to get their changes beyond the few faculty who were involved in planning the work. Discussions occurred about what was different, which led to a desire to more formally understand how and under what conditions attempts to adopt active learning principles in P2C2 courses might transform a department's culture related to teaching. Thus, a subset of the ALM RAC members decided to pursue an Improving Undergraduate STEM Education grant from the National Science Foundation to help other mathematics departments transform their teaching and learning in the calculus sequence. A five-year collaborative grant was awarded to the Association of Public and Land-grant Universities (APLU), University of Colorado Boulder, University of Nebraska–Lincoln, and San Diego State University in September 2016 (NSF DUE-1624643, 1624610, 1624628, and 1624639). *Student Engagement in Mathematics through an Institutional Network for Active Learning (SEMINAL)* is focused on answering the research question: What conditions, strategies, interventions, and actions at the departmental and classroom levels contribute to the initiation, implementation, and institutional sustainability of active learning in the undergraduate calculus sequence (precalculus through Calculus 2) across varied institutions?

The SEMINAL grant funds a two-phase research project, both collecting qualitative and quantitative data. Phase 1 (2016–2018) collected and analyzed data from six institutions, which had successfully implemented ALM in in the P2C2 sequence (Smith, Webb, Martin, Strom, & Voigt, forthcoming). Phase 2 (2018–2021) solicited nine additional institutions (through an open call for proposals), each seeking to implement active learning strategies in P2C2 courses and positively change their departmental cultures around teaching and learning.

Although built on a foundation of the ALM RAC work, the SEMINAL grant moves a step further, shifting from individuals implementing active learning to departmental transformation. In parallel to SEMINAL, the work of the ALM RAC continues. As the MTE-Partnership has shifted a focus from local teams participating in single RACs toward local teams engaging in multi-faceted program transformation efforts, the ALM RAC has gained members. The ALM RAC has thus been working recently to better organize findings and PDSA Cycles to date to optimize bringing new members up to speed. In the next section, four vignettes from a subset of ALM RAC member institutions are shared, to illustrate different aspects of the primary drivers, as enacted in different contexts.

ILLUSTRATING ALM RAC CHALLENGES

The following four vignettes from ALM RAC participating institutions tell portions of their change stories and illustrate the challenges encountered when pursuing the primary drivers in the driver diagram. Although each university's change story is unique, these stories are generally representative of the ongoing work of the ALM RAC.

First, Kennesaw State University's story depicts an example of initial change efforts, focusing on will-building. Next, Western Michigan University's story illustrates the development and enactment of a long-term vision. The University of Nebraska at Omaha's story shares ALM RAC efforts to develop appropriate curricular resources to support active student engagement. Finally, the University of Nebraska–Lincoln's story depicts how the ALM RAC attends to instructor capacities and student dispositions.

Will-Building and Initiating Change Efforts

Will-building is an important component for fostering instructional change, and is a focus of the work at Kennesaw State University. As a SEMINAL Phase 2 partner, the mathematics department is using a "top-down" and a "bottom-up" approach (NRC, 2012) and leveraging communities of practice (Henderson et al., 2012) to expand conversations about teaching and learning and provide supports to encourage faculty to integrate active learning practices into P2C2 courses. For its top-down and bottom-up approach, the mathematics department has: (a) worked with the administration to update the placement process to ensure that entering students are placed in mathematics courses best suited for them; (b) aligned the department's work with the goals of the College—to increase academic success, progression, retention, and graduation rates for students with different backgrounds and experiences and to foster faculty excellence in teaching; and (c) implemented a course coordination program for all sections of Precalculus and Calculus 1 to facilitate both horizontal and vertical alignment of course content. Layered with these efforts is the development of instructor communities of practice. To leverage the benefits of communities of practice to support instructional change, the department is holding monthly department-wide teaching conversations about research-based instructional practices that foster student success and engaging groups of P2C2 faculty in yearlong faculty learning communities focused on using active learning in mathematics. Faculty learning community participants are also provided with undergraduate learning assistants to help facilitate active learning during class and given teaching tools (e.g., student response systems) to support their efforts.

Although Kennesaw State University has had several successes, such as instructors accepting increased coordination efforts and the support of the Center for Excellence in Teaching and Learning for building communities of practice, the department is still challenged with finding ways to include part-time faculty in these conversations. Many part-time faculty are on campus to teach in the evenings and are not available to participate in departmental teaching conversations or faculty learning communities as these events occur during the day. The course coordinators are currently working to create electronic resources (e.g., active learning activities and related assessment items) that all faculty can use, but this does not address the value of engaging part-time faculty in discussions with peers about teaching.

Another challenge is helping students to shift their thinking about teaching and learning mathematics. Active learning classrooms involve students in taking more ownership for their learning and their peers' learning, and the thinking process may be quite different from what students have experienced in prior mathematics courses. As such, some students are resistant to engaging fully in active learning experiences. Although instructors share their reasons for using these practices and learning assistants offer support for students as they adjust to the new expectations, additional support is needed to fully transition the teaching and learning experience.

Participating in the ALM RAC has fostered the exchange of ideas with other mathematics departments working toward similar goals. Sharing ideas, resources, and learnings has been particularly valuable to Kennesaw State University so that they have strategies for will-building early in their efforts. Ongoing interactions with ALM RAC colleagues will help expand understandings for how to best support all faculty in using active learning and how to support students in learning mathematics in new ways. Thus, this vignette illustrates a program that is among the more recent members to the ALM RAC; early work has focused on the drivers of will-building, instructor supports, and improving student attitudes.

Coalition of the Willing and Building Long-Term Vision through Multiple Liaisons

The Department of Mathematics at Western Michigan University offers three different version of calculus: Calculus, Calculus for Science/Engineering, and Applied Calculus. The calculus and science/engineering versions are both two-semester sequences that feed into the same two-semester sequence of Multivariate Calculus and Matrix Algebra and Differential Equations and Linear Algebra. The department's initial improvement focus on the calculus sequence was a consequence of faculty dissatisfaction from numerous departments with the level of student success. For example, College of Engineering and Applied Sciences programs were successfully recruiting minority students; however, these students were failing to progress in their chosen degrees. The college attributed the lack of success to the students' inability to successfully pass through the first gateway course in their programs, Calculus 1. The Department of Mathematics chose to redesign the general track of Calculus 1 because this course serves as a gateway course to majors in five different colleges: College of Arts and Sciences, College of Engineering and Applied Sciences, College of Business, College of Aviation, and College of Education.

The efforts to examine and restructure Calculus 1 had many stakeholders whose participation and feedback were essential to success. Building a common vision among these stakeholders was a critical component early in the design and implementation of changes to the course and to developing the will to maintain the changes in the face of resistance inside and outside the department. Both institutional and extra-institutional stakeholders provided support to the department to investigate issues related to student success in calculus and then to develop interventions to address the lack of student progression through the calculus sequence.

The chair of the Department of Mathematics formed a calculus task force to investigate issues around student success in the calculus sequence in the fall of 2014. The calculus task force pulled membership broadly from partner disciplines in the Colleges of Arts and Sciences and Engineering and Applied Sciences (deans, chairs, and faculty); staff from the advising offices and Student Success Services; instructors from local community colleges; and faculty from other universities in Michigan. The calculus task force identified numerous issues that impacted students' abilities to successfully complete the P2C2 sequence: placement, student dispositions and preparedness, content organization, teacher-centered pedagogy, and lack of coordination. One of the findings of the calculus task force was that over the previous decade the DFW[3] rates in both Calculus 1 and 2 had been steadily climbing from 33% to 50%. While the aggregated data showed a clear upward trend, the disaggregated data showed a high degree of variability among sections in the same semester and across semesters. This variability was believed to be a natural consequence of the minimal footprint of coordination of the instruction of the course. Little effort was made to coordinate the instruction of the sections beyond requiring a common text and a common list of sections to cover. Typically, instructors for Calculus 1 were assigned after graduate courses and upper-level undergraduate courses because it was commonly believed that any mathematics faculty member had the content knowledge needed to teach the course.

Collaborative coordination was used as an avenue to redesign the curriculum, pedagogy, and assessments and to develop a common vision of the course among the mathematics department faculty. A team of mathematics and mathematics education faculty who were willing to teach Calculus 1 every semester for three years was identified, and the chair of the department committed to staffing Calculus 1 first. The initial changes established a common sequencing of content, emphasis in coverage of topics, methods of teaching and assessment, and grading scale and point distribution of types of assessments. The teaching team met prior to the start of the semester to create a common syllabus. During the semester, the team met twice weekly to discuss techniques of teaching, to develop common assessments and rubrics, to discuss different methods of teaching particular content and to discuss issues that students may be struggling with in the course. The first year, the cohesiveness of the team was further enhanced by the participation of the co-coordinators in a Faculty Learning Community on Student Engagement in the Classroom through the Office of Faculty Development, funded by the dean of the College of Arts and Sciences. The Faculty Learning Community met biweekly for an academic year, and the conversations in that context cross-pollinated the efforts of the calculus team. Each of the co-coordinators has taught one section of the course since the redesign of the course began. Each year the team has brought in additional graduate students and faculty members to work on the implementation of the redesigned curriculum. This expansion of the team is also a critical part of building a common long-term vision.

Two extra-institutional agencies—MTE-Partnership and the Gardner Institute—provided key support to the redesign team. The provost of Western Michigan University had spearheaded the department's involvement with MTE-Partnership and supported the redesign team's attendance to the MTE-Partnership annual conferences and to the ALM RAC summer conferences. The ALM RAC supported the team through the use of classroom activities, information on how to use learning assistants to support student learning, and research tools to study the effectiveness of the changes being implemented. In parallel with participation in MTE-Partnership, the dean of the College of Arts and Sciences spearheaded Western Michigan University becoming part of the Gardner Institute's Gateways to Completion (G2C) program. This program guides departments through a process of redesigning courses to promote student success. The Gateway to Success annual conference has pre-conference sessions for faculty teams to attend workshops on teaching, community building, and on the use of analytics to predict student success. Throughout the academic year, the Gardner Institute provides webinars on active teaching and assessment methods. As part of campus involvement in the Gateways to Completion effort, the dean budgeted for the incorporation of learning assistants into all sections of Calculus 1, and the chair provided funds to pay for graders for each instructor.

The work at Western Michigan is far from complete; the curriculum and pedagogical changes incorporated into the teaching of Calculus 1 have produced significant results in student progression in their programs. The DFW rate in 2014–2015 was 50.3%; in 2016–2017, the rate was down to 23.3%. The formation of the calculus task force played a key role in creating an understanding of the extent and nature of the problems causing high DFW rates in the P2C2 sequence and provided important connections between faculty in the department and stakeholders from other departments, colleges, student services, and local institutions. The calculus task force provided support and protection for those faculty who were tasked with designing and implementing changes to the calculus course. The task force helped to: (a) develop a common vision of the issues surrounding student success in the calculus sequence, (b) obtain buy-in from faculty on the proposed changes adopted to address those issues, (c) support and protect those faculty implementing the changes, and (d) elicit support from partner disciplines and administration (programmatically and monetarily) for efforts put in place. While the two extra-institutional agencies provided needed support for the team through training and resources, upper administration funneled financial resources to the team as a consequence of participation with national organizations that were focusing on the issue of student success in calculus. Involvement with national organizations placed the program's redesign efforts in a broader context and brought in expertise that was valued by the department faculty. This story of will-building and effectively using department leadership to advance change efforts illustrates the ALM RAC primary driver related to will-building, and the associated driver of building coalitions of stakeholders.

Developing Activities to Support Active Learning Mathematics

A few members of the ALM RAC from the Universities of Nebraska at Omaha and Colorado Boulder collaborated to develop some hands-on activities that could comprise challenging and cognitively demanding tasks. One such activity was developed for differential equations and slope fields. In this activity, the participants each received a plastic bag with laminated cards. The cards either had a differential equation on them or a drawing of a slope field. Participants worked in small groups to match each differential equation with the corresponding slope field.

This card sorting activity ended up being a springboard for the creation of active learning curricular materials that are now used in many mathematics courses around the country. These activities are called TACTivities—activities that include moving (or tactile) pieces. They are designed to encourage students to work collaboratively and use deep mathematical thinking while completing the TAC-Tivities. ALM RAC members from the University of Colorado Boulder and the University of Nebraska at Omaha worked together to develop several additional TACTivities for the P2C2 sequence. This team created several types of TACTiv-ites: (a) sorting, (b) matching, (c) ordering, (d) dominoes, (e) determining the error, and (f) games, and also wrote a set of course notes to help promote active learning in their respective calculus sequences. Some of these materials are available online: http://math.colorado.edu/activecalc/. To further support mathematics teacher development, the TACTivities are also used in mathematics methods courses for prospective secondary teachers and incorporated into professional development experiences for practicing teachers.

One example of a TACTivity that was created by the Boulder/Omaha Active Learning Alliance (BOALA) group is *Indefinite Integral Dominoes* (see Figure 6.3). This TACTivity is used to review indefinite integrals and to help students practice indefinite integrals while making deeper mathematical connections. Students work in groups of two to four with all of the dominoes cut out on the solid lines. Then they match the indefinite integral with the corresponding antiderivative. Some integrals are easier and can be determined by memorization, such as the integral (g) in Figure 6.3. Other integrals require students to use algebra or trigonometry to rewrite the integrand before matching it with its antiderivative, such as integral (m) in Figure 6.3. This TACTivity often takes between 30 and 45 minutes for students to complete and could be used as an exam review.

A barrier to implementing active learning strategies is the extensive time it takes to create new instructional materials; thus, a set of materials for P2C2

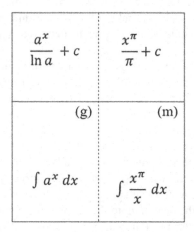

FIGURE 6.3. Sample TACTivity. Sample TACTivity, "Indefinite Integral Dominoes", in which students match small (paper) dominoes with indefinite integrals with antiderivatives.

courses made available to ALM RAC members helped others make progress on improving their P2C2 curricula. The successes at the universities using active learning courses has been phenomenal. For example, at the University of Nebraska at Omaha, the rates of success for students enrolled in two active learning classes (grades of C or higher in Calculus 1 and Calculus 2) was over 90% while those successful in Calculus 2 without an active learning class was a little over 77%. Both universities are still employing active learning methodologies, even though there have been some faculty members who have moved to other universities. Overall, this vignette is an example of ALM RAC members collaborating to address the first primary driver by creating mathematical tasks that support active learning pedagogies.

Attending to Student Attitudes and Capacities of Instructors

The ALM reforms at the University of Nebraska–Lincoln (UNL) began around 2012 as part of an effort by the Department of Mathematics to address low pass rates for Precalculus courses. Department leaders could see that university administrators were beginning to think more about freshman retention and six-year graduation rates. Local data showed two-thirds of freshmen enroll in a mathematics course in their first semester, with no other department having even one-third of the freshmen class enrolled. Thus, mathematics grades correlate strongly with freshmen retention. UNL developed its own theory of change that included more details than the overall ALM RAC driver diagram. Precalculus was (and still is) largely taught by graduate teaching assistants and is the first course they teach as instructors of record. One key dimension to the change process was that the Department of Mathematics wanted to develop a better understanding of the students who enrolled in entry-level mathematics courses. In particular, the Department wanted to know students' reasons for taking Precalculus and their beliefs about mathematics so that they could identify ways to better engage students in the learning process.

The ALM RAC developed the Collegiate Active Learning in Calculus Survey (CALCS) to use as one of the common data collection instruments administered near the beginning and end of the semester to capture student attitudes toward mathematics (usefulness of mathematics, flexible orientation toward mathematics, non-productive beliefs about mathematics, active learning), as well as student perceptions of the pervasiveness of active learning strategies in math class and intent to take future math courses. This survey was adapted from the University of Colorado's Colorado Learning Attitudes about Science Survey (CLASS; Adams, Perkins, Dubson, Finkelstein, & Wieman, 2004). Some sample items are shown in Table 6.2. Beliefs from the beginning to the end of each semester often follow a monotonic decreasing trajectory (Grover, 2015). Thus, seeing statistically flat belief scores or even less negative belief scores would be evidence of the effectiveness of ALM on student attitudes. By 2016, students at UNL were exhibiting beliefs that were statistically the same, from the beginning of the semester to

TABLE 6.2.   Sample Items from the Collegiate Active Learning in Calculus Survey.

Usefulness of Mathematics

- I think about the math in my everyday life.
- I study math to learn knowledge that will be useful in my life outside of school.
- Mathematical formulas express meaningful relationships among variables.

Flexible Orientation Toward Mathematics

- There are times I solve a math problem more than one way to help my understanding.
- When studying math, I relate the important information to what I already know rather than just memorizing it the way it is presented.
- I am not satisfied until I understand why something works the way it does.

Non-Productive Beliefs About Mathematics

- A significant problem in learning math is being able to memorize all the information I need to know.
- Knowledge in math consists of many disconnected topics.
- In doing a math problem, if my calculation gives a result very different from what I'd expect, I'd trust the calculation rather than going back through the problem.

Active Learning Mathematics

- I can learn from hearing other people's mathematical thinking, even if their thinking is not correct.
- Understanding math basically means being able to communicate your reasoning with others.
- When a question is left unanswered in math class, I continue to think about it afterward.

the end of the semester. Thus, in addition to the rising pass rates for the targeted courses, the goal of improving student attitudes (stopping the decline) has been achieved and maintained.

In addition to better understanding students, the UNL focused on instructor training; UNL knew it would not be enough to simply tell people to teach in different ways. Since the target instructors comprised graduate students and a few experienced lecturers, the training focused on graduate students. The pre-semester workshop was expanded to one week, and the newly hired director of first-year mathematics created and taught a yearlong pedagogy course to second-year graduate students during their first year as an instructor of record. Of note, first-year graduate students were most frequently put in charge of calculus recitation sections. To create the time for the second-year graduate students to take this teaching seminar, they were assigned a 1–1 load instead of the typical 2–1 load. To offset the reduction in instructional capacity, class sizes were increased from around 32 to around 40. At the same time, learning assistants were hired for these larger classes to help facilitate group work. Lai et al. (2016) noted in a report on the teaching seminar and graduate students' experiences that the seminar has been well received and includes a focus on teaching instructors how to effectively facilitate group work and other ALM strategies.

Sharing the results of the efforts to understand students and train instructors with the ALM RAC members has helped others to make similar changes. As part of the NIC efforts, a UNL researcher helps ALM RAC members collect and analyze CALCS data. The UNL instructor training has been shared with the ALM RAC and with the broader MTE-Partnership community. The aspects of UNL's work shared here address the ALM RAC drivers related to instructor training and coming to better understand students, in order to improve student attitudes toward mathematics.

FOCUS ON IMPROVEMENT

The vignettes on the previous pages provide small glimpses into the ALM RAC, addressing challenges via the work of local partnership teams to enact active learning in P2C2 courses. As a first step for local partnership teams within the ALM RAC, each assembled a group of stakeholders. The stakeholders could share their own perspectives of the contexts and together developed a vision for effective mathematics teaching and learning. In the sites where changes have encompassed whole departments and have been sustained, the mathematics department chair was an active proponent of change efforts. The efforts could be tied to university strategic plans, promoting interest and commitment from deans and other administrators. These successful efforts also included faculty seen as leaders (with or without formal leadership positions), and the P2C2 coordinator(s), when such a position existed. The successful teams worked through cycles of change: build a common vision, assess the context (local data), select change strategies, test and refine change strategies, plan next steps, disseminate results, and ensure sustainability. Throughout the cycles of change, teams kept an understanding that improvement is a process rather than a destination (see Figure 6.4). Although sustainability of changes seems far in the future, the local teams that included plans for sustainability from the beginning—particularly to handle turnover in key personnel—seems to be a key to successful change initiatives.

The Mathematical Association of America's *Instructional Practices Guide* (2018) describes a potential vision for effective mathematics teaching and learning at the undergraduate level. Building a common vision is not accomplished through a single meeting, but often takes several months of meeting with different groups of stakeholders. Along with building a common vision, each team gained access to local data, including not just overall DFW rates, but also data broken down by various student demographics (e.g., gender, ethnicity, first-generation college students, Pell-eligible, transfer status). ALM RAC members found examining course-taking trajectories and student majors to be helpful. Some campuses were able to demonstrate growth by breaking down student progress through mathematics courses based on grades in previous mathematics courses (e.g., what happens to students who get a grade of B in Precalculus—how many of them go on to pass Calculus 1 in the next two semesters?). Other campuses have used their data to find the inequities in student outcomes; developing specific plans to address demonstrated inequities helped some faculty convince administrators to provide resources for ALM change efforts.

FIGURE 6.4. AMTE Standards' Improvement Cycle. Improvement cycle from AMTE's (2017) Standards for Preparing Teachers of Mathematics.

Finally, members of the ALM RAC have recognized the strength of working collaboratively as a NIC. Members working together across different contexts helped everyone accelerate change efforts. Learning what did and did not work in other contexts (and knowing what those contexts are), allowed everyone to make more informed choices for individual change strategies and related efforts. Working together with a NIC provided not only resources but also ongoing motivation to continue working on change efforts over time.

The field knows more now about how people learn; students need to be actively engaged in learning, which is hard for most students to do while listening to lectures (e.g., Laursen et al., 2011; Laursen, Hassi, Kogan, & Weston, 2014). Attempting to change the culture of a mathematics department so that active learning is the status quo is not an easy task, but it can be done. The ALM RAC and MTE-Partnership have benefited from following the Carnegie approach to educational improvement (see Bryk, Gomez, Grunow, & LeMahieu, 2015). The ALM RAC also welcomes new members; see website (http://mtep.info/ALM) for more information about how to join this community.

READING LIST

The following annotated bibliography provides recommended readings that can help immerse someone in ideas about institutional change, effective teaching practices, and effective calculus programs.

1.   Elrod, S., & Kezar, A. (2016). *Increasing student success in STEM: A guide to systemic institutional change.* Washington, DC: Association of American Colleges and Universities.

– Practical handbook for transforming institutions (including colleges) and thinking about change comprehensively and systemically.

2. Bryk, A. S., Gomez, L., Grunow, Al., & LeMahieu, P. (2015). *Learning to improve: How America's schools can get better at getting better.* Boston, MA: Harvard Education Publishing.
   – This book from the Carnegie Foundation focuses on K–12 school systems and how to apply improvement science to efforts.

3. Freeman, S., Eddy, S. L., McDonough, M., Smith, M. K., Okoroafor, N., Jordt, H., & Wenderoth, M. P. (2014). Active learning increases student performance in science, engineering, and mathematics. *Proceedings of the National Academy of Sciences, 111*(23), 8410–8415.
   – This article demonstrates positive student outcomes at the undergraduate level, connected with the implementation of active learning strategies in science, mathematics, and engineering courses.

4. Bressoud, D., Carlson, M. P., Mesa, V., & Rasmussen, C. (2013). The calculus student: Insights from the Mathematical Association of America national study. *International Journal of Mathematical Education in Science & Technology, 44*(5), 685–698.
   – Summary document of seven characteristics of undergraduate calculus programs with positive student outcomes.

5. Kezar, A. (2014). *How colleges change: Understanding, leading, and enacting change.* New York, NY: Routledge.
   – Longer book than the Elrod & Kezar handbook, detailing how colleges and approach change systemically.

6. Hayward, C. N., Kogan, M., & Laursen, S. L. (2015). Facilitating instructor adoption of inquiry-based learning in college mathematics. *International Journal of Research in Undergraduate Mathematics Education, 1,* 1–24.
   – Article discussing inquiry-based learning and how college mathematics departments can help instructors adopt such teaching practices.

7. Laursen, S. L., Hassi, M.-L., Kogan, M., & Weston, T. J. (2014). Benefits for women and men of inquiry-based learning in college mathematics: A multi-institution study. *Journal for Research in Mathematics Education, 45*(4), 405–418.
   – This research article shows the positive impact of inquiry-based teaching practices on student outcomes, particularly for female students.

8. *MAA Instructional Practices Guide.* (2018). Retrieved from https://www.maa.org/programs-and-communities/curriculum%20resources/instructional-practices-guide
   – This handbook targets mathematics instructors who want practical tips for how to improve instructional practices, including task design, course design, and assessment.

As follows are additional selected publications by ALM RAC members about their ALM RAC work.

1. Lawler, B. R., Ronau,R. N., & Mohr-Schroeder, M. J. (Eds.). (2016). *Proceedings of the fifth annual Mathematics Teacher Education Partnership conference.* Washington, DC: Association of Public and Land-grant Universities.

    Smith, W. M., Lawler, B. R., Bowers, J., & Augustyn, L. C. (Eds.) (2017). *Proceedings of the sixth annual Mathematics Teacher Education Partnership.* Washington, DC: Association of Public and Land-grant Universities.

    Smith, W. M., Lawler, B. R., Strayer, J., & Augustyn, L. C. (Eds.) (2018). *Proceedings of the seventh annual Mathematics Teacher Education Partnership.* Washington, DC: Association of Public and Land-grant Universities.

    – The MTE-Partnership proceedings include a summary of ALM RAC work of the year, and chapters/abstracts of research work completed by ALM RAC members. Each of the proceedings includes multiple chapters about ALM RAC and its work.

2. Apkarian, N., Bowers, J., O'Sullivan, M., & Rasmussen, C. (2018). A case study of change in the teaching and learning of Precalculus to Calculus 2: What we're doing with what we have. *PRIMUS, 28*(6), 528–549.

    – This article describes the changes at one university, which has been a member of ALM RAC.

3. *PRIMUS* (2017). *PRIMUS, 17*(7, special issue). Retrieved from https://www.tandfonline.com/doi/abs/10.1080/10511970.2017.1393846

    – This special issue of PRIMUS is dedicated to active learning strategies in the first two years of collegiate mathematics.

4. Ernst, D. C., Hodge, A., Yoshinobu, S. (2017). What is inquiry-based learning? *Notices of the AMS, 64*(6), 570–574. Retrieved from https://www.ams.org/journals/notices/201706/rnoti-p570.pdf

    – This article describes active learning strategies in more detail, including work done at one ALM RAC member university.

5. Lai, Y., Smith, W. M., Wakefield, N. P. Miller, E. R., St. Goar, J., Groothuis, C. M., & Wells, K. M. (2016). Characterizing mathematics graduate student teaching assistants' opportunities to learn from teaching. In J. Dewar, P. Hsu, & H. Pollatsek (Eds.), *Mathematics education: A spectrum of work in mathematical sciences departments* (pp. 73–88). Association for Women in Mathematics Series. Cham, Switzerland: Springer International Publishing.

    – This chapter describes the way one ALM RAC member university provides opportunities for graduate students to learn how to teach in ways that engage students actively.

6. Wakefield, N., & Smith, W. M. (2016). Enriching student's online home-work experience in pre-calculus courses: Hints and cognitive supports. In T. Fukawa-Connolly (Ed.), *Proceedings of the 19th Research in Undergraduate Mathematics Education Conference* (pp. 1393–1394). Pittsburgh, PA.
    - This chapter describes how one ALM RAC member university changed the hint feature in an online homework system to better support active learning goals.

For current information on the ALM RAC, please visit mtep.info/ALM.

CONCLUSION

A summary of the key lessons learned by the ALM RAC include:

- To increase the quality and quantity of secondary mathematics teachers, freshmen-level mathematics courses need to be taught in ways that increase (rather than decrease) student engagement and interest in mathematics;
- Approach change systematically and with an explicit plan;
- Understand the local contexts (including beliefs, knowledge, and experiences of stakeholders, from students to provosts);
- Active leaders help efforts scale up and become sustained;
- Plan for sustainability, including turnover in personnel; and
- See improvement as cycles rather than as a destination.

These key lessons are drawn from the experiences of all the ALM RAC members, including efforts that encountered barriers and have not yet been successfully scaled up to encompass entire departments.

Moving forward, the ALM RAC members continue to share lessons learned and to use their collective wisdom to move local efforts forward. In keeping with NIC principles, ALM RAC members have adopted stances of continuous improvement: transformation is not a destination, but an unending process of improvement. Findings from the SEMINAL grant related to key change levers to effect departmental transformations will be utilized by ALM RAC members to aid in their own local efforts. The ALM RAC members are also working to better organize a dynamic way to share mathematical activities, which can include instructor notes and comments and revisions added by other users of the materials (e.g., notes about other local contexts and how activities were adapted to fit those contexts). In these efforts, the ALM RAC is hoping to learn from the Transformation Working Group's knowledge generation and management system research (see Chapter 2 of this book).

By improving lower-level mathematics courses and embodying active-learning pedagogies that have the potential to enhance K–16 mathematics teaching and learning, the ALM RAC contributes to the MTE-Partnership's ultimate goal of increasing the quality and quantity of mathematics teachers. Ten percent to 25% of freshmen STEM majors at research-intensive universities switch away from any STEM major after their first year of college, often citing bad experiences in their freshman mathematics course (e.g., Rasmussen, Ellis, & Bressoud, 2015). Improving freshman mathematics courses is a promising avenue to reduce such negative experiences and subsequent exit from STEM career pathways. The Association of Mathematics Teacher Educators' (2017) *Standards for the Preparation of Teachers of Mathematics* reference high-quality mathematics courses for future teachers; pedagogies that actively engage students and focus on reasoning are well-aligned with this vision for high-quality mathematics courses. Additionally, since teachers' default pedagogy tends to be how they themselves experienced mathematics as learners (e.g., Feiman-Nemser, 2012), having future teachers experience active learning pedagogies as undergraduates will help to reinforce such pedagogies in their own future practices as mathematics teachers.

ENDNOTES

1. Work on this chapter was supported in part by a grant from The Helmsley Charitable Trust. All findings and opinions are those of the authors, and not necessarily those of the funding agency.
2. The Plan-Do-Study-Act (PDSA) Cycle is an iterative four-step process to plan and document the viability/efficacy of a proposed change to a system by planning a small change, implementing the change, observing the results, and then acting on what is learned.
3. In addition to grades of D, F, and Withdraw (DFW), at Western Michigan, the DFW rate includes grades of Incomplete.

REFERENCES

Adams, W. K., Perkins, K. K., Dubson, M., Finkelstein, N. D., & Wieman, C. E. (2004). The design and validation of the Colorado Learning Attitudes about Science Survey. *2004 Physics Education Research Conference, 790,* 45–48.

Aguirre, J. M., Mayfield-Ingram, K., & Martin, D. B. (2013). *The impact of identity in K–8 mathematics; Rethinking equity-based practices.* Reston, VA: National Council of Teachers of Mathematics.

Association of Mathematics Teacher Educators. (2017). *Standards for preparing teachers of mathematics.* Raleigh, NC: Author. Retrieved from http://amte.net/standards

Astin, A. (1999). Student involvement: A developmental theory for higher education. *Journal of College Student Development, 40*(5), 518–529.

Bank, B. J. Slavings. R. L., & Biddle, B. J. (1990). Effects of peer, faculty, and parental influences on students' persistence. *Sociology of Education, 63*(3), 208–225.

Battey, D., Neal, R., Leyva, L. A., & Adams-Wiggins, K. (2016). The interconnectedness of relational and content dimensions of quality instruction: Supportive teacher-student relationships in urban elementary mathematics classrooms. *The Journal of Mathematical Behavior, 42,* 1–19.

Boaler, J., & Brodie, K. (2004, October). The importance, nature and impact of teacher questions. In D. E. McDougall & J. A. Ross (Eds.), *Proceedings of the 26th annual meeting of the North American chapter of the International Group for the Psychology of Mathematics Education* (vol. 2, pp. 773–781). Toronto, Ontario, Canada.

Bonwell, C. C., & Eison, J. A. (1991). *Active learning: Creating excitement in the classroom.* ASHE-ERIC Higher Education Report No. 1. (ERIC Document Reproduction Service No. 336 049)

Bressoud, D., Carlson, M. P., Mesa, V., & Rasmussen, C. (2013). The calculus student: Insights from the Mathematical Association of America national study. *International Journal of Mathematical Education in Science & Technology, 44*(5), 685–698.

Bressoud, D., Mesa, V., & Rasmussen, C. (Eds.). (2015). *Insights and recommendations from the MAA national study of college calculus.* Washington, DC: MAA Press.

Bryk, A. S., Gomez, L., Grunow, A., & LeMahieu, P. (2015). *Learning to improve: How America's schools can get better at getting better.* Boston, MA: Harvard Education Publishing.

Cai, J., & Cirillo, M. (2014). What do we know about reasoning and proving? Opportunities and missing opportunities from curriculum analyses. *International Journal of Educational Research, 64,* 132–140.

Callahan, K. M. (2016a). The nature of mathematical conversations among prospective middle school teachers' in a mathematics content course. *Proceedings of the 43rd Annual Meeting of the Research Council on Mathematics [RCML],* (pp. 75–82). Orlando, FL.

Callahan, K. M. (2016b). Prospective middle school teachers' generalizing actions as they reason about algebraic and geometric representations of even and odd numbers. *Teacher Education & Practice, 29*(4), 630–641.

Callahan, K. M., & Hillen, A. F. (2012). Prospective teachers' transition from thinking arithmetically to thinking algebraically about even and odd numbers. In L. R. Van Zoest, J. J. Lo, & J. L. Kratky (Eds.), *Proceedings of the 34th Annual Meeting of the North American Chapter of the International Group for the Psychology of Mathematics Education,* (pp. 584–590). Kalamazoo, MI: Western Michigan University.

Chamberlin, M. T. (2005). Teachers' discussions of students' thinking: Meeting the challenge of attending to students' thinking. *Journal of Mathematics Teacher Education, 8,* 141–170.

Cobb, P., & Hodge, L. (2002). A relational perspective on issues of cultural diversity and equity as they play out in the mathematics classroom. *Mathematical Thinking and Learning, 4,* 249–284

Cobb, P., Boufi, A., McClain, K., & Whitenack, J., (1997). Reflective discourse and collective reflection. *Journal of Research in Mathematics Education, 28*(3), 258–277.

Conference Board of the Mathematical Sciences. (2012). *The mathematical education of teachers II.* Providence, RI and Washington, DC: American Mathematical Society and Mathematical Association of America.

Cuoco, A., Goldenberg, E. P., & Mark, J. (1996). Habits of mind: An organizing principle for mathematics curricula. *Journal of Mathematical Behavior, 15,* 375–402.

Davidson, N. (1971). The small group-discovery method as applied to calculus instruction. *The American Mathematical Monthly, 78*(7), 789–791.

Dee, J. R., Henkin, A. B., & Hearne, J. L. (2011). Enabling initiative and enterprise: Faculty-led course redesign in a STEM discipline. *Educational Research Quarterly, 35*(1), 34–64.

Driscoll, M., & Moyer, J. (2001) Using students' work as a lens on algebraic thinking. *Mathematics Teaching in the Middle School, 6*(5), 282–287.

Ellis, A. B. (2011). Generalizing-promoting actions: How classroom collaborations can support students' mathematical generalizations. *Journal for Research in Mathematics Education, 42*(4), 308–345.

Elrod, S., & Kezar, A. (2016). *Increasing student success in STEM: A guide to systemic institutional change.* Washington, DC: Association of American Colleges and Universities.

Empson, S. B., & Jacobs, V. R. (2008). Learning to listen to children's mathematical thinking. In D. Tirosh & T. Wood (Eds.), *The international handbook of mathematics teacher education: Tools and processes in mathematics teacher education* (vol. 2, pp. 257–281). Rotterdam, The Netherlands: Sense Publishers.

Feiman-Nemser, S. (2012). *Teachers as learners.* Cambridge, MA: Harvard University Press.

Freeman, S., Eddy, S. L., McDonough, M., Smith, M. K., Okoroafor, N., Jordt, H., & Wenderoth, M. P. (2014). Active learning increases student performance in science, engineering, and mathematics. *Proceedings of the National Academy of Sciences, 111*(23), 8410–8415.

Grover, R. (2015). *Student conceptions of functions: How undergraduate mathematics students understand and perceive functions.* Unpublished doctoral dissertation, Boulder, CO.

Hayward, C. N., Kogan, M., & Laursen, S. L. (2016). Facilitating instructor adoption of inquiry-based learning in college mathematics. *International Journal of Research in Undergraduate Mathematics Education, 2,* 59–82.

Henderson, C., Dancy, M., & Niewiadomska-Bugaj, M. (2012). Use of research-based instructional strategies in introductory physics: Where do faculty leave the innovation-decision process? *Physical Review Special Topics-Physics Education Research, 8*(2), 020104. Retrieved from: https://doi.org/10.1103/PhysRevSTPER.8.020104

Herbel-Eisenmann, B. A., & Phillips, E. D. (2005). Using student work to develop teachers' knowledge of algebra. *Mathematics Teaching in the Middle School, 11*(2), 62–66.

Hiebert, J., Carpenter, T. P., Fennema, D., Fuson, K. C., Wearne, D., Murray, H., Olivier, A., & Human, P. (1997). *Making sense: Teaching and learning mathematics with understanding.* Portsmouth, NH: Heinemann.

Hobson, E. H. (1996). Encouraging self-assessment: Writing as active learning. *New Directions for Teaching and Learning. Using Active Learning in College Classes: A Range of Options for Faculty, 67,* 45–58.

Howell, K. M. (2006). *An examination of the relationship between participation in academic-centered peer interactions and students' achievement and retention in mathematics-based majors.* Unpublished Doctoral Dissertation, University of Maryland.

Izsák, A., & Sherin, M. G. (2003). Exploring the use of new representations as a resource for teacher learning. *School Science and Mathematics, 103*(1), 18–27.

Jacobsen, M. (1997). *Bridging the gap between early adopters' and mainstream faculty's use of instructional technology.* (ERIC Document Reproduction Service No. ED 423 785)

Jansen, A., Cooper, B., Vascellaro, S., & Wandless, P. (2016). Rough-draft talk in mathematics classrooms. *Mathematics Teaching in the Middle School, 22*(5), 304–307.

Jardine, R. (1997). Active learning mathematics history. *Problems, Resources, and Issues in Mathematics Undergraduate Studies, 7*(2), 115–122.

Kazemi, E., & Franke, M. L. (2004). Teacher learning in mathematics: Using student work to promote collective inquiry. *Journal of Mathematics Teacher Education, 7*, 203–235.

Kerrigan, J. (2018). Active learning strategies in the mathematics classroom. *College Teaching, 66*(1), 35–36.

Kezar, A. (2014). *How colleges change: Understanding, leading, and enacting change.* New York, NY: Routledge.

Krathwohl, D. R. (2002). A revision of Bloom's taxonomy: An overview. *Theory into practice, 41*(4), 212–218.

Ladson-Billings, G. (1997). It doesn't add up: African American students' mathematics achievement. *Journal for Research in Mathematics Education, 25*(6), 697–708.

Lai, Y., Smith, W. M., Wakefield, N. P. Miller, E. R., St. Goar, J., Groothuis, C. M., & Wells, K. M. (2016). Characterizing mathematics graduate student teaching assistants' opportunities to learn from teaching. In J. Dewar, P. Hsu, & H. Pollatsek (Eds.), *Mathematics education: A spectrum of work in mathematical sciences departments* (pp. 73–88). Association for Women in Mathematics Series. Cham, Switzerland: Springer International Publishing.

Laursen, S., Hassi, M. L., Kogan, M., Hunter, A. B., & Weston, T. (2011). *Evaluation of the IBL mathematics project: Student and instructor outcomes of inquiry-based learning in college mathematics.* Boulder, CO: Colorado University.

Laursen, S. L., Hassi, M.-L., Kogan, M., & Weston, T. J. (2014). Benefits for women and men of inquiry-based learning in college mathematics: A multi-institution study. *Journal for Research in Mathematics Education, 45*(4), 405–418.

Leatham, K. R., Peterson, B. E., Stockero, S. L., & Van Zoest, L. R. (2015). Conceptualizing mathematically significant pedagogical opportunities to build on student thinking. *Journal for Research in Mathematics Education, 46*(1), 88–124.

Legrand, M. (2001). Scientific debate in mathematics courses. In D. Holton (Ed.), *The teaching and learning of mathematics at university level: An ICMI study* (pp. 127–135). Dordrecht, The Netherlands: Klumer Academic Publishers.

Lipnevich, A. A., MacCann, C., Krumm, S., Burrus, J., & Roberts, R. D. (2011). Mathematics attitudes and mathematics outcomes of U.S. and Belarusian middle school students. *Journal of Educational Psychology, 103*(1), 105–118.

Liu, R., & Liu, E. (2000). *Institutional integration: An analysis of Tinto's theory.* (Report number HE033302). Cincinnati, Ohio: Paper presented as the Annual Forum of the Association for Institutional Research. (ERIC Document Reproduction Service No. ED44629)

Loo, C. M., & Rolison, G. (1986). Alienation of ethnic minority students at a predominantly white university. *Journal of Higher Education, 57*(1), 58–77.

Lubienski, S. T. (2002). A closer look at black-white mathematics gaps: Intersections of race and SES in NAEP achievement and instructional practices data. *The Journal of Negro Education, 71*(4), 269–287.

Ma, X., & Kishor, N. (1997). Assessing the relationship between attitude toward mathematics and achievement in mathematics: A meta-analysis. *Journal for Research in Mathematics Education, 28,* 26–47.

Mathematics Association of America (MAA). (2018). *Instructional practices guide.* Retrieved from https://www.maa.org/programs-and-communities/curriculum%20resources/instructional-practices-guide

McGuire, S. Y. (2015). *Teach students how to learn: Strategies you can incorporate into any course to improve student metacognition, study skills, and motivation.* Sterling, VA: Stylus Publishing.

Meyers, C., & Jones, T. B. (1993). *Promoting active learning: Strategies for the college classroom.* San Francisco, CA: Jossey-Bass Publishers.

Millett, K. C. (2001). Making large lectures effective: An effort to increase student success. In D. Holton (Ed.), *The teaching and learning of mathematics at university level: An ICMI Study* (pp. 137–152). Dordrecht, The Netherlands: Klumer Academic Publishers.

National Council of Teachers of Mathematics. (1989). *Curriculum and evaluations standards for mathematics.* Reston, VA: Author.

National Council of Teachers of Mathematics (2000). *Principles and standards for school mathematics.* Reston, VA: Author.

National Council of Teachers of Mathematics. (2014). *Principles to actions: Ensuring mathematical success for all.* Reston, VA: Author.

National Governors Association Center for Best Practices, Council of Chief State School Officers. (2010). *Common core state standards: Mathematics.* Washington, DC: Author.

National Research Council (1989). *Everybody counts: A report to the nation on the future of mathematics education.* Washington, DC: Author.

National Research Council, Committee on the Status, Contributions, and Future Directions of Discipline Based Education Research, Board on Science Education. (2012). *Discipline-based education research: Understanding and improving learning in undergraduate science and engineering.* Washington, DC: The National Academies Press.

Olson, S., & Riordan, D. G. (2012). *Engage to excel: Producing one million additional college graduates with degrees in science, technology, engineering, and mathematics.* Report to the President. Washington, DC: Executive Office of the President.

Pascarella, E. T., & Terenzini, P. T. (1991). *How college affects students: Findings and insights from twenty years of research.* San Francisco, CA: Jossey Bass.

Pierson, J. L. (2008). *The relationship between patterns of classroom discourse and mathematics learning.* Ann Arbor, MI: ProQuest.

Prince, M. (2004). Does active learning work? A review of the research. *Journal of Engineering Education, 93*(3), 223–231.

Pourdavood, R., Carignan, N., Martin, B., & Sanders, M. (2005). Culture, social interaction, and mathematics learning. *Focus on Learning Problems in Mathematics, 27*(1&2), 38–62.

Rasmussen, C., Ellis, J., & Bressoud, D. (2015). Who are the students who switch out of calculus and why? *Proceedings of the Korean Society of Mathematics Education: 2015 International Conference on Mathematics Education* (pp. 1–27). Seoul National University, South Korea.

Richardson, V. (1996). The role of attitudes and beliefs in learning to teach. In J. P. Sikula, T. J. Buttery, & E. Guyton (Eds.), *Handbook of research on teacher education* (2nd ed., pp. 102–119). New York, NY: Macmillan.

Russell, S. J. (1999). Mathematical reasoning in the elementary grades. *Developing Mathematical Reasoning in Grades K–12, 61*, 1.

Schraw, G., & Dennison, R. S. (1994). Assessing metacognitive awareness. *Contemporary Educational Psychology, 19*, 460–475.

Sherman, H. J., & Christian, M. (1999). Mathematics attitudes and global self-concept: An investigation of the relationship. *College Student Journal, 33*(1). Retrieved from http://www.projectinnovation.com/college-student-journal.html

Smith, M. S., & Stein, M. K. (1998). Selecting and creating mathematical tasks: From research to practice. *Mathematics Teaching in the Middle School, 3*(5), 344–350.

Smith, M. S., & Stein, M. K. (2011). *Five practices for orchestrating productive mathematics discussions.* Reston, VA: National Council of Teachers of Mathematics and Corwin Press.

Smith, M. S., & Stein, M. K. (2018). *Five practices for orchestrating productive mathematics discussions* (2nd ed.). Reston, VA: National Council of Teachers of Mathematics and Corwin Press.

Smith, W. M., & Webb, D. C. (2017). Active learning in mathematics research action cluster. In W. M. Smith, B. R. Lawler, J. Bowers, & L. Augustyn (Eds.), *Proceedings of the Sixth Annual Mathematics Teacher Education Partnership* (pp. 49–53). Washington, DC: Association of Public and Land-grant Universities.

Smith, W. M., Webb, D. C., Martin, W. G., Ström, A., & Voigt, M. (forthcoming). *Student engagement in mathematics through an institutional network for active learning: Recommendations to mathematics departments.* Conference Board of Mathematical Sciences. Washington, DC: MAA Press.

Stein, M. K., Grover, B. W., & Henningsen, M. (1996). Building student capacity for mathematical thinking and reasoning: An analysis of mathematical tasks used in reform classrooms. *American Educational Research Journal, 33*(2), 455–488.

Stylianides, G. J. (2008). An analytic framework of reasoning-and-proving. *For the Learning of Mathematics, 28*(1), 9–16.

Stylianides, G. J. (2009). Reasoning-and-proving in school mathematics textbooks. *Mathematical thinking and learning, 11*(4), 258–288.

Taraban, R., Box, C., Myers, R., Pollard, R., & Bowen, C. W. (2007). Effects of active-learning experiences on achievement, attitudes, and behaviors in high school biology. *Journal of Research in Science Teaching, 44*(7), 960–979.

Vygotsky, L. S. (1978). *Mind in society.* Cambridge, MA: Harvard University Press.

Vygotsky, L. (1986). *Thought and language.* Cambridge, MA: The MIT Press.

Wahlberg, M. (1997). Lecturing at the bored. *The American Mathematical Monthly, 104*(6), 551–556.

Weissglass, J. (1993). Small-group learning. *The American Mathematical Monthly, 100*(7), 662–668.

White, D. Y. (2003) Promoting productive mathematical classroom discourse with diverse students. *The Journal of Mathematical Behavior, 22*(1), 37–53.

Wieman, C. (2017). *Improving how universities teach science: Lessons from the Science Education Initiative.* Cambridge, MA: Harvard University Press.

Wilson, D., Jones, D., Bocell, F., Crawford, J., Kim, M. J., Veilleux, N., ... & Plett, M. (2015). Belonging and academic engagement among undergraduate STEM students: A multi-institutional study. *Research in Higher Education, 56*(7), 750–776.

Yackel, E., & Cobb, P. (1996). Sociomathematical norms, argumentation, and autonomy in mathematics. *Journal for Research in Mathematics Education, 27*(4), 458–477.

Zhao, N., Wardeska, J. G., McGuire, S. Y., & Cook, E. (2014). Metacognition: An effective tool to promote success in college science learning. *Journal of College Science Teaching, 43*(4), 48–54.

SECTION III

OPPORTUNITIES TO LEARN IN CLINICAL SETTINGS

This section provides a review of literature and current research endeavors focused on clinical experiences provided to secondary mathematics teacher candidates. Chapter 7 provides an initial framing of research in this area by the MTE-Partnership. Chapter 8 describes the general organization of the Clinical Experiences Research Action Cluster (CERAC), which was formed to conduct research on improving clinical experiences in secondary mathematics teacher preparation. The next three chapters provide summaries of the subgroups working on particular aspects of clinical experiences: early clinical experiences during methods classes; the use of co-planning and co-teaching strategies in clinical experiences, particularly student teaching; and the paired placement model for student teaching, in which two teacher candidates are placed with one mentor teacher. Finally, Chapter 12 outlines suggestions for how secondary mathematics teacher preparation programs can improve their clinical experiences, based on the research of the CERAC.

CHAPTER 7

CLINICAL EXPERIENCES FOR SECONDARY MATHEMATICS TEACHER CANDIDATES

Marilyn E. Strutchens[1,] David Erickson,
Ruthmae Sears, & Jeremy Zelkowski

Clinical Experiences is the third driver identified by the Mathematics Teacher Educators Partnership (MTE-Partnership) as part of its theory of improvement described at the beginning of this book. This chapter provides a background for the development of the Clinical Experience Research Action Cluster (RAC), which initially began as a working group. The first section of the chapter reviews research and recommendations related to clinical experiences for secondary mathematics teacher candidates. These fall into three categories: mentor teachers and mentoring; internships and student teaching; and institutional partnerships or infrastructures that simultaneously support interns and their mentors. The next section of the chapter gives an overview of internship practices at universities and colleges. The final section of the chapter describes the MTE-Partnership's strategy for improving teacher candidates' clinical experiences at its institutions.

REVIEW OF LITERATURE

Traditionally, student teaching was a short component that occurred at the end of academic coursework for teacher preparation. Now, it is more frequently the last

The Mathematics Teacher Education Partnership: The Power of a Networked Improvement Community to Transform Secondary Mathematics Teacher Preparation, pages 179–198.
Copyright © 2020 by Information Age Publishing

of a series of school-based experiences that begin early in a teacher candidate's preparation program. According to Wilson, Floden, and Ferrini-Mundy (2001), clinical experiences vary both within and across institutions and may be designed to show what the job of teaching is like, develop skills in instruction and classroom management, and provide practical reality to concepts encountered in university coursework. Many of these experiences occur during prospective teachers' methods courses and may include teaching small groups of students in a teacher's classroom, working in an after-school program as a tutor, teaching a few random lessons to a whole class, and other activities that give teacher candidates the opportunity to experience some aspect of teaching. Later in the program are clinical experiences in which teacher candidates begin to take on formal instructional roles.

As stated in the introduction to this book, improving clinical experiences through partnerships with mentor teachers and other stakeholders is a primary driver toward the MTE-Partnership's improvement target of creating a gold standard for secondary mathematics teacher education candidates and increasing the number of graduating secondary mathematics teachers coming from a variety of racial or ethnic, cultural, sociolinguistic, or other backgrounds. The related problem of preparing and supporting mentor teachers arose from the need for practicing teachers to adjust to the shifts in teaching required by the *Common Core State Standards for Mathematics* (*CCSS-M*; National Governors Association Center for Best Practices, Council of Chief State School Officers, 2010) and other college and career ready standards. The need to shore up the skills of practicing teachers contributes to the inadequate supply of high-quality mentor teachers who can mentor prospective teachers. Furthermore, teacher preparation programs and the schools in which teacher candidates' clinical experiences occur may not share a common vision for teacher preparation and understanding of the *CCSS-M*.

These are important considerations when looking at the kind of field experiences that prospective teachers will have. Student teaching in which experienced teachers mentor prospective teachers is an especially important type of clinical experience. As a summary of research on teacher preparation concludes:

> Study after study shows that experienced and newly certified teachers alike see clinical experiences (including student teaching) as a powerful—sometimes the single most powerful—component of teacher preparation. Whether that power enhances the quality of teacher preparation, however, may depend on the specific characteristics of the field experience. (Wilson, Floden, & Ferrini-Mundy, 2001, p. 17)

During these experiences, prospective teachers develop *the craft of teaching—* the ability to design lessons that involve important mathematical ideas, design tasks that will help students to access those ideas, and to successfully carry out the lessons. This craft may include effectively launching lessons, facilitating student engagement with tasks, orchestrating meaningful mathematical discussions, and helping to make explicit the mathematical understandings that students construct (Leatham & Peterson, 2009, p. 115).

Principles to Actions: Ensuring Mathematical Success for All (National Council of Teachers of Mathematics [NCTM], 2014) describes the craft of mathematics teaching as involving eight mathematics teaching practices:

1. Establish mathematics goals to focus learning,
2. Implement tasks that promote reasoning and problem solving,
3. Use and connect mathematical representations,
4. Facilitate meaningful mathematical discourse,
5. Pose purposeful questions,
6. Build procedural fluency from conceptual understanding,
7. Support productive struggle in learning mathematics, and
8. Elicit and use evidence of student thinking.

Clinical experiences should provide teacher candidates with opportunities to develop these practices under the guidance of expert mentors, their mentor teachers, and university supervisors.

Mentor Teachers and Mentoring

As noted in Chapter 1 of this book, only about half of secondary mathematics teachers report using instructional practices and goals that promote the *CCSS-M* (Banilower et al., 2013; Markow, Macia, & Lee, 2013). Moreover, many secondary teachers prepared before the era of the *CCSS-M* need opportunities to study and learn mathematics and statistics that they have not previously taught (Conference Board of the Mathematical Sciences [CBMS], 2012). The need to shore up the skills of practicing teachers contributes to the inadequate supply of high-quality teachers who can mentor teacher candidates. This need may be especially acute for the case of middle grades teachers because few preparation programs address the mathematics of middle grades. University faculty must provide support to mentor teachers through professional development, sharing of relevant research, engaging in teaching students in the school, and other support as defined through the relationship.

Professional Development for Secondary Teachers. *The Mathematical Education of Teachers II* (CBMS, 2012) recommends three types of professional development for all practicing secondary teachers: further study of specific topics (e.g., graduate courses); experiences designed to help teachers understand the nature of doing mathematics and statistics (e.g., math teachers' circles[2], study groups focused on further mathematics[3], or immersion experiences[4]); and opportunities for discussing teaching and learning of mathematics (e.g., lesson study, observation and discussion of demonstration lessons, study groups focused on school mathematics, or other activities of professional learning communities).

Many of these activities involve mathematics or education faculty members offering in-service teachers opportunities to learn about the constraints and affordances of secondary school environments as well as become acquainted with individual teachers. Findings from research with elementary and middle grades teachers indicate that, although lesson study *per se* may not result in measurable changes, lesson study with use of "toolkits" for fractions, proportional relationships, or area of polygons has yielded increases in student learning and teachers' mathematical knowledge and mathematical knowledge for teaching (Gersten, Taylor, Keys, Rolfhus, & Newman-Gonchar, 2014; Lewis & Perry, 2014, 2015). Teachers' math circles are aimed primarily at middle grades teachers and oriented toward the nature of doing mathematics rather than teaching it. Despite this orientation, teachers in math circles have shown increases in the number concept and operation subsection of the Learning Mathematics for Teaching instrument (White, Donaldson, Hodge, & Ruff, 2013). See Chapters 4 and 5 in this book for discussions of mathematical knowledge for teaching.

Identifying Mentor Teachers. Two leading organizations offer ideas on how to identify mentor teachers. The American Association of Colleges for Teacher Education (2010) recommends that clinical teachers (a) have at least three years of teaching experience, be matched to their novice teachers by subject and grade level, and be selected jointly by preparation program and school faculty; (b) be determined by excellent supervisor and peer evaluations as well as outstanding performance on a teacher performance assessment; and (c) be trained in helping and supporting novice teachers, modeling excellent teaching practice and exercising positive problem-solving skills. The Council for the Accreditation of Educator Preparation (2015) requires that all clinical educators, including mentor teachers "demonstrate a positive impact on candidates' development and P–12 student learning and development" (Standard 2.1).

Examples of how mentor teachers can be nurtured, identified, and supported occur in projects funded by the National Science Foundation's Math and Science Partnership (MSP) program. For more than a decade, this research and development effort has worked to build capacity and integrate the work of science, technology, engineering and mathematics (STEM) disciplines and faculty with that of K–12 to strengthen and reform mathematics and science education. Many of the partnerships have groomed teacher leaders who work both in and out of the classroom and who can and do serve as mentor teachers for mathematics teacher candidates. The TEAM-Math Secondary Mathematics Teacher Leader Academy was one project in which in-service teachers (fellows) worked on advanced degrees and received professional development focused on teacher leadership responsibilities, such as mentoring new teachers and facilitating the growth of veteran teachers in implementing the mathematics teaching practices (NCTM, 2014) and serving as mentor teachers for teacher candidates. The fellows in the program also received preparation for providing professional learning opportunities for the teachers in their building, district, state, and at the national level (Strutchens & Martin, 2017). Another

MSP program known for its work in developing teacher leaders is the Math in the Middle Institute Partnership. The Math in the Middle Institute Partnership educated and supported teams of middle-level (grades 5–8) mathematics teachers to become leaders in their schools, districts, and Educational Service Units through developing the teachers' mathematics content knowledge, pedagogical practices, leadership skills, and the ability to conduct action research about their teaching practices and leadership skills, as well as the ability to apply what they had learned in their classrooms, schools, and districts (Augustyn & Lewis, 2011).

Mentoring. Mentoring—as opposed to simply supervising and evaluating—is encouraged as an approach to support the development of novice teachers, both teacher candidates and practicing teachers. The assumption is that novice teachers will learn how to teach as they engage in authentic teaching activities supported by a more experienced, knowledgeable, and effective teacher (Ball & Cohen, 1999; Lave & Wenger, 1991). Mentoring is also encouraged as a means to retain teachers in the profession (Brill & McCartney, 2008; Darling-Hammond, 2000; Ingersoll & Strong, 2011; and Chapter 13 of this book). The expectation is that the mentor will work within the mentee's *zone of proximal development*—guiding the mentee to greater levels of effectiveness by working with him or her to accomplish what can be achieved with the mentor's help and cannot be achieved by the mentee alone. Odell, Huling, and Sweeney (2000) assert that high-quality mentoring is a professional practice that must be learned and developed over time; includes careful selection, preparation, and ongoing professional development for new mentors; and involves experienced teachers as mentors and includes them in program design and evaluation. Overall, mentoring is viewed as a support for teacher learning across all stages of the professional continuum, including early field experiences, student teaching, induction, and professional development.

Preparation for Mentoring. Although effective teachers are often identified as possible mentor teachers, it is important to recognize that being an effective teacher is not sufficient for being a good or skillful mentor (Hobson et al., 2009). Skills associated with mentoring teachers are different from those used in teaching secondary students. Researchers have found that mentors' influence on novices' performance is greater when mentors have received specific professional development over time for the role (Evertson & Smithey, 2000; Giebelhaus & Bowman, 2002). To provide high-quality mentoring, a teacher must know what to observe and how to provide feedback, be skilled with collaboration and inquiry, understand how to keep communication open and resolve conflicts, be able to study his or her own teaching and to communicate the thought processes that go into planning a lesson, questioning students, and other pedagogical strategies so others can learn from them, as well as provide appropriate challenges for the novice.

Research shows that mentors who have had professional development related to mentoring prospective teachers are better able to assist their mentees with classroom management, problem solving, and lesson planning expertise (Evertson & Smithey, 2000). "Well-prepared mentor teachers combine the knowledge and skills of a competent classroom teacher with the knowledge and skills of a teacher of teaching" (Feiman-Nemser, 2001, p. 1037).

The Roles of Mentor Teachers and Teacher Candidates. Much like teaching, mentoring is a practice (Feiman-Nemser, 2001) that relies on the professional knowledge and judgment of the mentor to develop and support a novice in ways that move him or her toward greater levels of understanding and effectiveness. The mentor teacher identifies tasks of teaching (planning, enacting lessons, examining, and assessing students' thinking and work, and reflecting on teaching practices and their outcomes, reexamining practice in light of new evidence) and plans experiences, manages learning opportunities, and guides the novice teacher toward greater understanding. In addition, Schwille (2008) reported that effective mentor teachers deliver demonstration lessons that model specific teaching moves and engage the teacher candidate in thinking about the lesson and its objectives, reflecting on instructional observations, and considering what should occur next. Other traits of effective mentor teachers are co-planning and co-teaching lessons and providing support and feedback while the mentee is teaching such as conveying teaching moves that support lesson implementation (Schwille, 2008).

Even though these aforementioned traits of effective mentoring suggest a collaborative relationship between the mentor teacher and the mentee, Ambrosetti and Dekkers (2010) found that most definitions of mentoring suggested "a hierarchical relationship in which the mentor is more experienced than the mentee, or that the mentor has or can provide knowledge and skills that the mentee wants or needs" (p. 43). Based on their review of the literature, they identified roles and actions of mentor and novice teachers. The mentor serves as a supporter, role model, facilitator, assessor, collaborator, friend, trainer or teacher, protector, colleague, evaluator, and communicator. Although the literature provided little clarity about the role of the mentee, three studies that Ambrosetti and Dekkers found suggest that the novice's role includes becoming involved in the day-to-day routine of the classroom, observing the mentor in action, teaching lessons, evaluating, and reflecting.

In contrast to the hierarchical view, another view acknowledges the possibility that the novice teacher may have knowledge (e.g., about the use of technology to support learning) that the more experienced teacher may not have (Ambrosetti & Dekkers, 2010; Danielson, 2002). This bi-directional mentoring relationship provides a mutually beneficial relationship in which the novice and mentor have opportunities to learn and grow.

Various formats have been used to provide mentoring. These include: face-to-face interaction away from class, scheduled whole group meetings, phone follow-up, online forums or discussion groups, and in-class modeling (Dempsey & Christenson-Foggett, 2011). Regardless of the type of mentoring received, researchers (Hobson et al., 2008; Smith & Ingersoll, 2004) found the experience was more beneficial when the mentor teacher and novice teacher taught the same subject. In addition, Smith and Ingersoll (2004) found that beginning teachers who were mentored by an individual in the same field and who had participated in an induction program that included regular collaboration with other teachers were more likely to stay in the profession.

Internships and Student Teaching

Internships vary widely; for example, in 2012 state requirements for minimum length varied from five weeks in Virginia to 20 weeks in Maryland[5]. Other variations include the number of teacher candidates placed with the same mentor teacher, and length of time a candidate spends with a given mentor teacher. This section reviews what is known about these and other variations.

Traditional Internships. In the traditional apprentice-type internship, a teacher candidate is assigned to a mentor teacher's classroom for eight to 15 weeks. During this period, the candidate gradually assumes responsibility for all the mentor teacher's classes, then gradually returns responsibility to the mentor teacher. While the candidate is in charge of teaching, the university supervisor and the mentor teacher provide feedback to the student teacher about his or her growth over time based on classroom observations and artifacts (e.g., lesson plans or student work). This type of internship can either be highly beneficial or futile for teacher candidates, depending on the alignment between the candidates' university coursework, previous field experiences, the mentors' instructional practices, the mentors' beliefs about the purposes of the internship and their roles, and whether the emphasis of the internship is on the craft of teaching (Leatham & Peterson, 2009; NCTM, 2014) as described earlier, classroom management, or some combination of the two.

Leatham and Peterson (2010) list five reasons for dissatisfaction with traditional internships:

1. In some instances, the purposes of student teaching may not be explicit;
2. If candidates find themselves focusing more on classroom management than student thinking and orchestrating meaningful lessons then teaching becomes focused on learning how to survive in the classroom;
3. The experience could cause some candidates to focus more on their preparation and knowledge instead of on student thinking;
4. Candidates may be left alone without much collaboration with their mentors or other teachers in the school building; and
5. The internship may be perceived as instructorless if the mentor teacher has not been briefed on his or her responsibilities as mentor or the mentor teacher and university supervisor do not coordinate to support the intern. Traditional internships may also be problematic if there are large numbers of interns compared to the number of available mentor teachers.

On the other hand, some programs have dealt with some of these problems, making traditional internships more useful. For example, Strutchens and Martin (2013) have created a course syllabus that clearly defines the roles of the mentor teacher, university supervisor, and the intern. These roles are further explicated in a workshop each semester for mentor teachers and an orientation session for interns.

Among the MTE-Partnership secondary mathematics programs, traditional internships are the most prevalent. When surveyed in 2013, 28 of 30 respondents stated that they place one intern per teacher.

Multiple Interns with One Mentor. In order to address some of the disadvantages of traditional internships, Leatham and Peterson (2010) implemented a *paired-placement internship*. A pair of teacher candidates worked daily with an experienced mathematics mentor/coach who was devoted full time to helping the teacher candidates address the craft of teaching, plan lessons jointly, and teach those same lessons while actively observing, reflecting, and revising. The teacher candidates quickly came to realize not everyone learns the way they learn, and their focus shifted to the learning of their students rather than their own learning. For over eight years, Peterson and Leatham (2018) have continued to work on the structure of the paired placement model by putting mechanisms in place, such as focused observations, daily journals, student interviews, and specific responsibilities for each member of the teaching team which assist them in focusing on students' mathematical reasoning and sense making.

Nokes, Bullough, Egan, Birrell, and Hansen (2008) studied paired-placement internships of prospective secondary teachers, reporting that the interns learned through "tensions, dialog, and reflections" due to "being placed with a peer" (p. 2168). Results indicated students in the interns' classrooms benefited from the collaboration of the teaching team. The push of the interns to work through problems, that perturbation that comes from differences of opinion, led to better student understanding.

In summarizing research related to paired-placement internships, Mau (2013) reported that paired interns engaged in more frequent and varied communication, increased their willingness to take pedagogical risks, improved their levels of reflection, found methods for collaboration and cooperation in the teaching action, found ways to increase K–12 student learning, had better classroom management, and found strategies to handle tensions in perspective and performance (p. 54). Based on these findings, Mau (2013) recommended placing pairs of student teachers with one cooperating teacher and implementing a model of learning to teach that encourages collaboration, pedagogical risk taking, increased reflection, and better classroom management.

In agreement with Mau (2013), some members of the MTE-Partnership have found similar benefits to implementing paired placement. One respondent to the MTE-Partnership survey said:

> In part-time student teaching (the prospective teachers are not in the schools all day) we have two student teachers with a partner mentor. We would like to continue this model into full-time student teaching but have hit some district roadblocks. We see an advantage to having student teachers clustered in schools and with select mentors.

As follows is an example from an MTE-Partnership institution.

Paired-Placement Internship with Co-Teaching. In the fall of 2013, the University of Montana piloted the use of paired-placement internships at a local high school designed to involve co-teaching similar to that described by Bacharach, Heck, and Dahlberg (2010)[6]. Throughout the internship, all three teachers (the mentor and teacher candidates) were responsible for teaching, learning, and reflecting on student learning. Each intern taught one-third of the classes as the lead teacher, with the others observing, assisting as needed, and reflecting on student learning. This arrangement was beneficial to the interns and the school. The interns had the advantage of a highly experienced, well-versed mentor teacher who turned instruction over to them for most of the semester. The mentor was available to work with other teachers at the school as a part-time mathematics coach. Since this initial implementation by one of the MTE-Partnership institutions, several others have occurred, which are discussed in Chapter 11 in this section of this book.

In addition to the paired placement model, a group of nine interns were placed with the same mentor teacher as part of a professional learning community at an institution in Sydney, Australia, as described in the following section.

Professional Learning Experience Community with Coordinated Methods Workshop and Co-Teaching. Cavanagh and Garvey (2012) describe a one-year fast-track program in metropolitan Sydney for holders of bachelor's degrees in mathematics or a related field. Nine mathematics teacher candidates were assigned to one of two eighth-grade classes taught by the same mentor teacher. Thus, at each lesson there was a maximum of six observers (four or five candidates and the university supervisor). This experience had two phases:

- Semester 1: observe Grade 8 problem-solving lessons every two weeks, with follow-up activities at the university methods workshop taught by the supervisor.
- Semester 2: co-teach two or three Grade 8 problem-solving lessons, each time with a different teacher candidate.

Each of these lessons was followed by a discussion led by the mentor teacher. Afterward, each teacher candidate posted a reflection in an online forum. In anonymous survey responses, the teacher candidates were "uniformly positive in their evaluation of the learning community." They reported seeing connections between university coursework and lesson observations, also "many different ways" and "more options" to "enrich the teaching of mathematics."

The examples featured in this section provide a rationale for placing more than one teacher candidate with a qualified and effective mentor teacher who fosters their growth and development. Moreover, the examples highlight the collaboration, reflection, and positive risk taking that can happen in clinical experiences with more than one teacher candidate.

Yearlong Internships. In recent years, the notions of teacher quality and "well-qualified teacher" have been increasingly highlighted in national discussions of school reform. In these conversations, clinically based teacher education practices have been examined for their role in preparing well-qualified teachers. Quantitative and qualitative research supports the perception that increasing the amount of field experience and mentoring opportunities can have positive effects on prospective teachers' abilities to deal with the complex realities of today's classrooms (Spooner, Flowers, Lambert, & Algozzine, 2008). Consistent with these perceptions, some universities have extended the length of student teaching experiences. One trend has been to replace the traditional one-semester internship with a yearlong internship as recommended by the American Association of Colleges for Teacher Education (2010).

There are many types of yearlong internships, but they typically share key goals and features. Yearlong experiences allow prospective teachers to begin their placements at the very beginning of the school year, witnessing the first days and weeks of school. During the fall and spring semesters, these experiences often involve different phases that combine differing amounts of university coursework and placement responsibilities. Placements may or may not extend through the school's last day of classes. In most cases, student teachers spend the entire year with one mentor teacher, but variations exist that allow student teachers to spend time with a few different mentor teachers in the same school over the course of the year. Sometimes the yearlong experience occurs in the fourth year of an undergraduate program. Other programs stipulate a mandatory fifth year. Three examples are given as follows.

Senior Year Internship at Professional Development School. At the University of Maryland, students complete their yearlong student teaching experience during their senior year[7]. The teacher candidates begin their internships with teachers' meetings at the beginning of the school year. During the first few weeks, the teacher candidates attend the placement school on selected days, later settling into a schedule of two days at the school and two days in university classes with related coursework. During the second semester, the teacher candidates work with the same mentor teacher every day of the week. The four phases are:

- Phase 1 (fall): collaborative teaching planned by mentor;
- Phase 2 (fall and spring): collaboratively planned teaching;
- Phase 3 (spring): collaborative teaching planned by the candidate, responsibility devolves to candidate; and
- Phase 4 (spring): collaborative teaching, responsibility devolves to mentor.

Fifth Year Internship. At Michigan State University, education majors complete their yearlong internships during a required fifth year[8]. The student teachers attend their placement schools full time in both the fall and spring, with occasional release days on Fridays for online or live classes. As at the University of

Maryland, the Michigan State internships have several phases; however, the two fall phases are not consecutive. The five phases are:

- Co-teaching (fall): one "focus class";
- Guided lead teaching I and II (fall): increased responsibility for one or two classes other than focus class (2 to 3 weeks);
- Lead teaching (spring): responsibility for several classes (10 weeks); and
- April transition: responsibility reduced to the focus class while the candidate completes necessary coursework and job searching activities.

Practica and Residency Year for Grades 6–8. As part of a newly developed middle school teacher education program, the University of South Florida, in collaboration with its partner school district, moved from a semester-long to a yearlong internship with co-teaching[8]. This internship is intended to build on early field experiences (Practica 1 and 2) in which teacher candidates visit mathematics classes and engage in the planning and implementation of enacted lessons for the entire class, sub-groups, or particular individuals:

- Practicum 1: candidates are assigned to Grade 6;
- Practicum 2: candidates are assigned to Grade 7; and
- Yearlong internship: candidates are assigned to Grade 8.

Among other things, this approach addresses pragmatic issues that include the reluctance of teachers to "give up" their classes due to new teacher evaluation systems in which teacher ratings are closely linked to student test scores.

Impacts. What impacts do full-year internships have on teacher candidates? Ronfeldt and Reininger (2012) surveyed research on this question, finding that most research on the effects of lengthening internships relies upon rather simplistic comparisons of one versus two semesters of student teaching, or some versus no student teaching. Their analysis of entry and exit survey responses from 1,057 teacher candidates at 36 institutions between 2008 and 2010 found no significant differences between those who completed one versus two semesters of student teaching in perceptions of instructional preparedness, efficacy, and career plans. However, for candidates with 13 or fewer weeks of student teaching, perceptions of the quality of the experience were more strongly associated with perceptions of instructional preparedness. Moreover, teacher candidates who reported better quality student teaching experiences tended to report feeling more prepared and efficacious. They planned to stay longer in teaching than less satisfied peers. Some earlier studies found differences in perceptions of preparedness related to length. Spooner, Flowers, Lambert, and Algozzine (2008) analyzed survey responses from 119 candidates from one university. Some engaged in a traditional semester of student teaching while others completed a yearlong experience. Those in the yearlong internships reported better relationships with their supervising teachers, greater knowledge of school policies and procedures, and higher scores for the perceived adequacy of time spent in school than did the candidates in the semester-long experience. The two groups did not differ in perceptions of their teaching ability, which were generally favorable. Silvernail and Costello (1983) analyzed survey responses from 60 elementary education majors in four different programs. They found that teacher candidates in yearlong internships demonstrated a significant reduction in anxiety as compared to those in the traditional one-semester experience. This reduction occurred at the end of the first semester of the internship, again suggesting that the quality of a student teaching experience may be more important than its length.

Summary. We have described several types of internships. In each, the mentor teacher has a critical role. The mentor teacher's classroom serves as the environment in which a teacher candidate learns the craft of teaching (Leatham & Peterson, 2009). If mentor teachers understand their roles and have the pedagogical and content knowledge that will enable them to mentor the teacher candidates well, teacher candidates have the potential to have a successful experience. However, a mentor teacher is only one of the stakeholders in the education of teacher candidates. University supervisors, other university faculty members, and the candidates themselves also affect whether they develop knowledge and skills needed in teaching. University supervisors and other faculty members can play important roles such as coordinating coursework and field experiences or supporting mentor teachers. These roles can occur as parts of a larger partnership.

Interns and Mentors Supported by Comprehensive Partnerships

Internships—and the field and clinical experiences that precede them—necessarily involve cooperation between a university and one or more schools. As discussed in Chapter 1 of this book, cooperation involves building mutual trust and a common vision among participating individuals, as well as structural arrangements among and within participating institutions. Two types of partnerships with promise in both improving the quality of the field experiences of teacher candidates and supporting continuous growth for mentor teachers are professional learning communities (PLCs) and professional development schools (PDSs). In addition to other activities, these partnerships may incorporate professional development activities described earlier (e.g., lesson study).

Professional Learning Communities. PLCs are generally defined as groups of teachers, who, together with other stakeholders outside of the school, come together for the purpose of studying and learning (Putnam & Burke, 1992/2006). PLCs within schools have the end goal of improving teacher practice and, thus, student learning. Typically, they involve collective inquiry, collaborative teams, action orientation and experimentation, continuous improvement, and a focus on student learning (DuFour, 2004; Hord, 2008).

As described earlier in this chapter, a PLC formed by a group of teacher candidates, mentor teacher, and university supervisor can be used to support clinical experiences (Cavanagh & Garvey, 2012). Such an arrangement can benefit the mentor as well as the teacher candidates. The mentor in Cavanagh and Garvey's study reported being inspired to adapt his teaching of some problem-solving lessons by the teacher candidates' "innovative and creative lesson introductions and the ways they used technology to motivate and engage students" (p. 66). Post-lesson discussions were also useful because the candidates had "a perception of things that's really worthwhile" (p. 66).

Professional Development Schools. PDSs are schools that have entered into special relationships with the university regarding teacher training, induction, and professional development opportunities. Among other things, a PDS may create programs or provide an environment that help practicing teachers evolve into mentor teachers or teacher educators (Beaty-O'Ferrall & Johnson, 2010; Schussler, 2006).

The yearlong internships at the University of Maryland described earlier are examples of the supports and benefits to teachers that can be provided by PDSs. Student teachers are placed with mentor teachers who work at PDSs. Higher education faculty members provide mentor teachers with weekly foci for activities and discussion points to use with their interns and with professional development to enhance their own teaching.

Summary. PLCs and PDSs provide mechanisms for bidirectional relationships to be built between colleges of education and school partners around clinical experiences. Both approaches afford prospective teachers and mentor teachers the opportunities to grow at the same time: prospective teachers develop the craft of teaching, and mentor teachers hone their leadership and mentoring skills.

THE EMERGENCE OF THE CLINICAL
EXPERIENCE RESEARCH ACTION CLUSTER

As stated in the introduction for this book, improving clinical experiences through partnerships with mentor teachers and other stakeholders to develop instructional practices that promote student success toward the goals of the *CCSS-M* is one of the primary drivers for the target of the MTE-Partnership. This driver emerged from an initial focus on one of the MTE-Partnership's *Guiding Principles for Secondary Mathematics Teacher Preparation Programs* (2014), concerning mentor teachers:

> Effective mechanisms needed to prepare and support mentor teachers as they ensure teacher candidates receive field experiences that will help them progress as teachers of mathematics able to teach the *CCSS-M*.

In August of 2012, the MTE-Partnership's working group on mentor teachers began a discussion of what these effective mechanisms might be. Several challenges were identified in the initial discussion of preparing and supporting mentor teachers. Most teams stated that in the wake of the *CCSS-M* many of the mentor teachers may be similar to teacher candidates with regard to pedagogical content knowledge for *CCSS-M* objectives such as Grades 6–12 statistics and transformational geometry. Thus, some mentor teachers may not be able to help teacher candidates to unpack the mathematics well—that is, make tacit mathematical knowledge explicit. In addition, team leaders stated that the number of interns in most cases outnumber the number of high-quality mentor teachers. Another concern was the need for mentor teachers to have professional development for mentoring teacher candidates with respect to mathematical and pedagogical knowledge, and students' cultures and backgrounds. Finally, we agreed that education and mathematics faculty members need to come together with state and district personnel to define what we mean by effective teaching and provide professional development for both mentors and teacher candidates.

Along with challenges, working group members also discussed possible solutions. At least two teams stated that they place two interns with one mentor teacher to capitalize on the knowledge and skills of the mentor teacher and to place the interns in a collaborative learning situation. Another team discussed the possibility of developing the mentoring skills of an entire high school mathematics department so that the university does not depend on one teacher to mentor all interns and risk the loss of the sole mentor. Professional learning communities (PLCs) were also mentioned as viable mechanisms for simultaneously developing the mentor teachers' and the teacher candidates' knowledge. PLCs were also discussed as mechanisms that in the long run could become cost-effective and capacity-sustaining mechanisms through which partners could continue to develop effective practices related to the teaching and learning of mathematics.

A survey of the MTE-Partnership teams was conducted in the spring of 2013. The survey responses revealed that there is much work to do for some MTE-Partnership teams in developing relationships between faculty and school partners focused on field experiences. Only a few MTE-Partnership teams mentioned comprehensive partnerships, like PLCs or PDSs (see this chapter's appendix for statistical details). The next chapter gives an update on the changes in partnership programs related to field experiences.

Theory of Improvement for Clinical Experiences

This section describes a theory of improvement developed to address the problem of improving the clinical experiences of secondary mathematics teacher candidates. First, factors posited to address the problem are presented in a driver diagram (see Figure 7.1). Second, suggested measures and data useful in measuring progress relative to the problem are discussed. Lastly, potential interventions addressing the problem are presented.

Driver Diagram. Using the literature review and survey responses from MTE-Partnership institutions, a working group posed the clinical experiences problem illustrated by the initial driver diagram (see Figure 7.1). The general improvement target is given in the left-most column; this target is a primary driver identified by the MTE-Partnership as a whole and refined by the working group to focus more

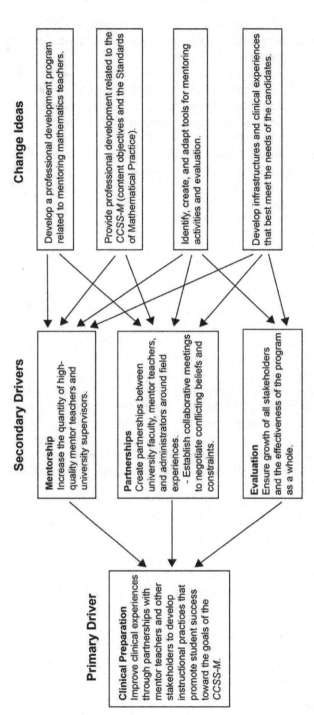

FIGURE 7.1. Driver Diagram for Clinical Experiences. The driver diagram for Clinical Experiences presents the factors posited to address the problem of improving the clinical experiences of secondary mathematics teacher candidates.

concretely on improving clinical experiences through partnerships with mentor teachers and other stakeholders to develop instructional practices that promote student success toward the goals of the *CCSS-M*.

The primary drivers are elements that are likely to create movement toward the improvement target. MTE-Partnership members said all stakeholders needed professional development and stronger commitments to the changes needed to modify clinical experiences, in order for teacher candidates to develop the knowledge and skills needed to implement the *CCSS-M* well.

Next steps: Research Action Clusters. Based on the literature reviewed and information from the MTE-Partnership survey, a working group developed three possible Research Action Clusters (RACs). Each addressed one of the following primary drivers shown in the driver diagram:

- Create professional development to support mentoring teacher candidates.
- Professional development for the *CCSS-M* via learning circles.
- Develop effective clinical experiences.

During the 2014 MTE-Partnership annual meeting, the three themes were discussed in detail by MTE-P members who were interested in clinical experiences. These members were given a survey that provided the leaders of the working group with information that helped the group to decide that creating a RAC that had elements of all three themes would best serve its purposes. Thus, Developing Effective Clinical Experiences became the title of the RAC that focuses on developing effective clinical experiences across the continuum from methods courses through student teaching and providing professional development to mentor teachers around mentoring and implementing the *CCSS-M*. Currently, 26 MTE-Partnership teams are participating in this RAC. Teams have been categorized by the intervention that they propose to carry out within their programs. In the next chapter, the MTE-Partnership shares how the Clinical Experiences RAC has evolved over the past five years.

APPENDIX A. STATISTICS FROM THE MTE-PARTNERSHIP SURVEY

This appendix reports on responses to the survey sent to MTE-Partnership teams in 2013. At that time, there were 35 teams. The MTE-Partnership has since expanded. Typically, the lead institution responded on behalf of its team. Occasionally more than one institution responded, leading to a total of 39 institutions responding.

What professional development do you provide for mentor teachers prior to them supervising interns?

Course or multi-session workshop	13.3%
Workshop	36.7%
Nothing organized	50%

Do you provide any continuing professional development for mentor teachers?

Yes	36.7%
No	63.3%

What relationships exist between the program faculty and mentor teachers?

Ongoing professional relationships	37%
Acquainted	50%
No real relationship	16%

Which type of internship do you use?

One intern per teacher with university representative supervisor	93.3%
Other	6.7%

Who from the university supervises interns or practicum students?

Retired teachers or adjuncts	most
Graduate students	few
Faculty members	3

ENDNOTES

1. Other contributors to this chapter include: Gladis Kersaint, University of Connecticut, Dana Franz, Mississippi State University, Adam Poetzel, University of Illinois at Urbana-Champaign, and Johannah Maynor.

2. For information about math teachers' circles, including a list of circles by state, see www.mathteacherscircle.org.

3. For information about study groups, see www.focusonmath.org/programs.

4. Some examples are listed at https://pcmi.ias.edu/affiliated.

5. http://www.edcounts.org/createtable/step1.php.

6. Description from chapter author David Erickson, The University of Montana.

7. Description based on March 2013 access of www.education.umd.edu/EDCI/info/internship.html.

8. Description from chapter author Gladis Kersaint, University of South Florida.

REFERENCES

Augustyn, L., & Lewis, W. J. (Eds.) (2011). *Math in the middle institute partnership: Making an impact, final report.* Lincoln, NE: University of Nebraska-Lincoln.

Ambrosetti, A., & Dekkers, J, (2010). The interconnectedness of the roles of mentor teachers and preservice teachers. *Australian Journal of Teacher Education, 35*(6), 42–55.

American Association of Colleges for Teacher Education. (2010). *Teacher preparation: Who needs it? The clinical component.* Washington, DC: Author.

Bacharach, N., Heck, T. W., & Dahlberg, K. (2010). Changing the face of student teaching through coteaching. *Action in Teacher Education, 32*(1), 3–14.

Ball, D., & Cohen, D. (1999). Developing practice, developing practitioners: Toward a practice-based theory of professional education. In L. Darling-Hammond & G. Sykes (Eds.), *Teaching as the learning profession* (pp. 3–32). San Francisco, CA: Jossey-Bass.

Banilower, E., Smith, S., Weiss, I., Malzahn, K., Campbell, K., & Weis, A. (2013). *Report of the 2012 national survey of science and mathematics education.* Chapel Hill, NC: Horizon Research.

Beaty-O'Ferrall, M., & Johnson, F. W. (2010). Using supportive team building to promote improved instruction, student achievement, and collaboration in an urban professional development school. *School-University Partnerships, 4*(1), 56–64.

Brill, S., & McCartney, A. (2008). Stopping the revolving door: Increasing teacher retention. *Politics & Policy, 36*(5), 750–774.

Cavanagh, M. S., & Garvey, T. (2012). A professional experience learning community for pre-service secondary mathematics teachers. *Australian Journal of Teacher Education, 37*(12), 57–75.

Conference Board of the Mathematical Sciences. (2012). *The mathematical education of teachers II.* Providence, RI and Washington DC: American Mathematical Society and Mathematical Association of America.

Council for the Accreditation of Educator Preparation. (2015). *CAEP accreditation standards.* Washington, DC: Author.

Danielson, L. (2002), Developing and retaining quality classroom teachers through mentoring. *The Clearing House, 75*(4), 183–85.

Darling-Hammond, L. (2000). How teacher education matters. *Journal of Teacher Education, 51*, 166–173.

Dempsey, I., & Christenson-Foggett, J. (2011). External mentoring support for early career special education teachers. *Australasian Journal of Special Education, 35*(1), 61–71.

DuFour, R. (2004). What is a "professional learning community"? *Educational Leadership, 61*(8), 6–11.

Evertson, C. M., & Smithey, M. W. (2000). Mentoring effects on protégés' classroom practice: An experimental field study. *The Journal of Educational Research, 93*(5), 294–304.

Feiman-Nemser, S. (2001). From preparation to practice: Designing a continuum to strengthen and sustain teaching. *Teachers College Record, 103*(6), 1013–1055.

Gersten, R., Taylor, M. J., Keys, T. D., Rolfhus, E., & Newman-Gonchar, R. (2014). *Summary of research on the effectiveness of math professional development approaches* (REL 2014-010). Washington, DC: U.S. Department of Education, Institute of Education Sciences.

Giebelhaus, C. R., & Bowman, C. L. (2002). Teaching mentors: Is it worth the effort? *The Journal of Educational Research, 95*(4), 246–254.

Hobson, A. J., Ashby, P., Malderez, A., & Tomlinson, P. D. (2008). Mentoring beginning teachers: What we know and what we don't. *Teaching and Teacher Education, 25*, 207–216.

Hord, S. M. (2008). Evolution of the professional learning community: Revolutionary concept is based on intentional collegial learning. *Journal of Staff Development, 29*(3), 10–13.

Ingersoll, R. M., & Strong, M. (2011). The impact of induction and mentoring programs for beginning teachers: A critical review of the research. *Review of Educational Research, 81*(2), 201–233.

Lave, J., & Wenger, E. (1991). *Situated learning: Legitimate peripheral participation.* Cambridge, UK: Cambridge University Press.

Leatham, K. R., & Peterson, B. E. (2009). Secondary mathematics cooperating teachers' perceptions of the purpose of student teaching. *Journal of Mathematics Teacher Education, 13*, 99–119.

Leatham, K. R., & Peterson, B. E. (2010). Purposefully designing student teaching to focus on students' mathematical thinking. In J. Lott & J. Luebeck (Eds.), *Mathematics teaching: Putting research into practice at all levels* (AMTE Monograph 7, pp. 225–239). San Diego, CA: Author.

Lewis, C., & Perry, R. (2014). Lesson study with mathematical resources: A sustainable model for locally-led teacher professional learning. *Mathematics Teacher Education and Development, 16*(1), 22–42.

Lewis, C., & Perry, R. (2015). A randomized trial of lesson study with mathematical resource kits: Analysis of impact on teachers' beliefs and learning community. In J. Middleton, J. Cai, & S. Hwang (Eds.), *Large-scale studies in mathematics education* (pp. 133–158). Cham, Switzerland: Springer International Publishing AG Switzerland.

Markow, D., Macia, L., & Lee, H. (2013). *The MetLife survey of the American teacher: Challenges for school leadership.* New York, NY: Metropolitan Life Insurance Company.

Mathematics Teacher Education Partnership. (2014). *Guiding principles for secondary mathematics teacher preparation.* Washington, DC: Association of Public and Land-grant Universities. Retrieved from: mtep.info/guidingprinciples

Mau, S. (2013). Letter from the editor: Better together? Considering paired-placements for student teaching. *School Science and Mathematics, 113*(2), 53–55.

National Council of Teachers of Mathematics. (2014). *Principles to actions: Ensuring mathematical success for all.* Reston, VA: Author.

National Governors Association Center for Best Practices, Council of Chief State School Officers. (2010). *Common core state standards: Mathematics.* Washington, DC: Author.

Nokes, J. D., Bullough, R. V. Jr., Egan, W. M., Birrell, J. R., & Hansen, J. M. (2008). The paired-placement of student teachers: An alternative to traditional placements in secondary schools. *Teaching and Teacher Education, 24*, 2168–2177.

Odell, S. J., Huling, L., & Sweeney, B. (2000). Conceptualizing quality mentoring: Background information. In S. J. Odell & L. Huling (Eds.), *Quality mentoring for novice teachers* (pp. 3–15). Indianapolis, IN: Kappa Delta Pi.

Peterson, B. E., & Leatham, K. R. (2018). The structure of student teaching can change the focus to students' mathematical thinking. In M. E. Strutchens, R. Huang, L. Losano, & D. Potari (Eds.), *Educating prospective secondary mathematics teachers. Monograph Series Edited by Kaiser, G.* (pp. 9- 26.) Cham, Switzerland: Springer.

Putnam, J., & Burke, J. (1992/2006). *Organizing and managing classroom learning communities.* Boston,MA: McGraw-Hill.

Ronfeldt, M., & Reininger, M. (2012). More or better student teaching? *Teaching and Teacher Education, 28*(8), 1091–1106.

Schussler, D. L. (2006). The altered role of experienced teachers in professional development schools: The present and its possibilities. *Issues in Teacher Education, 15*(2), 61–75.

Schwille, S. A. (2008). The professional practice of mentoring. *American Journal of Education, 115*(2), 139–167.

Silvernail, D., & Costello, M. (1983). The impact of student teaching and internship programs on preservice teachers' pupil control perspectives, anxiety levels, and teaching concerns. *Journal of Teacher Education, 34*(4), 32–36.

Smith, T. M., & Ingersoll, R. M. (2004). What are the effects of induction and mentoring on beginning teacher turnover? *American Educational Research Journal, 41*, 681–714.

Spooner, M., Flowers, C., Lambert, R., & Algozzine, B. (2008). Is more really better? Examining perceived benefits of an extended student teaching experience. *The Clearing House, 81*(6), 263–269.

Strutchens, M. E., & Martin, W. G. (2013). *CTSE 4920: Internship in secondary mathematics* [course syllabus]. Auburn, AL: Auburn University.

Strutchens, M. E., & Martin, W. G. (2017). The Transforming East Alabama Mathematics teacher leader academies. In N. Rigelman & M. McGatha (Eds.), *Elementary mathematics specialists: Developing, refining, and examining programs that support mathematics teaching and learning* (pp 77–84). Charlotte, NC: Information Age Publishing

White, D., Donaldson, B., Hodge, A., & Ruff, A. (2103). Examining the effects of Math Teachers' Circles on aspects of teachers' mathematical knowledge for teaching. *International Journal of Mathematics Teaching and Learning.* Published online September 26, 2013, at http://www.cimt.plymouth.ac.uk/journal/.

Wilson, S. M., Floden, R. E., & Ferrini-Mundy, J. (2001). *Teacher preparation research: Current knowledge, gaps, and recommendations.* Seattle, WA: Center for the Study of Teaching and Policy.

CHAPTER 8

IMPROVING CLINICAL EXPERIENCES FOR SECONDARY MATHEMATICS TEACHER CANDIDATES

Marilyn E. Strutchens, Ruthmae Sears, and Jeremy Zelkowski

The Clinical Experiences Research Action Cluster (CERAC) was formed by the working group on clinical experiences to address a two-fold problem:

1. There is an inadequate supply of quality mentor teachers to oversee clinical experiences. Too few teachers are well-versed in implementing the *Common Core State Standards for Mathematics* (*CCSS-M*; National Governors Association Center for Best Practices, Council of Chief State School Officers, 2010), and teachers are especially inexperienced with embedding the standards for mathematical practice into their teaching of content standards on a daily basis. Further, many veteran teachers do not implement the mathematics teaching practices as discussed in *Principles to Actions: Ensuring Mathematical Success for All* (National Council of Teachers of Mathematics [NCTM], 2014) on an ongoing basis.

2. Bidirectional relationships between the teacher preparation programs and school partners in which clinical experiences take place are rare. Such relationships that reflect a common vision and shared commitment

The Mathematics Teacher Education Partnership: The Power of a Networked Improvement Community to Transform Secondary Mathematics Teacher Preparation, pages 199–209.
Copyright © 2020 by Information Age Publishing
199

to the vision of the *CCSS-M* and other issues related to mathematics teaching and learning are critical to the development and mentoring of new teachers.

The work of CERAC encompasses a number of the principles and principle indicators from the MTE-Partnership's *Guiding Principles for Secondary Mathematics Teacher Preparation Programs* (Mathematics Teacher Education Partnership, 2014), including fostering partnerships between institutions of higher education, schools, and districts, and other stakeholders, such as state departments of education, and is focused on preparing teacher candidates who promote student success in mathematics, as described in *CCSS-M* and other college and career ready standards. In addition, the CERAC specifically addresses the standards in Figure 8.1 from the Association of Mathematics Teacher Educators (2017) *Standards for the Preparation of Teachers of Mathematics* (AMTE Standards).

In the CERAC, higher education faculty and partner school districts and schools work together to actively recruit, develop, and support in-service master secondary mathematics teachers who can serve as mentors across the teacher development continuum from prospective to beginning teachers. Moreover, the CERAC helps to ensure that teacher candidates have the knowledge, skills, and dispositions needed to implement educational practices found to be effective in supporting all secondary students' success in mathematics as defined in the *CCSS-M* and other college and career ready standards.

In 2015, five members of the CERAC participated in the Carnegie Foundation for the Advancement of Teaching's Networked Improvement Community (NIC) Learning Lab, which was geared toward supporting a small number of groups build the capacity and products to launch an NIC. These members attended four sessions over a nine-month period. The Learning Lab helped the leadership team to relaunch the RAC. The team went through the process as if it were just beginning to learn about NICs, which truly helped to ground its work better in terms of improvement science. A *fishbone diagram* was developed, based on the original working group's white paper, and is "a tool that visually represents a group's casual systems analysis and is also known as a cause and effect diagram" (Bryk, Gomez, Grunow, & LeMahieu, 2015, p. 198). The white paper also provided the initial driver diagram, which can be seen in Chapter 7 of this volume, and was revised during the Learning Lab to better reflect a measurable aim. Figure 8.2 contains the CERAC revised aim and driver diagram.

The CERAC's aim states: "During student teaching, teacher candidates will use each of the eight Mathematics Teaching Practices (NCTM, 2014) at least once a week during full time teaching." The eight research-based mathematics teaching practices (NCTM, 2014) promote learning aligned with new college and career ready content standards: 1) establish mathematics goals to focus learning; 2) implement tasks that promote reasoning and problem solving; 3) use and connect mathematical representations; 4) facilitate meaningful mathematical discourse;

Standard C.2. Pedagogical Knowledge and Practices for Teaching Mathematics
Well-prepared beginning teachers of mathematics have foundations of pedagogical knowledge, effective and equitable mathematics teaching practices, and positive and productive dispositions toward teaching mathematics to support students' sense making, understanding, and reasoning.
    C.2.1. Promote Equitable Teaching
    C.2.2. Plan for Effective Instruction
    C.2.3. Implement Effective Instruction
    C.2.4. Analyze Teaching Practice
    C.2.5. Enhance Teaching Through Collaboration with Colleagues, Families, and Community Members

Standard C.4. Social Contexts of Mathematics Teaching and Learning
Well-prepared beginning teachers of mathematics realize that the social, historical, and institutional contexts of mathematics affect teaching and learning and know about and are committed to their critical roles as advocates for each and every student.
    C.4.1. Provide Access and Advancement
    C.4.2. Cultivate Positive Mathematical Identities
    C.4.3. Draw on Students' Mathematical Strengths
    C.4.4. Understand Power and Privilege in the History of Mathematics Education
    C.4.5. Enact Ethical Practice for Advocacy

P.3. Opportunities to Learn to Teach Mathematics
An effective mathematics teacher preparation program provides candidates with multiple opportunities to learn to teach through mathematics-specific methods courses (or equivalent professional learning experiences) in which mathematics, practices for teaching mathematics, knowledge of students as learners, and the social contexts of mathematics teaching and learning are integrated.
    P.3.1. Address Deep and Meaningful Mathematics Content Knowledge
    P.3.2. Provide Foundations of Knowledge About Students as Mathematics Learners
    P.3.3. Address the Social Contexts of Teaching and Learning
    P.3.4. Incorporate Practice-Based Experiences
    P.3.5. Provide Effective Mathematics Methods Instructors

P.4. Opportunities to Learn in Clinical Settings
An effective mathematics teacher preparation program includes clinical experiences that are guided on the basis of a shared vision of high-quality mathematics instruction and have sufficient support structures and personnel to provide coherent, developmentally appropriate opportunities for candidates to teach and to learn from their own teaching and the teaching of others.
    P.4.1. Collaboratively Develop and Enact Clinical Experiences
    P.4.2. Sequence School-Based Experiences
    P.4.3. Provide Teaching Experiences with Diverse Learners
    P.4.4. Recruit and Support Qualified Mentor Teachers and Supervisors

FIGURE 8.1.   Selected Standards from the Standards for Preparing Teachers of Mathematics (Association of Mathematics Teacher Educators, 2017).

5) pose purposeful questions; 6) build procedural fluency from conceptual understanding; 7) support productive struggle in learning mathematics; and 8) elicit and use evidence of student thinking. Learning about the mathematics teaching practices and other equity-based instructional strategies must be at the core of teacher preparation coursework and reflected in clinical experiences. However, as mentioned in our problem statement, there are not enough mentor teachers at

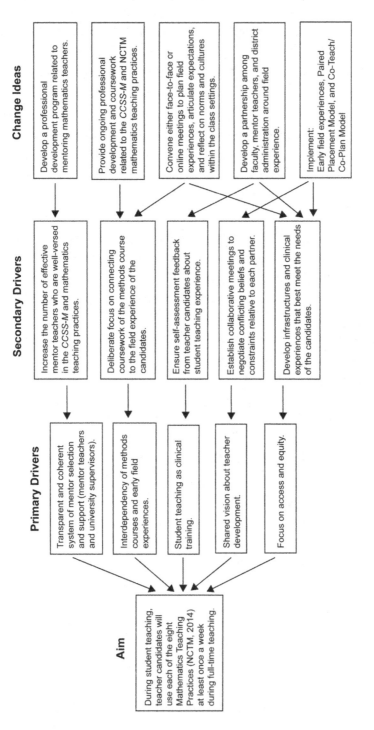

FIGURE 8.2. CERAC Driver Diagram. The revised aim and driver diagram for the Clinical Experiencs RAC contains primary drivers that represent the community's hypothesis about the main areas of influence necessary to advance the improvement aim (Bryk et al., 2015). The secondary drivers are system components that are hypothesized to activate each primary driver change (Bryk et al., 2015). Change ideas help move the primary drivers forward; change ideas are also action areas that the CERAC might address.

the secondary mathematics level prepared to foster the growth of teacher candidates, due to a lack of proficiency with this new approach to teaching, which is in alignment with the NCTM (1989, 1991, 1995, 2000, 2014, 2018) standards and position statements and the AMTE Standards. While university coursework can provide knowledge about content and about teaching strategies, it is during clinical experiences that prospective teachers develop the craft of teaching—for instance, the ability to design lessons that involve important mathematical ideas, design or select tasks that will help students to access those ideas, and implement instructional strategies to successfully execute the lesson (Leatham & Peterson, 2010a). Consequently, teacher candidates often find it difficult to learn to implement the mathematics teaching practices advocated by NCTM (2014).

Primary drivers represent the community's hypothesis about the main areas of influence necessary to advance the improvement aim (Bryk et al., 2015). As follows is a description of each of the primary drivers:

- Transparent and coherent system of mentor selection and support (mentor teachers & university supervisors): Organize mentor selection and support around deepening expertise with math content, math standards, mathematics teaching practices (NCTM, 2014), and mentoring strategies.
- Interdependency of methods course and early field experiences: Structure methods course assignments with a focus on the mathematics teaching practices (NCTM, 2014) and *CCSS-M* such that they include engagement of mentor teachers.
- Student teaching as clinical training: Ensure that requirements for student teaching and feedback during student teaching emphasize the responsibility of teacher candidates to advance mathematics learning among secondary students through collaboration with more expert mentors in use of MTPs (AMTE, 2017).
- Shared vision about teacher development: Ensure mutual agreement between district(s) and university about what quality teaching of secondary mathematics looks like and how to further skills of all teachers (including teacher candidates) and see mentor teaching as part of career ladder.
- Focus on access and equity: Disrupt long-standing teaching practices that contribute to inequities in learning outcomes of students.

The secondary drivers are system components that are hypothesized to activate each primary driver change (Bryk et al., 2015). Figure 8.2 shows the relationship between the primary drivers and the secondary drivers. Also included in Figure 8.2 are change ideas that can help move the primary drivers forward; change ideas are also action areas that the CERAC might address.

Currently, the CERAC consists of 26 university-led teams, each consisting of at least one mathematics teacher educator, mathematician, and school partner. The CERAC is divided into three sub-RACs, which focus on three types of field experiences: Methods, Paired Placement, and Co-Planning and Co-Teaching.

Each sub-RAC is implementing Plan-Do-Study-Act (PDSA) Cycles based on their goals and objectives. Teams work together via conference calls, email, and a communication platform. They use Dropbox, Google Drive, and the communications platform as a way of sharing files and materials. Additionally, they have held face-to-face meetings as a RAC that included breakout meetings for sub-RACs.

The sub-RACs also have overlap areas that drive and focus the RAC, such as the emphasis on the mathematics teaching practices (NCTM, 2014) and other equitable teaching practices, PD for mentors related to the *CCSS-M*, mentoring mathematics teacher candidates, and outcome measures. There are also specific goals to be attained within each of the sub-RACs, and each sub-RAC has developed its own specific research questions.

In 2017, the CERAC submitted a proposal and received funding for the Engaged Student Learning, Design or Development and Implementation (level 2) of IUSE of the National Science Foundation. The project is led by principal investigators from Auburn University, the University of South Florida, and the Association of Public and Land-grant Universities (APLU). The NSF Improving Undergraduate STEM Education (IUSE) grant, *Collaborative Research: Attaining Excellence in Secondary Mathematics Clinical Experiences with a Lens on Equity*[1], is implementing an improvement science study to answer the following question: How does a continuum of collaborative and student-focused clinical experiences, including co-planning /co-teaching and paired placement fieldwork models, impact prospective teachers' equitable implementation of the mathematics teaching practices (NCTM, 2014) across multiple institutional contexts?

The CERAC contends that the mathematics teaching practices are representative of equitable pedagogy as defined by Banks and Banks (1995) and asset-focused strategies as defined by Boykin (2014). According to Banks and Banks (1995), "equity pedagogy is teaching strategies and classroom environments that help students from diverse racial, ethnic, and cultural groups attain the knowledge, skills, and attitudes needed to function effectively within, and help create and perpetuate, a just, humane, and democratic society" (p. 152). In addition, Boykin (2014) suggested that asset-focused strategies should be used in classrooms to increase students' engagement that leads to desired outcomes. Boykin (2014, p. 513) posited that the following strategies have been most empirically verified as asset-based strategies:

- Teacher-student relationship quality: entails the provision of a socially and emotionally supportive yet demanding and high-expectations classroom learning environment (Ladson-Billings, 1995; NCTM, 2014, 2018);
- Collaborative learning: entails collaborative intellectual exchanges among students and ensures that all classroom participants are actively involved in the learning process (Boaler, 2006; Cohen et al., 1999);

- Meaningful learning: conveys building on students' past experience and prior knowledge and making connections in school to significant events in students' lives (Ladson-Billing, 1995);
- Cultural resources: captured as building proactively on the cultural, family, and community assets, values, and practices students bring with them to the classroom (Gonzalez, Andrade, Civil, & Moll, 2001; NCTM, 1991, 2000, 2014, 2018); and
- Information processing quality: involves directly teaching students problem-solving and learning strategies, and promoting higher-order thinking and critical understanding with respect to subject matter (NCTM, 1989, 1991, 1995, 2000, 2014, 2018).

The funding from the grant with the emphasis of the equity lens to the work is propelling the CERAC's work forward in a positive and comprehensive direction.

MEASURES

The CERAC is using the following measures to track the teacher candidates' development of proficiency with the mathematics teaching practices (NCTM, 2014) with attention to equity across the clinical experience continuum:

- *The Mathematics Teaching Practices Survey* is designed to monitor the extent to which teacher candidates have read, discussed, observed, planned, enacted, and received feedback on each MTP across the continuum. The teacher candidates are asked to complete the survey multiple times throughout the continuum. A subscale including MTPs particularly relevant to equitable teaching practices is used.
- *The Mathematics Classroom Observation Protocol for Practices (MCOP²)* is a K–16 mathematics classroom instrument designed to measure the degree of alignment of the mathematics classroom with the standards for mathematical practice from the *CCSS-M* (National Governors Association Center for Best Practices, Council of Chief State School Officers, 2010); the NCTM (2000) process standards; and recommendations for undergraduate mathematics instruction. The instrument contains 17 items intended to measure three primary constructs (student engagement, lesson content, and classroom discourse) (Gleason, Livers, & Zelkowski, 2015). To accommodate the group's focus on equity, the CERAC has highlighted the indicators that particularly focus on equitable pedagogical practices within the protocol. Members of the research team have received training to ensure reliable ratings when administering the MCOP² at key points throughout the clinical experiences' continuum. See Figure 8.3 for a crosswalk of the MCOP² with the mathematics teaching practices (NCTM, 2014) and the standards for mathematical practice (National Governors Association Center for Best Practices, Council of Chief State School Officers, 2010).

Mathematical Classroom Observation Protocol for Practices - MCOP² Crosswalk with the Mathematics Teaching Practices and the Standards for Mathematical Practice

MCOP² Item	Mathematics Teaching Practice(s) (MTPs)	Standards for Mathematical Practice (SMP)
1. Students engaged in exploration/ investigation/problem solving	2. Implementing tasks that promote reasoning and problem solving 1. Establishing mathematics goals to focus learning	1. Make sense of problems and persevere in solving them 7. Look for and make use of structure 8. Look for and express regularity in repeated reasoning
2. Students used a variety of means (models, drawings, graphs, concrete materials, manipulatives, etc.) to represent concepts	3. Use and connect mathematical representations	5. Use appropriate tools strategically 1. Make sense of problems and persevere in solving them
3. Students were engaged in mathematical activities	2. Implementing tasks that promote reasoning and problem solving	1. Make sense of problems and persevere in solving them
4. Students critically assessed mathematical strategies	4. Facilitate meaningful mathematical discourse	3. Construct viable arguments and critique the reasoning of others 1. Make sense of problems and persevere in solving them 8. Look for and express regularity in repeated reasoning
5. Students persevered in problem solving	7. Support productive struggle in learning mathematics	1. Make sense of problems and persevere in solving them 2. Reason abstractly and quantitatively 3. Construct viable arguments and critique the reasoning of others 5. Use appropriate tools strategically
6. The lesson involved fundamental concepts of the subject to promote relational/conceptual understanding	6. Build procedural fluency from conceptual understanding	8. Look for and express regularity in repeated reasoning 7. Look for and make use of structure
7. The lesson promoted modeling with mathematics	2. Implementing tasks that promote reasoning and problem solving	4. Model with mathematics 2. Reason abstractly and quantitatively
8. The lesson provided opportunities to examine mathematical structure (Symbolic notation, patterns, generalizations, conjectures, etc.)	1. Establishing mathematics goals to focus learning	8. Look for and express regularity in repeated reasoning 7. Look for and make use of structure

MCOP²	MTP	SMP
9. The lesson included tasks that have multiple paths to a solution or multiple solutions	2. Implementing tasks that promote reasoning and problem solving	1. Make sense of problems and persevere in solving them
10. The lesson promoted precision of mathematical language	4. Facilitate meaningful mathematical discourse	6. Attend to precision 3. Construct viable arguments and critique the reasoning of others
11. The teacher's talk encouraged student thinking	5. Pose purposeful questions 8. Elicit and use evidence of student learning	1. Make sense of problems and persevere in solving them
12. There were a high proportion of students talking related to mathematics	4. Facilitate meaningful mathematical discourse	3. Construct viable arguments and critique the reasoning of others
13. There was a climate of respect for what others had to say	4. Facilitate meaningful mathematical discourse	3. Construct viable arguments and critique the reasoning of others
14. In general, the teacher provided wait-time	8. Elicit and use evidence of student learning	1. Make sense of problems and persevere in solving them
15. Students were involved in the communication of their ideas to others (peer-to-peer)	4. Facilitate meaningful mathematical discourse	3. Construct viable arguments and critique the reasoning of others
16. The teacher uses student questions/comments to enhance conceptual mathematical understanding	4. Facilitate meaningful mathematical discourse 5. Pose purposeful questions 8. Elicit and use evidence of student learning	1. Make sense of problems and persevere in solving them 3. Construct viable arguments and critique the reasoning of others

FIGURE 8.3.   Mathematical Classroom Observation Protocol for Practices - MCOP² Crosswalk with the Mathematics Teaching Practices and the Standards for Mathematical Practice. A crosswalk of the MCOP², MTPs, and SMP. The SMP column lists the primary SMP first tied to the MCOP² item. See Gleason, Livers, & Zelkowski (2017) Appendix B from the MCOP² external validation study. Depending on the lesson being implemented and the design of the lesson for what students do, the MTP and SMP columns may vary slightly from these suggested listings.

- *MTE-Partnership Program Completer Survey* is designed for program completers to self-assess their success in developing the craft of teaching, based on the MTE-Partnership's *Guiding Principles for Secondary Mathematics Teacher Preparation Programs* (2014) and the mathematical teaching practices (NCTM, 2014). The survey also asks them to assess the success of their preparation program in alignment with the *Guiding Principles*, including their clinical experiences. Responses to the *Guiding Principles* addressing equity will receive particular focus.
- Members of the RAC will be asked to participate in *focus groups* at common events, such as RAC/MTE-Partnership meetings and/or professional development experiences. In these focus groups, participants will be asked to describe strengths, challenges, and complexities of the use of the various clinical experiences' models at their respective institutions. They also provide insight into what constructs should be preserved, revised, or deleted from the models for future PDSA Cycles. At least one focus group with each of the sub-RACs will be conducted annually.

As stated earlier in the chapter, sub-RACs also use PDSA Cycles to refine modules and other tools developed by the group to aid in the implementation of the different clinical experiences' models and teaching strategies.

In the next three chapters, CERAC shares the work of each of the sub-RACs and how they are working to meet the overall aim of the RAC. The sub-RACs are presented in the following order: Methods, Co-Planning and Co-Teaching, and Paired Placement. Each of the sub-RACs is unique in their approach to working toward the aim of the CERAC. The Methods sub-RAC is focusing on designing modules and piloting them across the MTE-Partnership to ensure that the modules are effective in providing the mentor teacher and teacher candidates with shared experiences that move teaching and learning forward. The Co-Planning and Co-Teaching sub-RAC focuses on developing professional learning experiences around co-planning and co-teaching models for mentor teachers and the teacher candidates placed in their classrooms. The Paired Placement sub-RAC develops protocols and workshops for mentor teachers, university supervisors, and teacher candidates to be able to implement the paired placement student teaching models across multiple contexts.

ENDNOTE

1. Funded by the National Science Foundation Directorate for Education & Human Resources, Division of Undergraduate Education (DUE)— Improving Undergraduate STEM Education (IUSE), Development & Implement I & II: Engage Student Learning—Grant ID#s: 1726998, 1726362, and 1726853. Any opinions, findings, and conclusions or recommendations expressed in this material are those of the author(s) and do not necessarily reflect the views of the National Science Foundation.

REFERENCES

Association of Mathematics Teacher Educators. (2017). *Standards for preparing teachers of mathematics*. Raleigh, NC: Author. Retrieved from http://amte.net/standards

Banks, C. A., & Banks, J. A. (1995). Equity pedagogy: Component of multicultural education. *Theory into Practice, 34*(3), 152–158.

Boaler, J. (2006). How a detracked mathematics approach promoted respect, responsibility, and high achievement. *Theory into Practice, 45*(1), 40–46.

Boykin, A. W. (2014). Human diversity, assessment in education, and the achievement of excellence and equity. *The Journal of Negro Education, 83*(4), 499–521.

Bryk, A. S., Gomez, L. M., Grunow, A., & LeMahieu, P. G. (2015). *Learning to improve: How America's schools can get better at getting better*. Cambridge, MA: Harvard Education Press.

Cohen, E. G., Lotan, R. A., Scarloss, B. A., & Arellano, A. R. (1999). Complex instruction: Equity in cooperative learning classrooms. *Theory into Practice, 38*(2), 80–86.

Gleason, J., Livers, S. D., & Zelkowski, J. (2015). *Mathematics classroom observation protocol for practices MCOP²: Descriptive manual booklet*. Tuscaloosa, AL: The University of Alabama. Retrieved from http://bit.ly/MCOP2-descriptive-manual

Gleason, J., Livers, S. D., & Zelkowski, J. (2017). Mathematics Classroom Observation Protocol for Practices (MCOP²): Validity and reliability. *Investigations in Mathematical Learning, 9*(3), 111–129.

Gonzalez, N., Andrade, R., Civil, M., & Moll, L. (2001). Bridging funds of distributed knowledge: Creating zones of practices in mathematics. *Journal of Education for Students Placed at Risk, 6*(1/2), 115–132.

Ladson-Billings, G. (1995). Making mathematics meaningful in multicultural contexts. In W. Secada, E. Fennema, & L. B. Adjian (Eds.), *New directions for equity in mathematics education* (pp. 126–145). New York, NY: Cambridge University Press.

Mathematics Teacher Education Partnership. (2014). *Guiding principles for secondary mathematics teacher preparation*. Washington, DC: Association of Public and Land-grant Universities. Retrieved from mtep.info/guidingprinciples

National Council of Teachers of Mathematics. (1989). *Curriculum and evaluation standards for school mathematics*. Reston, VA: Author.

National Council of Teachers of Mathematics. (1991). *Professional standards for teaching mathematics*. Reston, VA: Author.

National Council of Teachers of Mathematics. (1995). *Assessment standards for school mathematics*. Reston, VA: Author.

National Council of Teachers of Mathematics. (2000). *Principles and standards for school mathematics*. Reston, VA: Author.

National Council of Teachers of Mathematics (2014). *Principles to actions: Ensuring mathematical success for all*. Reston, VA: Author.

National Council of Teachers of Mathematics. (2018). *Catalyzing change in high school mathematics: Initiating critical conversations*. Reston, VA: Author.

National Governors Association Center for Best Practices, Council of Chief State School Officers. (2010). *Common core state standards: Mathematics*. Washington, DC: Author.

CHAPTER 9

ENGAGING MENTOR TEACHERS WITH TEACHER CANDIDATES DURING METHODS COURSES IN CLINICAL SETTINGS[1]

Jeremy Zelkowski[2], Jan Yow, Mark Ellis and Patrice Waller

The Methods sub-Research Action Cluster (sub-RAC) has focused its work on the two-fold problem in mathematics teacher education discussed in Chapter 8 of this book. The focus of Methods is to develop knowledge and understanding of critical components with teacher candidates that ultimately improve student outcomes when put into practice. Once this knowledge and understanding begins to take shape, the sub-RAC engages bidirectional learning between *teacher candidate* and their *mentor teacher* in a clinical school setting. To accomplish this endeavor to ultimately improve teacher candidate quality, mentoring, and student outcomes, this sub-RAC develops, tests, revises, and finalizes modules for secondary mathematics methods coursework. These modules offer opportunities to support teacher candidates with developing knowledge and skills needed to plan and implement lessons that (a) engage students in the Standards for Mathematical Practice (SMP) in the *Common Core State Standards for Mathematics* (*CCSS-M*; National Governors Association Center for Best Practices, Council of Chief State School Officers, 2010) through the use of the eight mathematics teaching prac-

The Mathematics Teacher Education Partnership: The Power of a Networked Improvement Community to Transform Secondary Mathematics Teacher Preparation, pages 211–234.
Copyright © 2020 by Information Age Publishing
211

tices (MTPs) (National Council of Teachers of Mathematics [NCTM], 2014) and (b) facilitate conversations and activities between mentor teachers and teacher candidates about the SMP and MTPs reflecting a bidirectional relationship of professional learning in preparing the next generation of mathematics teachers.

This work supports the overall aim of the entire Clinical Experiences Research Action Cluster (CERAC) to have teacher candidates use each of the eight mathematics teaching practices at least once a week during the full-time student teaching and, more specifically, addresses the primary driver calling for attention to the interdependency of methods courses and field experiences. The Methods sub-RAC focuses on two secondary drivers that call for a deliberate focus on:

1. connecting coursework of the methods course to the field experience of the teacher candidates, and
2. efforts to increase the number of effective mentor teachers who are well versed in their state standards, the SMP, and MTPs (see Figure 9.1).

This work is well aligned with the four *Guiding Principles for Secondary Mathematics Teacher Preparation Programs* (MTE-Partnership, 2014). Most specifically, the Methods sub-RAC highlights the second guiding principle, which focuses on enhancing communication among the partners involved in a secondary mathematics teacher preparation program, in order to: clarify program goals, to assess the effectiveness of the program, and to guide program development and revision. This work most closely addresses two main elements of the Association of Mathematics Teacher Educators' (Association of Mathematics Teacher Educators [AMTE], 2017) *Standards for the Preparation of Teachers of Mathematics* (AMTE Standards): Standard C.2. Pedagogical Knowledge and Practices for Teaching Mathematics and P.4. Opportunities to Learn in Clinical Settings. Included within C.2 and P.4 is the recognition of the importance of embedding equity in the work; the Methods sub-RAC is conscious of and attentive to the importance of Standard C.2.1 Promoting Equitable Teaching, which states, "Well-prepared beginning teachers of mathematics structure learning opportunities and use teaching practices that provide access, support, and challenge in learning rigorous mathematics to advance the learning of every student" (AMTE, 2017, p. 13).

This chapter shares insights into the process of applying an improvement science approach through a Networked Improvement Community (NIC) to analyze problems of practice related to clinical experiences associated with secondary mathematics methods coursework and to develop, test, and revise modules that address these problems. Faculty members from multiple secondary teacher preparation programs who are part of three sub-RACs for Clinical Experiences do this work collaboratively, paying particular focus to each of our areas of the CERAC driver diagram (see Figure 9.1). Methods sub-RAC members use Plan-Do-Study-Act (PDSA) Cycles to pilot module materials and generate data to inform revisions (Deming, 1950, 1993). This process, described in-depth as follows, requires

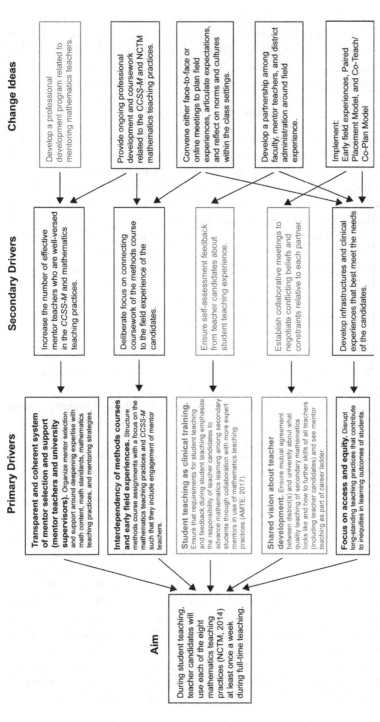

FIGURE 9.1.   CERAC Driver Diagram. The Methods sub-RAC focuses its work on all of these drivers, with the exceptions of the boxes that are faded, and mainly addresses the primary driver calling for attention to the interdependency of methods courses and field experiences.

the articulation of expected outcomes as well as the intentional and non-intrusive collection of actionable data about implementation (e.g., time spent, clarity of instructions, quality of resources) to inform revisions. The work allows adaptations that acknowledge local contexts while adhering to a core set of activities and strategies.

REVIEW OF LITERATURE

Clinical experiences for teacher candidates in partner schools that are well-tied to mathematics teacher preparation coursework are essential (Ball, Sleep, Boerst, & Bass, 2009; Forzani, 2014). Furthermore, offering clinical experiences integrated within methods courses is even more beneficial to teacher candidates' development (Darling-Hammond, 2014). Strutchens (2017), in a review of research that examined field experiences connected to methods courses for teacher candidates of mathematics, found evidence of the importance of explicitly connecting what teacher candidates are learning in their coursework to actual classroom practice (Cavey & Berenson, 2005; Nguyen, Dekker, & Goedhart, 2008; Ricks, 2011). We review the various components of the aspects of importance with respect to making connections for teacher candidates for integrating university methods coursework, clinical experiences, mentor teachers, and engaging students.

Variations in Methods Courses

The CERAC primary driver that emphasizes *the interdependency of methods course(s) and field experiences* is the primary focus driver of the Methods sub-RAC. Methods courses provide teacher candidates the opportunity to learn about mathematics pedagogical strategies, as well as to examine the role of the mathematics teacher for student learning (Grossman, Hammerness, & McDonald, 2009). Research has found, however, that there is great variation among mathematics methods courses across institutions in terms of the learning goals, key assignments, and connection to fieldwork (Kidd, 2008; Taylor & Ronau, 2006; Yee, Otten, & Taylor, 2017). While most methods courses emphasize pedagogical content knowledge and concurrent engagement with (though not necessarily a direct connection to) field experiences (Yee, Otten, & Taylor, 2017; Youngs & Hong, 2013), in their survey asking faculty to rate the importance of a set of key topics (so-called "touchstones" in their article), Yee, Otten, and Taylor (2017) noted significant variance in importance for two touchstones in particular— "issues of equity, status, fairness, and social justice" and "needs of underrepresented populations"—with methods faculty in departments of education valuing these more highly (approximately 4.3/5.0 versus 3.6/5.0) than methods faculty in departments of mathematics. Yee and colleagues' findings point to a larger issue at hand. Alignment of program curriculum, design, goals, and visions across both education and mathematics faculty is critical. Tatto (1998) has long advocated for this critical role of shared faculty vision and enactment of teacher preparation.

Mentor Teacher Preparation

Recent calls ask teacher preparation programs to focus more attention on the mentor teacher's preparation for supervision (Gareis & Grant, 2015; Hoffman et al., 2015). Given that mentor teachers' critique of teacher candidates is influenced by their own beliefs about effective teaching (Goodwin, Roegman, & Reagan, 2016), and teacher candidates' confidence in their teaching style can be positively and negatively affected by their mentor teachers (Izadinia, 2015), it is important that mentor teachers model practices aligned with what is being learned in methods coursework and engage in discussions with teacher candidates about their growth in ways that reinforce and refine methods coursework content (Lawley, Moore, & Smajic, 2014). It is also imperative that mentor teachers and methods instructors have shared visions and responsibilities in preparing teacher candidates. This shared experience with methods course assignments provides a rich context for mentor teacher and teacher candidate co-learning (Capraro, Capraro, & Helfeldt, 2010; Wood & Turner, 2015). Common tasks experienced by mentor teachers and teacher candidates can support a joint development of understanding of effective teaching (Gunkel & Wood, 2016).

Supporting Teachers

Knowing the importance of clinical experiences embedded within methods courses, as well as the importance of work between mentor teachers and teacher candidates in secondary classrooms, the methods sub-RAC module development is focused on the SMP of the *CCSS-M* and MTPs (NCTM, 2014), further supported by the AMTE Standards (2017). Prior research has shown teachers struggle to promote meaningful student engagement with the SMP in their classrooms (Hiebert et al., 2005) and even have difficulty understanding and interpreting the SMP (Olson, Olson, & Capen, 2014). Just as students need extended, deliberate, and focused opportunities to unpack the SMP (Mateas, 2016), so do teachers and teacher candidates (Bostic & Matney, 2013).

Engaging Students

Additionally, the MTPs (NCTM, 2014) provide direction for helping teachers implement lessons that engage students in the SMP (Leinwand, Huinker, & Brahier, 2014). Similar to the SMP, the MTPs call for ambitious teaching with rigorous content. Therefore, teacher candidate preparation must include deliberate opportunities for engagement with both. Rogers and Steele (2016) shared results from a study of graduate teaching assistants (GTAs) who taught elementary mathematics content courses. This work provides an example this sub-RAC uses with the CERAC's driver diagram and intended outcomes. Rogers and Steele spoke to the need for teacher candidates to engage in tasks during the courses that are "consistent with the effective mathematics teaching practices outlined in *Principles to Actions* (NCTM, 2014)" (Rogers & Steele, 2016, p. 375). Their analysis of GTA interview data and pedagogical decisions indicated different types of pedagogical support for GTAs (and one may argue faculty of teacher candidates) that is important for teacher candidates' development, "including supports that help develop (a) content-specific pedagogical knowledge and practices, (b) pedagogical strategies for facilitating mathematical discussions, and (c) critically reflective teaching practices" (Rogers & Steele, 2016, p. 409). More and more, as reflected in the MTPs, teachers, including teacher candidates and mentor teachers, are being encouraged to engage students in "authentic, non-routine problem-solving tasks and dialogic discourse intended to develop students' understandings of concepts and practices in ways that foster their abilities to solve problems, reason, and communicate mathematically" (Swars et al., 2018, p. 127). Finally, Boston et al. (2017) speak to four key points in their deep dive into the eight MTPs:

1. The MTPs are a coherent and connected set of practices that support mathematics teaching and learning,
2. Thoughtful and thorough lesson planning must precede ambitious teaching,
3. Deliberate reflection and judicious adjustments must occur and required time for those processes, and
4. Instruction must be equitable—each and every student must have the opportunity to learn mathematics with understanding. (pp. 213–214)

The need to help teacher candidates and mentor teachers build a strong, common understanding of the SMP and MTPs is driven by the research and supports the CERAC secondary driver, increase the number of effective mentor teachers who are well versed in their state standards and MTPs, which closely aligns to the AMTE Standards (2017) P.4.1 and P.4.4. See additional discussion in Chapter 7.

Integrating Equitable Teaching Practices

Effective methods courses need to integrate strong field experiences (AMTE Standard P.4.2). Mentor teachers impact teacher candidates' development, and mentor teachers can be impacted by co-learning alongside teacher candidates as a form of professional development. Opportunities for mentor teachers and teacher candidates to unpack, better understand, and experience the SMP and MTPs are imperative. These opportunities also need to be embedded in equitable teaching practices. Bartell et al. (2017) called for explicit and deliberate attention to how SMP are enacted to ensure equitable teaching and learning for all students. They further offered nine equitable practices for helping teachers think about equity as they engage in the SMP, such as establish norms for participation, position students as capable, and press for academic success. Similarly, Aguirre, Mayfield-Ingram, and Martin (2013), shared five equity-based practices in mathematics classrooms to provide teacher candidates with opportunities to engage with diverse learners in clinical settings (AMTE Standard P.4, indicator P.4.3). These equity-based practices are to:

1. go deep with mathematics,
2. leverage multiple mathematical competencies,
3. affirm mathematics learners' identities
4. challenge spaces of marginality, and
5. draw on multiple sources of knowledge (math, culture, language, family, and community).

The work of the Methods sub-RAC was undergirded by the research as well as the need to elevate and embed equitable practices within the modules. With this vision established, the next section includes a description of the work of the Methods sub-RAC as it relates to module development to further address the CERAC drivers.

DESCRIPTION OF ACCOMPLISHED WORK

An important goal of the Methods sub-RAC's modules is to engage mentor teachers in specific, productive interactions with teacher candidates around the SMP and MTPs. This goal is meant to closely align mentor teacher practices with credential program coursework and address the often heard criticism that teacher preparation coursework and fieldwork are disjointed (Capraro, Capraro, & Helfeldt, 2010). In the module design, there are specific opportunities for teacher candidates to engage with mentor teachers in ways that are not overly burdensome for the mentor teachers (e.g., watch a five-minute video clip and have a 10- to 15-minute discussion guided by a rubric). Thus, providing mentor teachers the opportunity to share their experience and expertise to support teacher candidates learning about the SMP and the MTPs gives mentor teachers a greater sense of buy-in with the program's vision of mathematics teaching and learning. Additionally, these short but intense opportunities for targeted conversations about course-related topics give mentor teachers the chance to reflect on their own knowledge and implementation of teaching practices that support students' engagement with the SMP and demonstrate the MTPs; in a sense, this provides an informal professional learning opportunity for mentor teachers.

The three activities within each module follow a similar sequence and structure. That is, Activity 1 asks methods course instructors to engage and focus on teacher candidates' preconceived beliefs and knowledge about the module's focus while beginning to advance understanding of the specific pedagogical knowledge domain for teacher candidate development. Activity 2 follows by introducing teacher candidates to research in the knowledge domain so that they begin to understand effective strategies for teaching and learning to apply it in their methods course assignments. Then, Activity 3 culminates each module by engaging the mentor teacher with the teacher candidate in a practicum setting with respect to the module focus. Providing the mentor teacher and teacher candidate an opportunity to learn bidirectionally is critical to the final activity of each module.

Module 1—Standards for Mathematical Practice

The first module focuses on the Standards for Mathematical Practice (SMP). The Methods sub-RAC decided to focus its first module on the SMP given both the research and anecdotal data that teacher candidates, mentor teachers, and methods faculty all needed to better familiarize themselves with the SMP. The Methods sub-RAC also answered the call for continued ways to support mentor teachers' continued learning (Graybeal, 2013) by engaging the mentor teachers and teacher candidates with co-learning opportunities. Module 1 consists of three activities designed to be implemented consecutively. Some methods faculty implement the three activities in three back-to-back course meetings, while other methods faculty may implement the activities across the timeline of an entire course depending on program needs, which are discussed in detail later.

Activity 1 begins with teacher candidate generating a list of mathematical habits of typical 10th grade students. Once that list is generated, which usually involves entries such as *ask for the rule to get the answer* and *avoid word problems*, teacher candidates then review the SMP and generate a list of what practices (i.e. habits) these seem to be proposing. Next, teacher candidates compare the two lists to discover differences between the habits of mind that 10th grade mathematics students typically have and to what the SMP intend.

Activity 2 within Module 1 engages teacher candidates in an inquiry lesson centered on the properties of quadrilaterals. Teacher candidates complete the lesson as students, with the methods faculty serving the role as the teacher facilitator. Next, teacher candidates reflect on the lesson, focusing on what SMP were demonstrated during the lesson and how the methods faculty's *pedagogical moves* helped to facilitate growth in those SMP. Finally, the teacher candidates watch a video excerpt of the same lesson being taught by another teacher with high school students in a diverse public high school and are asked, again, to reflect on the lesson to determine how the teacher facilitated the lesson in order to invoke student engagement in the SMP. The diversity of the students in the video provides an entry point for teacher candidates and mentor teachers to discuss issues of equity, access, and empowerment. High expectations and high levels of instruction are seen enacted and attained with a group of students who look like the students in the classrooms of teacher candidates and mentor teachers. The video offers an exemplar of what productive mathematical discussions can occur in their classrooms.

Activity 3 of Module 1 asks teacher candidates to watch the same video excerpt alongside their mentor teacher. Using a set of discussion prompts such as: *What were the students doing?*, *What content were they discussing?*, and *How would you describe their level of engagement?*, the teacher candidates and mentor teachers discuss what SMP they see in the video excerpt and how what they see may impact their own instruction. Teacher candidates and mentor teachers use the *Engaging in the Mathematical Practices Look-Fors* rubric (Elementary Mathematics Specialists & Teacher Leader Project, 2012) to review the indicators for each of the SMP and discuss whether they characterize any of the interactions from the video. Teacher candidates then write a reflection based on the discussion with their mentor teachers.

Given that the module on SMP was the first module developed by the Methods sub-RAC, it has been implemented more than the other two modules and has gone through many PDSA Cycles of iterative development. For example, Activity 1 was initially planned (Plan phase) with teacher candidates working in groups to review the SMP and the reporting out to the larger group (Do phase). When survey and exit-slip data were reviewed (Study phase), feedback indicated that teacher candidates needed more engagement across groups before sharing out about the SMP. Therefore, a Gallery Walk was included (Act phase) to allow teacher candidates to read, review, and comment on the SMP summaries before groups reported back out to the group. For more details and some findings based on implementation of Module 1, see Yow, Waller, and Edwards (2018).

Module 2—Lesson Planning

The second module created by the Methods sub-RAC focuses on ambitious lesson planning through the framework of the MTPs and the Mathematics Classroom Observation Protocol for Practice (MCOP^2) (Gleason, Livers, & Zelkowski, 2015, 2017). The lesson planning module is designed to be implemented in a mathematics methods course that is centered on developing the ideas of equitable instructional practices that promote mathematics learning, through making explicit connections to the MTPs and SMP.

Module 2 is composed of three activities. Activity 1 begins with the assumption that this activity is teacher candidates' first attempt at writing a lesson plan, prior to any specific or formal mathematics lesson planning instruction. This design choice is intended to challenge and expose teacher candidates' preconceived beliefs about teaching and learning mathematics. Teacher candidates are given a mock assessment and a contextual setting(s) (e.g., an eighth-grade mathematics class or a first-year algebra course) with appropriately aligned state standards and given 45 minutes in which to write a lesson plan that prepares learners for the given assessment, the best they can. The purpose of this initial lesson plan draft is to provide an opportunity both for faculty teaching methods courses to uncover teacher candidates' beliefs and knowledge about lesson planning and quality instructional practices, so that teacher candidates can begin to realize the importance and challenge of designing ambitious lessons that engage and support all students in making sense of mathematics. The expectation is that teacher candidates will create initial lesson plans that overwhelmingly are focused on procedures and examples, give little attention to learners' needs and strategies for engagement, and consist mostly of teacher-led *instruction* followed by student *practice* work.

MCOP² Item #16

16) The teacher uses student questions/comments to enhance conceptual mathematical understanding.	
Score	Description
1	There is some evidence the teacher does use student questions/comments to enhance conceptual mathematical understanding.
0	There is no evidence the teacher ever uses student questions/comments to enhance conceptual mathematical understanding. (No anticipation of struggles, misconceptions, or questions by students).

FIGURE 9.2. Mathematics Classroom Observation Protocol for Practice (MCOP²). This figure depicts an example of the MCOP² item #16 modified for assessing lesson plans.

After writing this initial lesson plan, teacher candidates provide a copy to their methods faculty and, before the next class meetings, teacher candidates are asked to self-evaluate using the MCOP² Lesson Planning Rubric that is a modification of the MCOP² observation instrument designed to examine aspects of a lesson plan that can be explicitly documented during planning (see Figure 9.2). The faculty will also generate MCOP² Lesson Planning Rubric scores with feedback for each lesson plan. This rubric serves as an initial checklist for many of the 16 indicators that are found on the MCOP² observation instrument and encourages teacher candidates to think about several aspects of lesson planning ensuring evidence in their plans. Some of these indicators include evidence that (a) students will engage in exploration, investigation, or problem solving; (b) students will use a variety of means to represent concepts; (c) students will critically assess mathematical strategies; and (d) the lesson (plan) involves fundamental concepts of the subject to promote relational/conceptual understanding; and so forth (see Figure 9.3 for an MTP, SMP, and MCOP² crosswalk).

In addition to focusing on teacher facilitation and student engagement, the MCOP² also encourages a focus on equity. "Equitable teaching ensures that all students are able to have entry into the mathematics experience, construct their own knowledge, and share their reasoning, strategies, and solutions with others thus creating a positive classroom community" (Gleason et. al., 2015, p. 21). Several indicators in the MCOP² can examine power dynamics in the classroom. Specifically, student engagement indicator number 13 asks if there is a climate of respect for what others had to say. This indicator encourages the teacher to think, plan, and reflect on whether many students are sharing, questioning, and commenting during the lesson—including their struggles—or if only a few students are being called on by the teacher. Moreover, MCOP² items 2, 3, 9, and 12 share characteristics that indicate how well students are engaged in authentic rich mathematics experiences. Asking teacher candidates to intentionally think about these indicators as they learn to plan lessons is meant to increase the likelihood that they consider equitable teaching to be the bar to which they aspire to reach as an early career teacher.

The purpose of Activity 2 in Module 2 is for teacher candidates to have an opportunity to work with teacher candidate peers to reflect on their initial MCOP^2 Lesson Planning Rubric scores and consider ways they might improve the lesson plan to be more closely aligned with the MTPs. Following Activity 1, methods faculty are encouraged to (a) discuss the difference between procedural and conceptual questions and the difference between problem solving and skill building, (b) ask teacher candidates to brainstorm what is important to consider if their plans are to be revised to support student success with the assessment item, and (c) introduce the MCOP^2 indicators. With this additional information and structure that has been provided around lesson planning, teacher candidates at this point should be able to provide constructive feedback in a more informed manner. After receiving feedback from their peers, teacher candidates rewrite their lesson plans and receive additional feedback from their methods faculty.

The purpose of Activity 3 is to engage the mentor teacher with the teacher candidates in the lesson planning and self-assessment process through the lens of the MCOP^2. Activity 3 requires that teacher candidates plan, implement, and reflect on a lesson (20–45 minutes in length). Teacher candidates and mentor teachers are asked to collaboratively use the MCOP^2 Lesson Planning Rubric to co-plan the lesson. The activity is designed to impact the secondary driver, *increase the number of effective mentor teachers who are well versed in their state standards and MTPs,* and to respond to the call made in the literature around mentor teacher preparation and the influence the mentor teachers have on teacher candidates. The ultimate goal is to observe teacher candidates to produce high-quality lessons during formal observations, as well as how mentor teachers can evaluate teacher candidates with the MCOP^2. The Methods sub-RAC work has further demonstrated how methods faculty may use the MCOP^2 as a grade-bearing observation assessment (Zelkowski & Gleason, 2016).

We acknowledge the high likelihood that the iterative PDSA Cycles during the 2018–2019 academic year will result in modifications to the lesson planning module, given that we expect eight pilot testing sites to generate data that will inform revisions. As the team and pilot sites complete their PDSA Cycles, the final version of Module 2 is planned for release for the start of the 2019–2020 academic year.

Module 3–Student Feedback

The third module, the student feedback module, serves as a capstone for the Methods sub-RAC work. That is, the SMP and lesson plan modules put teacher candidates' mathematics-teacher developmental trajectory on the pathway to advance students' mathematical learning and that teacher candidates see demonstrated impact on student learning through feedback.

Activity 1 begins the module by establishing a start to understanding feedback to facilitate learning based on teacher candidates' self-beliefs about teacher feedback, but further relates well to Task 3 of the edTPA (Educative Teacher Perfor-

Mathematical Classroom Observation Protocol for Practices - MCOP²
Crosswalk with the Mathematics Teaching Practices and the Standards for Mathematical Practice

MCOP² Item	Mathematics Teaching Practice(s) (MTPs)	Standards for Mathematical Practice (SMP)
1. Students engaged in exploration/ investigation/problem solving	2. Implementing tasks that promote reasoning and problem solving 1. Establishing mathematics goals to focus learning	1. Make sense of problems and persevere in solving them 7. Look for and make use of structure 8. Look for and express regularity in repeated reasoning
2. Students used a variety of means (models, drawings, graphs, concrete materials, manipulatives, etc.) to represent concepts	3. Use and connect mathematical representations	5. Use appropriate tools strategically 1. Make sense of problems and persevere in solving them
3. Students were engaged in mathematical activities	2. Implementing tasks that promote reasoning and problem solving	1. Make sense of problems and persevere in solving them
4. Students critically assessed mathematical strategies	4. Facilitate meaningful mathematical discourse	3. Construct viable arguments and critique the reasoning of others 1. Make sense of problems and persevere in solving them 8. Look for and express regularity in repeated reasoning
5. Students persevered in problem solving	7. Support productive struggle in learning mathematics	1. Make sense of problems and persevere in solving them 2. Reason abstractly and quantitatively 3. Construct viable arguments and critique the reasoning of others 5. Use appropriate tools strategically
6. The lesson involved fundamental concepts of the subject to promote relational/ conceptual understanding	6. Build procedural fluency from conceptual understanding	8. Look for and express regularity in repeated reasoning 7. Look for and make use of structure
7. The lesson promoted modeling with mathematics	2. Implementing tasks that promote reasoning and problem solving	4. Model with mathematics 2. Reason abstractly and quantitatively
8. The lesson provided opportunities to examine mathematical structure (Symbolic notation, patterns, generalizations, conjectures, etc.)	1. Establishing mathematics goals to focus learning	8. Look for and express regularity in repeated reasoning 7. Look for and make use of structure

MCOP²	MTPs	SMP
9. The lesson included tasks that have multiple paths to a solution or multiple solutions	2. Implementing tasks that promote reasoning and problem solving	1. Make sense of problems and persevere in solving them
10. The lesson promoted precision of mathematical language	4. Facilitate meaningful mathematical discourse	6. Attend to precision 3. Construct viable arguments and critique the reasoning of others
11. The teacher's talk encouraged student thinking	5. Pose purposeful questions 8. Elicit and use evidence of student learning	1. Make sense of problems and persevere in solving them
12. There were a high proportion of students talking related to mathematics	4. Facilitate meaningful mathematical discourse	3. Construct viable arguments and critique the reasoning of others
13. There was a climate of respect for what others had to say	4. Facilitate meaningful mathematical discourse	3. Construct viable arguments and critique the reasoning of others
14. In general, the teacher provided wait-time	8. Elicit and use evidence of student learning	1. Make sense of problems and persevere in solving them
15. Students were involved in the communication of their ideas to others (peer-to-peer)	4. Facilitate meaningful mathematical discourse	3. Construct viable arguments and critique the reasoning of others
16. The teacher uses student questions/comments to enhance conceptual mathematical understanding	4. Facilitate meaningful mathematical discourse 5. Pose purposeful questions 8. Elicit and use evidence of student learning	1. Make sense of problems and persevere in solving them 3. Construct viable arguments and critique the reasoning of others

FIGURE 9.3.   Mathematical Classroom Observation Protocol for Practices - MCOP² Crosswalk with the Mathematics Teaching Practices and the Standards for Mathematical Practice. A crosswalk of the MCOP², MTPs, and SMP. The SMP column lists the primary SMP first tied to the MCOP² item. See Gleason, Livers, & Zelkowski (2017) Appendix B from the MCOP² external validation study. Depending on the lesson being implemented and the design of the lesson for what students do, the MTP and SMP columns may vary slightly from these suggested listings.

Self-Feedback	Task Feedback
Feedback that addresses evaluations about the learning (usually positive, as in praise)	Feedback that addresses how the learner has understood or completed the assigned task
Praise. Good job!	*Evaluative (correct or incorrect)*

Process Feedback	Self-Regulation Feedback
Feedback that addresses the primary process that the learner is using or may use to complete the task	Feedback that addresses the learner self-monitoring or regulating their own learning or progress
Descriptive	*Prescriptive*

FIGURE 9.4.   Four Levels of Student Feedback. The four levels of student feedback that improve mathematical learning goals (Hattie & Timperly, 2007).

mance Assessment) that is framed by Wiggins' (1998) *Educative Assessment: Designing Assessments to Inform and Improve Student Performance.* Furthermore, Module 3 centers the work of Black and William's (2010) conclusion that formative assessment is essential to improving instruction and learning to align with the MTPs (NCTM, 2014). Teacher candidates learn from a chosen assessment from their more recent coursework and use the lack thereof or strength of the feedback to make direct connections to the readings of the activity.

The purpose of Activity 2 extends teacher candidates' experiences from Activity 1, but focuses more on the product of quality feedback as opposed to their own experiences with feedback. Activity 2 extends to define and have teacher candidates understand the types of feedback. Feedback to any student should specify particular qualities and development toward the mathematical goals that have been set by teachers (MTP #1) while avoiding comparisons to other students. Furthermore, high-quality feedback, whether formative or summative, can improve outcomes for lower achievers more than other students, while raising achievement for all students. This student feedback module focuses on the development of teacher candidates' understanding of the four levels of feedback shown in Figure 9.4 (Hattie & Timperly, 2007):

1.   The *task*, in which feedback is about correct/incorrect with directions to acquire more information as needed;
2.   The *process*, which is used to create the performance or product aimed at the required learning processes;
3.   *Self-regulation*, which is designed to improve confidence through self-evaluation to continue and advance engagement in the task; and
4.   The *student as a person*, which focuses on the personal learning and not the performance or product.

Activity 3 then extends the first two activities to the mentor teacher's classroom with the teacher candidate. The activity asks teacher candidate and their mentor teacher to meet and determine an appropriate assessment in which the teacher candidate can engage the mentor teacher and provide feedback to students as it relates to the prior two activities. Mentor teachers engage the teacher candidate with an assessment where the teacher candidate can bring coursework knowledge on feedback to improve student learning in the clinical setting. The shared vision for improve student learning ultimately drives this team approach.

The Methods sub-RAC completed the first PDSA Cycle related to Module 3 during the 2018–2019 academic year at two sites. The lead developers will pilot an additional PDSA Cycle during the 2019–2020 academic year. The project goal focus includes a final product of readiness for implementation in the fall of the 2020–2021 academic year.

LESSONS LEARNED

From the start of the focused work of the three CERAC's sub-RACs, the Methods sub-RAC has gained many valuable insights to share with the mathematics teacher education community. This chapter has provided an overview of three modules and the developmental process. While the PDSA Cycles reflect a descriptive manner to enact revisions on a continual basis, the most relevant lessons learned have come about due to differences in the sub-RAC members' institutional commitments, program design, and resources. This finding has been the biggest contributor to module design, development, and refinement and provides mathematics teacher education programs the ability to implement the modules in differing contexts while not compromising the objectives for teacher candidates and clinical experiences as a whole. Next, we elaborate further on implementation in different contexts based on what we learned during these processes, including a diagram (see Figure 9.5) mapping our recommendations for implementation in various contexts.

Contexts for Implementing the Modules

During the developmental processes of creating, testing, revising, and implementing different versions of these modules, it became apparent the most challenging aspects of transforming mathematics teacher preparation are related to the reality that not all programs have the same resources. By resources we include the number of methods courses, the number of clinical experience hours, the number of faculty to support teaching, flexibility in pairing teacher candidates and mentor teachers, field observations, supervision, the number of specialized mathematics content courses for teachers, and administrative support to reach the MTE-Partnership gold-standard in preparation programs aligned to the *Guiding Principles* (MTE-Partnership, 2014) and the AMTE Standards (2017). The development teams for each module took these standards into serious consideration so that the modules can be implemented in a variety of programs with different designs and resources. We elaborate on the extreme cases and add descriptions for programs with resources that fall somewhere in the middle of the extremes. Each module includes instructions specifically for the implementation by methods faculty.

Minimally Resourced Programs. For programs that have minimal resources, we outline what our work has discovered. For the purposes of adhering to the *Guiding Principles* and the AMTE Standards, we acknowledge an important point: In situations in which there are inadequate resources for mathematics teacher preparation, just hoping that those minimal resources somehow work is not nearly enough. Together with the CERAC and the MTE-Partnership as a whole, we point to the justification for college and university administrators to increase resources for teacher preparation efforts to more closely align to the call in the AMTE Standards. We support the AMTE Standards and the call to action #5:

> Higher education administrators overseeing programs that prepare teachers of mathematics must ensure that commitment and focus is placed on those programs, including allocating qualified personnel and resources needed to achieve the vision of this document. A program that does not make clear and strong commitments to ensuring that its candidates are well-prepared as described in this volume is not justifiable. Mediocrity cannot be an option. (AMTE, 2017, p.166)

Programs with only a single mathematics methods course (or a general methods or STEM methods course) spanning one semester with a clinical experience are not likely to have teacher candidates who are ready for a full-time student teaching internship, particularly in light of the growing use of high-stakes, externally assessed teaching portfolios, such as edTPA (Educative Teacher Performance Assessment). We recognize the constraints of such programs may require an overlap of the Module 1 SMP and Module 2 Lesson Planning in one semester.

Module 1 focuses on taking teacher candidates from beginner to novice and emerging knowledge about the SMP, and we recommend implementing this project at the onset of the single methods course. It might be wise to collaborate with a mathematics faculty member who could implement the first two activities as part of a geometry course, for example, that is part of an undergraduate content preparation pathway. Activities 1 and 2 can be implemented early on, in the first two weeks of a semester. Activity 3 can be implemented later, after teacher candidates have started a corresponding clinical experience with a mentor teacher. In situations where teacher candidates do not have a clinical experience and a mentor teacher, our suggestion is to first seek to change the program structure to accommodate a mentor teacher in a clinical setting to align with best practices and the AMTE Standards. Adding time for a clinical experience into a teacher preparation program does not add instructional time to a university's budget. Thus looking for manners in which universities can improve clinical experiences is paramount. While there needs to be some coordination of a clinical experience, the module is embedded into a methods course in which the activities are coordinated by the methods instructor.

Module 2 focuses on moving from a novice/beginner lesson planner toward a competent teacher candidate. We recommend implementation of Activities 1 and 2 after the first two weeks of the course when Module 1's first two activities are finished. We recommend that methods faculty provide their own lesson planning instruction between the Module 2's Activities 1 and 2 as noted in the instructions. For Activity 3, we recommend waiting until lesson planning is at a competent level by teacher candidates before they can co-plan with their mentor teacher to implement a lesson or mini-lesson in a mathematics classroom. The co-planning with the mentor teacher and teacher candidate should take place after the Module 1 Activity 3 has been completed by the teacher candidate and mentor teacher.

Module 3 focuses on providing rich and appropriate feedback to students based on mathematical goals of lessons and activities. The student feedback module is well connected to and suited for preparation within Task 3 of the edTPA. Activities 1 and 2 should be implemented well after the completion of the activities in Modules 1 and 2. The student feedback Module 3's Activity 3 can be delayed until teacher candidates are embedded into their student teaching semester as well. However, we do recommend as part of the methods course instruction that Activities 1 and 2 of Module 3 are completed during the course, at a minimum before teacher candidates enter student teaching.

To adequately prepare the next generation of mathematics teachers to meet the *Guiding Principles* "gold standard" (MTE-Partnership, 2014) for well-prepared beginning teachers of mathematics, commitments are required to also increase the resources to meet the AMTE Standards. This work includes multiple mathematics methods courses and structured sequences of clinical experiences prior to or concurrent with student teaching. We further recommend the review of how sequenced methods courses can affect licensure exams such as edTPA for teacher candidates (see Zelkowski, Campbell, & Gleason, 2018; Zelkowski & Gleason, 2018).

Multiple Sequenced Methods Courses. For programs that have at least two sequential mathematics methods courses with clinical experiences prior to or concurrent with the student teaching internship, we recommend the following use of Modules 1, 2, and 3 spread across two semesters. In the first semester, our recommendation is to implement the Module 1's three activities before midterm while teacher candidates are learning about the foundations of teaching and learning mathematics. If possible, we also encourage methods faculty to collaborate with mathematics faculty of a geometry course where Activities 1 and Activity 2 could be jointly implemented, since Module 1 has a heavy focus on geometric content and the SMP.

After the midterm and completion of Module 1, we recommend implementing the Module 2 in an effort to have teacher candidates move from novice/beginner to competent (not proficient or expert) lesson planners for the start of the second methods course. During the second semester, our recommendation for Module 3 is to use local judgments on whether teacher candidates are ready to focus on student assessment and feedback. We envision that some teacher candidates will be ready for unit planning while others will still need time for lesson planning. However, we acknowledge that Module 3 should begin no later than the semester midterm as a means to engage the mentor teacher and teacher candidate in activities that connect the teacher preparation coursework to the clinical setting of the classroom. As a readiness measure for a portfolio requirement such as edTPA, moving through all three modules is important prior to the student teaching internship.

After completing the Modules 1 and 2 in the first methods course, our recommendation is that methods faculty determine where and how to best fit Module 3 on student feedback into their curriculum, during their lone methods course or at the start of the internship. Figure 9.5 depicts considering implementation of Module 3 in the middle half of the second methods course. Our intent is that when entering student teaching internships, teacher candidates will have experienced working with mentor teachers to: (a) make sense of the SMP, (b) integrate lesson planning for high leverage teaching practices (e.g. MTPs), and (c) understand the impact of student feedback related to mathematical goals (MTP #1). If all three modules are implemented, discussed, and tied to preparation curricula, student teaching internships should see improved readiness and internship outcomes.

For programs with additional methods courses and/or clinical experiences, our only recommendation is to use the information above and make local judgments about an appropriate timeline for implementing the sequence of the Modules 1, 2, and 3 prior to the student teaching internship. We include a three-semester sequenced set of methods course recommendation in Figure 9.5.

Future Plans

We first recognize many efforts outside of the MTE-Partnership and all of the CERAC regarding specific work on the design and implementation of mathematics methods courses (e.g., Kastberg, Tyminski, Lischka, & Sanchez, 2018) and the more recently published handbook on field-based clinical experiences (see Hodges & Baum, 2018). Moving forward, to link these two fields of professional work with the AMTE Standards, the Methods sub-RAC's future plans include: (a) national dissemination of these modules for implementation and research; (b) manuscripts on the development and results from the PDSA Cycles; and (c) the firm integration of mathematics methods courses with field-based clinical experiences, strategically engaging mentor teachers and teacher candidates to improve student teaching internship readiness. This work brings mentor teachers into the secondary teacher preparation program coursework and evaluation of teacher candidates by providing a collaborative set of guidelines aimed at erasing the stigma of disjointed teacher preparation coursework and clinical field experiences (Capraro, Capraro, & Helfeldt, 2010; National Research Council, 2010). To truly transform

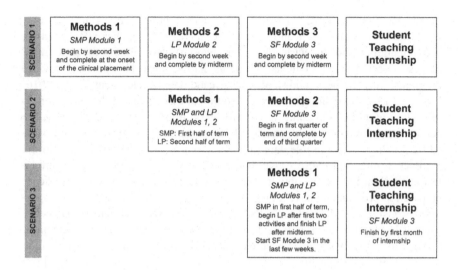

FIGURE 9.5.   Scenarios for Implementation of Methods Modules. Recommended or possible scenarios for implementation of Methods modules for engaging cooperating teachers in the development of teacher candidates during coursework.

mathematics teacher education and move toward an agreed upon shared set of experiences that develop teacher candidates into well-prepared beginning teachers, the disjointed and siloed nature of what we do as professional mathematics teacher educators in our own programs and methods courses needs to reach some point of consistency on a larger scale. The three modules of the MTE-Partnership Methods sub-RAC serve as starting points developed across many preparation programs that have been vetted and validated by many mathematics teacher educators through iterative cycles of feedback and teacher candidate/mentor teacher responses. We will seek to publish the final version of the three modules through a journal or publishing outlet (e.g., AMTE) so that mathematics teacher educators and their preparation programs across the nation have full access for implementation, practice, and research purposes.

SUMMARY

The Methods sub-RAC's modules described in this chapter provide the mathematics teacher education community with aspects of work that integrate important learning experiences for teacher candidates through structured and/or sequenced clinical experiences while meaningfully engaging the mentor teacher within each module. The qualities of each module address directly or partially the AMTE Standards generally (as previously mentioned) and, more specifically, standards for high school and middle grades mathematics teachers utilizing the MTPs.

The Methods sub-RAC's use of PDSA Cycles to improve and validate (Kane, 2012) the effective use of the modules on the standards for mathematical practice, lesson planning, and student feedback has demonstrated the need for an iterative process across different contexts, programs, and settings. This process provides programs with stronger validity of teacher candidate development (Kane, 2016; Lavery et al., 2018), though we expect continual assessment of the effectiveness of each module in local contexts. Our focus on improvement science using the Networked Improvement Community (NIC) of the MTE-Partnership has provided rich, valuable, and usable data to make strong iterative improvements to the modules during each PDSA Cycle. It is often very difficult within one or only a few institutions to capture enough data from small secondary mathematics teacher preparation programs to have a large impact on quality with researchable opportunities. Using the NIC and MTE-Partnership collaborative structures across upward of 15 institutions has provided a strong validation approach to the outcomes, and methods faculty should expect to accomplish effective implementation of the three modules in a structured sequence. Given this work is timely and has been intense over multiple years and institutions, we have kept a focus on improvement central to the iterations through PDSA Cycles rather than timeliness of releasing modules. Our work remains central to the focus on the MTE-Partnership and CE-RAC, as seen in the driver diagram. Moreover, by using secondary and primary drivers as our guiding tools, we are able to address the CERAC aim by specifically improving the literature on directly engaging the teacher candidate and mentor teachers in high-quality activities to address some of the criticisms of teacher preparation. Learning from multiple site implementations and iterations has and will continue to provide methods faculty with quality modules. These three modules, when fully completed, will have been well tested in multiple settings to support linking the work in methods courses to field-based clinical experiences by engaging the mentor teacher in meaningful ways with teacher candidates.

ENDNOTES

1. Work on this chapter was supported in part by a grant from the National Science Foundation (#1726998). All findings and opinions are those of the authors, and not necessarily those of the funding agency.
2. Other contributors to this chapter include: Justin Boyle, formerly of The University of Alabama and Belinda Edwards of Kennesaw State University.

REFERENCES

Aguirre, J. M., Mayfield-Ingram, K., & Martin, D. B. (2013). *The impact of identity in K–8 mathematics: Rethinking equity-based practices*. Reston, VA: National Council of Teachers of Mathematics.

Association of Mathematics Teacher Educators. (2017). *Standards for preparing teachers of mathematics.* Available online at http://amte.net/standards

Ball, D. L., Sleep, L., Boerst, T. A., & Bass, H. (2009). Combining the development of practice and the practice of development in teacher education. *The Elementary School Journal, 109*, 458–474.

Bartell, T., Wager, A., Edwards, A., Battey, D., Foote, M., & Spencer, J. (2017). Toward a framework for research linking equitable teaching with the Standards for Mathematical Practice. *Journal for Research in Mathematics Education, 48*(1), 7–21.

Black, P., & William, D. (2010). Inside the black box: Raising standards through classroom assessment. *Phi Delta Kappan, 92*(1), 81–90.

Bostic, J., & Matney, G. (2013). Overcoming a common storm: Designing PD for teachers implementing the common core. *Ohio Journal of School Mathematics, 67*, 12–19.

Boston, M., Dillon, F., Smith, M. S., & Miller, S. (2017). *Taking action: implementing effective mathematics teaching practices.* Reston, VA: National Council of Teachers of Mathematics.

Capraro, M. M., Capraro, R. M., & Helfeldt, J. (2010). Do differing types of field experiences make a difference in teacher candidates' perceived level of competence? *Teacher Education Quarterly, 37*(1), 131–154.

Cavey, L., & Berenson, S. (2005). Learning to teach high school mathematics: Patterns of growth in understanding right triangle trigonometry during plan study. *Journal of Mathematics Behavior, 24*(2), 171–190.

Darling-Hammond, L. (2014). Strengthening clinical preparation: The holy grail of teacher education. *Peabody Journal of Education, 89*(4), 547–561.

Deming, W. E. (1950). *Elementary principles of the statistical control of quality.* Tokyo, Japan: Japanese Union of Scientists and Engineers (JUSE).

Deming, W. E. (1993). *The new economics.* Cambridge, MA: MIT Press.

Elementary Mathematics Specialists & Teacher Leader Project. (2012). *Engaging in the mathematical practices (Look Fors).* Retrieved from http://www.nctm.org/Conferences-and-Professional-Development/Principles-to-Actions-Toolkit/Resources/5-SMPLookFors/

Forzani, F. (2014). Understanding "core practices" and "practice-based" teacher education: Learning from the past. *Journal of Teacher Education, 65*(4), 357–368.

Gareis, C. R., & Grant, L. W. (2014). The efficacy of training cooperating teachers. *Teaching and Teacher Education, 39*, 77–88.

Gleason, J., Livers, S. D., & Zelkowski, J. (2015). *Mathematics Classroom Observation Protocol for Practices MCOP²: Descriptive manual booklet.* Tuscaloosa, AL: The University of Alabama. http://bit.ly/MCOP2-descriptive-manual

Gleason, J., Livers, S. D., & Zelkowski, J. (2017). Mathematics Classroom Observation Protocol for Practices (MCOP²): Validity and reliability. *Investigations in Mathematical Learning, 9*(3), 111–129.

Goodwin, A. L., Roegman, R., & Reagan, E. M. (2016). Is experience the best teacher? Extensive clinical practice and mentor teachers' perspectives on effective teaching. *Urban Education, 51*(10), 1198–1225.

Graybeal, C. D. (2013). Learning to look for the Standards for Mathematical Practice. *SRATE Journal, 22*(2), 8–13.

Grossman, P., Hammerness, K., & McDonald, M. (2009). Redefining teaching: Re-imagining teacher education. *Teachers and Teaching: Theory and Practice, 15*(2), 273–290.

Gunckel, K. L., & Wood, M. B. (2016). The principle–practical discourse edge: Elementary preservice and mentor teachers working together on co-learning tasks. *Science Education, 100*(1), 96–121.

Hattie, J. & Timperly, H. (2007). The power of feedback. *Review of Educational Research, 77*(1), 81–112.

Hiebert, J., Stigler, J., Jacobs, J., Givvin, K., Garnier, H., Smith, M., Hollingworth, H., Manaster, A., Wearne, D., & Gallimore, R. (2005). Mathematics teaching in the United States today (and tomorrow): Results from the TIMSS 1999 Video Study. *Educational Evaluation and Policy Analysis, 27*(2), 111–132.

Hoffman, J. V., Wetzel, M. M., Maloch, B., Greeter, E., Taylor, L., DeJulio, S., & Vlach, S. K. (2015). What can we learn from studying the coaching interactions between cooperating teachers and preservice teachers? A literature review. *Teaching and Teacher Education, 52*, 99–112.

Hodges, T., & Baum, A. (2018). *Handbook of research in field-based teacher education.* Hershey, PA: IGI Global.

Izadinia, M. (2015). A closer look at the role of mentor teachers in shaping preservice teachers' professional identity. *Teaching and Teacher Education, 52*, 1–10.

Kane, M. (2012). All validity is construct validity. Or is it? *Measurement: Interdisciplinary Research and Perspectives, 10*(1–2), 66–70.

Kane, M. (2016). Validation strategies: Delineating and validating proposed interpretations and uses of test scores. In S. Lane, M. R. Raymond, T. M. Haladyna, S. Lane, M. R. Raymond, & T. M. Haladyna (Eds.), *Handbook of test development* (2nd ed., pp. 64–80). New York, NY: Routledge/Taylor & Francis Group.

Kastberg, S. E., Tyminski, A. M., Lischka, A. E., & Sanchez, W. B. (2018). *Building support for scholarly practices in mathematics methods.* Charlotte, NC: Information Age Publishing.

Kidd, M. (2008). A comparison of secondary mathematics methods courses in California. In P. M. Lutz (Ed.), *Secondary mathematics methods courses in California* (pp. 1–5). Bakersfield, CA: California Association of Mathematics Teacher Educators.

Lavery, M., Jong, C., Krupa, E., Bostic, J., & Carney, M. (2018). Developing an assessment with validity in mind. In J. Bostic, E. Krupa, & J. Shih (Eds.), *Assessment in mathematics education contexts* (pp. 12–39). New York, NY: Routledge.

Lawley, J. J., Moore, J., & Smajic, A. (2014). Effective communication between preservice and cooperating teachers. *The New Educator, 10*(2), 153–162.

Leinwand, S., Huinker, D., & Brahier, D. (2014). Principles to actions: Mathematics programs as the core for student learning. *MatheMatics teaching in the Middle school, 19*(9), 516–519.

Mateas, V. (2016). Debunking myths about the standards for mathematical practice. *Mathematics Teaching in the Middle School, 22*(2), 92–99.

Mathematics Teacher Education Partnership. (2014). *Guiding principles for secondary mathematics teacher preparation.* Washington, DC: Association of Public and Land-grant Universities. Retrieved from mtep.info/guidingprinciples

National Council of Teachers of Mathematics. (2014). *Principles to actions: Ensuring mathematical success for all.* Reston, VA: Author.

National Governors Association Center for Best Practices, Council of Chief State School Officers. (2010). *Common core state standards: Mathematics.* Washington, DC: Author.

National Research Council (2010). *Preparing teachers: Building evidence for sound policy*. Washington, DC: The National Academies Press.

Nguyen, T., Dekker, R., & Goedhart, M. (2008) Preparing Vietnamese student teachers for teaching with a student-centered approach. *Journal of Mathematics Teacher Education, 11*(1), 61–81.

Olson, T., Olson, M., & Capen, S. (2014). The common core standards for mathematical practice: Teachers' initial perceptions and implementation considerations. *Journal of Mathematics Education Leadership, 15*(2), 11–20.

Ricks, T. E. (2011). Process reflection during Japanese lesson study experiences by prospective secondary mathematics teachers. *Journal of Mathematics Teacher Education, 14*(4), 251–267.

Rogers, K. C., & Steele, M. D. (2016). Graduate teaching assistants' enactment of reasoning-and-proving tasks in a content course for elementary teachers. *Journal for Research in Mathematics Education, 47*(4), 372–419.

Strutchens, M. E. (2017). Current research on prospective secondary mathematics teachers' field experiences. In M. E., Strutchens, R. Huang, L. Losano, J. P.da Ponte, M. C. de Costa Trindade Cyrino, M. R., & Zbiek (Eds), *ICME-13 topic surveys: The mathematics education of prospective secondary teachers around the world* (pp. 33–44). Cham, Switzerland: Springer.

Swars, S. L., Smith, S. Z., Smith, M. E., Carothers, J., & Myers, K. (2018). The preparation experiences of elementary mathematics specialists: examining influences on beliefs, content knowledge, and teaching practices. *Journal of Mathematics Teacher Education, 21*(2), 123–145.

Tatto, M. T. (1998). The influence of teacher education on teachers' beliefs about purposes of education, roles, and practice. *Journal of Teacher Education, 49*, 66–77.

Taylor, M., & Ronau, R. (2006). Syllabus study: A structured look at mathematics methods courses. *AMTE Connections, 16*(1), 12–15.

Wiggins, G. (1998). *Educative assessment: Designing assessments to inform and improve student performance*. San Francisco, CA: Jossey-Bass.

Wood, M. B., & Turner, E. E. (2015). Bringing the teacher into teacher preparation: Learning from mentor teachers in joint methods activities. *Journal of Mathematics Teacher Education, 18*(1), 27–51.

Yee, S., Otten, S., & Taylor, M. (2017). What do we value in secondary mathematics teaching methods? *Investigations in Mathematics Learning, 10*(4), 187–201.

Youngs, P., & Hong, Q. (2013). The influence of university courses and field experiences on Chinese candidates' elementary mathematics knowledge for teaching. *Journal of Teacher Education, 64*(3), 244–261.

Yow, J. A., Waller, P., & Edwards, B. (2018). A national effort to integrate field experiences into secondary mathematics methods courses. In T. Hodges & A. Baum (Eds.), *Handbook of research in field-based teacher education*. Hershey, PA: IGI Global.

Zelkowski, J., Campbell, T. G., & Gleason, J. (2018). Programmatic effects of capstone math content and math methods courses on teacher licensure exams. In W. M. Smith, B. R. Lawler, J. F. Strayer, & L. Augustyn (Eds.), *Proceedings of the seventh annual Mathematics Teacher Education Partnership conference* (pp. 91–96). Washington, DC: Association of Public and Land-grant Universities.

Zelkowski, J., & Gleason, J. (2016). Using the MCOP^2 as a grade bearing assessment of clinical field observations. In B. R. Lawler, R. N. Ronau, & M. J. Mohr-Schroeder

(Eds.), *Proceedings of the fifth annual Mathematics Teacher Education Partnership conference* (pp. 129–138). Washington, DC: Association of Public and Land-grant Universities.

Zelkowski, J., & Gleason, J. (2018). Programmatic effects on high stakes measures in secondary math teacher preparation. In L. Venenciano & A. Redmond-Sanogo (Eds.), *Proceedings of the 45th annual meeting of the Research Council on Mathematics Learning*. Baton Rouge, LA.

CHAPTER 10

USING CO-PLANNING AND CO-TEACHING STRATEGIES TO TRANSFORM SECONDARY MATHEMATICS CLINICAL EXPERIENCES[1]

Maureen Grady, Ruthmae Sears,
Jamalee (Jami) Stone, and Stephanie Biagetti

During clinical experiences, prospective teachers should be encouraged to make connections to content addressed within their teacher preparation program coursework (Zeichner, 2010). Prospective teachers also need to be mentored by highly effective teachers who are able to provide appropriate guidance and feedback (Darling-Hammond, 2010); see Chapter 7 for further discussion of this literature. Therefore, using an improvement science research design (Bryk, Gomez, Grunow, & LeMahieu, 2015), the Co-Planning and Co-Teaching subgroup was formed as a part of the Clinical Experiences Research Action Cluster (CERAC) to study how the use of co-planning and co-teaching could support collaborative pairs and how this sub-RAC could develop resources to help teachers enact co-teaching within a secondary mathematics context (Strutchens, Iiams, & Sears, 2016).

The Mathematics Teacher Education Partnership: The Power of a Networked Improvement Community to Transform Secondary Mathematics Teacher Preparation, pages 235–256.
Copyright © 2020 by Information Age Publishing
235

In this chapter, we will focus on the work of the Co-Planning and Co-Teaching (CPCT) sub-RAC. We will discuss relevant literature that clarifies our theory of change (implementing co-planning and co-teaching during clinical experiences) and then describe how the CPCT sub-RAC supports the CERAC goal of increasing opportunities for teacher candidates to use each of the eight National Council of Teachers of Mathematics' (NCTM, 2014) mathematics teaching practices. Next, we will summarize initial efforts of the CPCT sub-RAC's Plan-Do-Study-Act (PDSA) Cycles across the years. We will conclude by sharing the lessons we have learned about how to support CPCT practices in clinical experiences.

The CPCT sub-RAC comprises mathematics educators from 11 institutions in six different states, within the continental United States, and has been in existence since 2012. The CPCT sub-RAC noted limitations of clinical experiences relative to a disconnect between coursework and clinical experiences and the lack of quality mentors to support prospective teachers in developing effective teaching practices (Sears et al., 2017b). The CPCT sub-RAC members believed that co-teaching, by providing better support for teacher candidates in field experiences, had the potential to address the needs of diverse learners and to strengthen relationships between university and school district partners. Hence, the mathematics teacher educators have made a commitment to implement co-planning and co-teaching at their respective institutions. Taking into account that programmatic structures vary (inclusive of residency models, post-baccalaureate certifications, number of methods courses, and practicum requirements), the sub-RAC was cognizant that the implementation of co-planning and co-teaching would need to complement the uniqueness of each institution's expectations and organizational structure for clinical experiences.

RELEVANT LITERATURE ABOUT CO-PLANNING AND CO-TEACHING

In considering how to improve clinical experiences, the Networked Improvement Community (NIC; Martin & Gobstein, 2015) began by considering the published research on co-planning and co-teaching. A review of the literature indicates that co-teaching has been used and researched within the special education context for a number of years (Friend, Cook, Hurley-Chamberlain, & Shamberger, 2010). Much of the theory and research about co-teaching focuses on its use with a general education teacher and a special education teacher in an inclusion classroom setting (Murawski & Swanson, 2001). Within this context, a variety of co-teaching strategies (see Table 10.1) have been used.

Within the special education context, co-teaching is perceived to be beneficial for teachers and K–12 students. For example, the Scruggs, Mastropieri, and McDuffie (2007) meta-synthesis of the literature on co-teaching between content area teachers and special educators in inclusion classrooms noted that the use of co-teaching strategies increased cooperation and collaboration between students, provided students with more teacher attention, and increased academic success

TABLE 10.1.   Co-Teaching Strategies[1]

Strategy	Definition
One teach, one observe	One teacher has primary instructional responsibility while the other gathers specific observational information on students or the (instructing) teacher. The key to this strategy is to focus the observation on specific behaviors. Both the teacher candidate and the cooperating teacher are able to take on either role.
One teach, one assist	One teacher has primary instructional responsibility while the other assists students with their work, monitors behaviors, or corrects assignments, often lending a voice to students or groups who hesitate to participate or add comments.
Station teaching	Station teaching occurs when the co-teaching pair divides the instructional content into parts. Each teacher instructs one of the groups. The groups then rotate or spend a designated amount of time at each station. Independent stations are often used along with the teacher-led stations.
Parallel teaching	Parallel teaching occurs when the class is divided, with each teacher instructing half the students. However, both teachers are addressing the same instructional material. Both teachers are using the same instructional strategies and materials. The greatest benefit to this method is the reduction of the student-to-teacher ratio.
Supplemental teaching	Supplemental teaching allows one teacher to work with students at their expected grade level while the other teacher works with those students who need the information or materials extended or remediated.
Alternative (differentiated) teaching	This teaching strategy provides two approaches to teaching the same information. The learning outcome is the same for all students; however, the avenue for getting there is different.
Team teaching	Team teaching incorporates an invisible flow of instruction with no prescribed division of authority. Using a team-teaching strategy, both teachers are actively involved in the lesson. From the students' perspective, there is no clearly defined leader–both teachers share the instruction, are free to interject information, and are available to assist students and answer questions.

[1]Bacharach, Heck, & Dahlberg, 2010

for students with disabilities. The researchers also found that teachers often reported substantial professional growth from their involvement in co-teaching settings (Scruggs, Mastropieri, & McDuffie, 2007).

The documented benefits of using co-teaching strategies within special education, coupled with the challenges of recruiting effective mentor teachers and adequately supporting prospective teachers during clinical experiences, have led to a growing body of literature about the use of co-teaching strategies during clinical experiences (Diana, Jr., 2014). Researchers found that when co-teaching strategies are used during clinical experiences, prospective teachers developed better classroom management (Bacharach, 2007; Heck, Bacharach, & Dahlberg, 2008); increased their collaboration skills (Bacharach, 2007; Darragh, Picanco, Tully, & Henning, 2011; Heck, Bacharach, & Dahlberg, 2008); exhibited better knowledge of the curriculum (Bacharach, 2007; Heck, Bacharach, & Dahlberg, 2008); and were afforded more time to be involved with instructional responsibilities

(Bacharach, 2007) when compared to prospective teachers in traditional placements. Prospective teachers reported that co-teaching with their mentor teachers helped them feel more like a "real teacher" (Bacharach, 2007; Heck, Bacharach, & Dahlberg, 2008); provided them with more access to resources and support (Bacharach, 2007); and increased opportunities to differentiate instruction (Darragh, Picanco, Tully, & Henning, 2011; Morton & Birky, 2015). Moreover, there is evidence that prospective teachers who participated in co-teaching settings continued to place high value on their experiences with using co-teaching strategies during their induction years as in-service teachers (Yopp & Young, 1999). Researchers found that, although prospective teachers initially missed the continual partnership with their co-teachers, they tended to reach out and form strong collaborative relationships in their new teaching settings (Guise & Thiessen, 2017; Wassell & LaVan, 2009).

In addition to the benefits for prospective teachers, co-teaching was found to be beneficial for both students and mentor teachers. Co-teaching provided mentor teachers with opportunities for professional growth as they planned and taught with their prospective teachers (Bacharach, 2007; Morton & Birky, 2015; Scruggs, Mastropieri, & McDuffie, 2007). Additionally, students in co-teaching classrooms experienced more attention from their teachers (Bacharach, 2007; Bacharach, Heck, & Dahlberg, 2010; Morton & Birky, 2015; Scruggs, Mastropieri, & McDuffie, 2007), obtained exposure to different teacher perspectives and instructional practices (Bacharach, 2007; Darragh, Picanco, Tully, & Henning, 2011; Morton & Birky, 2015), had fewer disruptions in classroom instruction from administrative tasks or student misbehavior (Bacharach, Heck, & Dahlberg, 2010; Morton & Birky, 2015), and engaged in lessons with greater depth of content and more opportunities for student engagement (Morton & Birky, 2015). Additionally, students in co-teaching classrooms demonstrated improved academic performance (Bacharach, 2007; Bacharach, Heck, & Dahlberg, 2010) and tended to be more cooperative and collaborative (Scruggs, Mastropieri, & McDuffie, 2007).

Co-teaching cannot occur in isolation, and co-planning is critical to effective co-teaching. Murawski and Lochner (2011) state, "Without co-planning, teachers are at best working together in a parallel or reactive manner" (p. 175). Nevertheless, research about co-planning and its role in co-teaching settings is limited. Many researchers have asserted that co-planning is essential to co-teaching and is often one of the largest challenges to successful implementation of co-teaching (e.g., Dieker, 2001; Howard & Potts, 2009; Magiera, Smith, Zigmond, & Gebauer, 2005). Despite the recognized importance of co-planning, most of the available literature about co-planning tends to focus on a call for more common planning time and to provide general advice about planning, such as plan for cooperative grouping (Bryant & Land, 1998); "select an appropriate environment without distractions" (Murawski, 2012, p. 9); and use co-planning time to "plan for instruction" (Howard & Potts, 2009, p. 3). Cayton and Grady (2016), concerned about the lack of specific strategies to help with co-planning, used the co-teaching strategies as inspiration to propose a set of six co-planning strategies, shown in Table 10.2.

Grady, Cayton, Preston, and Middleton (2018) found that secondary mathematics co-teaching collaborative pairs used a variety of these co-planning strategies and found co-planning helpful and rewarding. One mentor teacher remarked, "I think that it is the best way to grow an intern and to grow as a clinical teacher. Co-Planning has to be some of my favorite times with my intern, because I learn as much as they do" (p. 5).

Although the research to date on co-teaching and co-planning in student teaching settings is promising, there is little research available about the effectiveness of CPCT in secondary settings and, particularly, in secondary mathematics settings. Thus, the CPCT sub-RAC seeks to use improvement science to research the use of CPCT in secondary mathematics field experiences, investigate the supports necessary for successful implementations of CPCT in these settings, and study the ways that CPCT can support teacher candidates to better enact equitable teaching practices, such as those embodied in the mathematical teaching practices (NCTM, 2014) and the Association of Mathematics Teacher Educators' *Standards for Preparing Teachers of Mathematics* (AMTE, 2017).

Connecting Co-Planning and Co-Teaching to the CERAC Driver Diagram

The CPCT sub-RAC sought to explicate how co-planning and co-teaching could be used within secondary mathematics clinical experiences (Sears et al., 2017b). The CPCT sub-RAC members sought to address the primary drivers, secondary drivers, and change ideas identified in the CERAC driver diagram

TABLE 10.2. Co-Planning Strategies[1]

Strategy	Definition
One plans, one assists	Each co-teacher brings a portion of the lesson, although one clearly has the main responsibility. The team works jointly on final planning.
Partner planning	Co-teachers take responsibility for about half of the components of the lesson plan. Then they complete the plan collaboratively.
One reflects, one plans	One teacher thinks aloud about the main parts of the lesson, and the other teacher writes the plan.
One plans, one reacts	One co-teacher plans, and the other makes suggestions for improvement.
Parallel planning	Each member of the co-teaching team develops a lesson plan, and the two bring them together for discussion and integration.
Team planning	Both teachers actively plan at the same time and in the same space with no clear distinction of who takes leadership.

[1]Adapted from Grady, Cayton, Preston, & Middleton, 2018

(see Figure 10.1) to increase the likelihood that the overarching aim would be achieved. The primary sub-RAC change idea of implementing CPCT in clinical experiences, along with associated professional development and support, is embedded within the overall CERAC driver diagram and complements the work being done across the RAC.

Promoting Communication. The primary drivers for the CERAC advocate for transparency, interdependency, clinical training opportunities, shared vision, and attention to access and equity among teacher educators, district partners (inclusive of mentor teachers), and prospective teachers. Thus, it is important that effective modes of communication are used to ensure these drivers are addressed. To support transparent and coherent mentorship selection, and to support mentor teachers throughout clinical experiences, the CPCT sub-RAC collaborated with district partners to identify effective teachers, organized professional development training initiatives, and disseminated practical resources that could be used to help mentor teachers and prospective teachers conceptualize how co-planning and co-teaching could be used within a secondary mathematics context (Sears et al., 2017a). To foster alignment between methods courses and clinical experiences, prospective teachers were asked to describe the extent to which they attended to the *Common Core State Standards for Mathematics* (*CCSS-M*; National Governors Association Center for Best Practices, Council of Chief State School Officers, 2010) within their classroom settings. Prospective teachers were also asked to regularly reflect on the extent to which their instructional practices aligned with what they learned in their methods course(s).

Moreover, mathematics teacher educators in the CPCT sub-RAC also agreed to serve as university supervisors for clinical experiences and to teach practicum courses such that they could readily promote the interdependency between theoretical underpinnings addressed in the methods courses and actual events that transpire during clinical experiences. To promote student teaching as clinical training, the CPCT sub-RAC advocated that prospective teachers be treated as teachers from the onset. Additionally, to build a shared vison the CPCT sub-RAC attended monthly online meetings, communicated frequently via email and developed annual PDSA Cycles. Additionally, the sub-RAC made a commitment to promote access and equity within their individual university settings.

Collaborations Across the Mathematics Teacher Education Partnership. In order to make progress on the CERAC primary driver of increasing the number of highly qualified mentor teachers, the sub-RAC addressed the CERAC change idea of providing professional development to mentor teachers (see Figure 10.1). This professional development was done in partnership with districts, which identified highly qualified teachers who might serve as mentors. The focus of the professional development included both co-planning and co-teaching strategies as well as the vision for mathematics education in national standards documents (e.g., AMTE, 2017; NCTM, 2014). To connect coursework to clinical experiences, assessment measures for clinical experiences placed value on the use

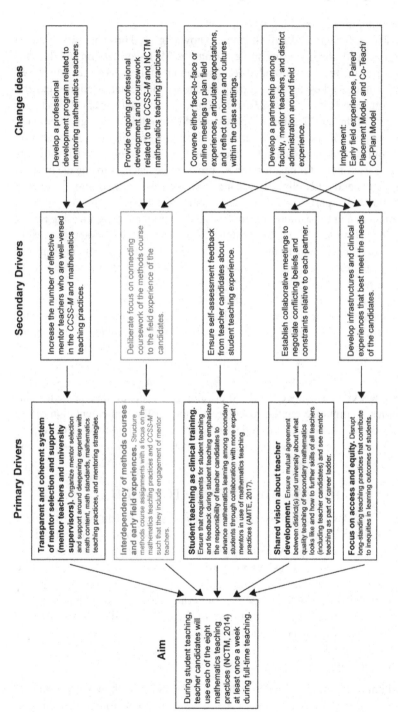

FIGURE 10.1. CERAC Driver Diagram. The Co-Planning and Co-Teaching sub-RAC focuses its work on all of these drivers, with the exceptions of the boxes that are faded as those are a main focus of the Methods sub-RAC.

of co-planning and co-teaching strategies. The data gathered from collaborative pairs over time documented growth in their knowledge and possible shifts in their perspectives. Additionally, using surveys from the MTE-Partnership and the sub-RAC, prospective teachers provided feedback about their experiences when they used co-planning and co-teaching during secondary mathematics clinical experiences that attended to the *CCSS-M* and its Standards for Mathematical Practice.

Members of the CPCT sub-RAC participated in monthly meetings where they obtained support for challenges faced, considered how to progress given university constraints, and gained insights about the variance among their institutions relative to the amount of time allocated for prospective teachers to engage in clinical experiences and other clinical teaching opportunities. Furthermore, to develop infrastructure to support the adoption of co-planning and co-teaching during clinical experiences and to promote student learning we used an apprenticeship model for learning (Brosnan, Jaede, Brownstein, & Stroot, 2014). The apprenticeship model for learning considers the intersectionality of using co-planning and co-teaching strategies and cognitive coaching to meet the needs of diverse learners in order to support student thinking.

Building Professional Development. For the change idea of implementing co-planning and co-teaching during clinical experiences to address the aim statement, there existed a need for professional development training for the mentor teachers and teacher candidates who worked together in the co-planning and co-teaching model. Moreover, to account for the variance across programs, and to gain insight into site-specific needs, the feedback from the mathematics teacher educators was vital. Thus, we sought to develop professional development training modules, to organize regular meetings to monitor the implementation of co-planning and co-teaching, and to reflect on changes that should be made to promote sustainability of the change idea with consideration to institutional norms and cultures.

THE EVOLUTION OF THE CPCT MODEL

The use of PDSA Cycles allowed the NIC of mathematics educators to rapidly test and evaluate the nature of the implementation of co-planning and co-teaching during clinical experiences. Across the years, the CPCT sub-RAC PDSA Cycles focused on promoting students' learning, implementing and measuring effective teaching practices with a lens of equity, and disseminating resources to help others conceptualize how co-planning and co-teaching could be used within a secondary mathematics setting. The members of the CPCT sub-RAC used findings from each cycle to create the subsequent cycle.

Before the First Sub-RAC PDSA Cycle: Getting Started

Initial implementation of CPCT in the sub-RAC began with two of its member institutions who were already in the early years of implementing co-teaching in their secondary mathematics internships. The Ohio State University was using co-teaching as a tool to change the focus of the internship experience away from a focus on teacher actions to an apprenticeship model where both mentor and mentor teacher focus together on student learning. Planning and instruction in this model center on three questions (Brosnan, Jaede, Brownstein, & Stroot, 2014):

1. What do students need to learn?
2. How will you know if they learn?
3. In what tasks will students engage to ensure learning happens?

One of the mentor teachers described the difference this way, "CPCT really changed my thinking. I could have given you a thousand examples of good teaching. Then the facilitator would say, 'Where is your evidence of student learning?' So, it really made me think." She continued, "I've started to work through that process with the student teacher, so it has made me rethink: 'Why I am [teaching] this, and is it leading to student learning?'"

Meanwhile, East Carolina University began using CPCT to transform clinical experiences to an apprenticeship model. Their focus was on building teacher identity and a sense of the classrooms as a shared space with the mentor teacher and teacher candidate. They were moving from a model in which the teacher candidate often moved directly from observing to being the sole teacher of one class, increasing their responsibility throughout the internship by increasing the number of classes that they were "in charge of." By the end of the internship the teacher candidate had full responsibility for all classes and the mentor teacher's only role was to provide feedback to the teacher candidate. The co-teaching strategies, by providing multiple models for how two teachers could work together, provided a new vision of the internship as one where the teacher candidate acts as and is perceived by students as being a teacher from the first day of the internship. Figure 10.2 depicts the changing level of responsibility of the teacher candidate and mentor during the internship. Note that the teacher candidate has some responsibility for instruction from the beginning and that the mentor teacher continues to have some responsibility for instruction throughout the internship. Implementation of this model of shared responsibility also addressed concerns that potential mentor teachers had about the effect on student learning of "handing their classes over" to a novice teacher by keeping mentor teachers engaged at some level throughout the internship.

Both universities provided training to mentor teachers and teacher candidates on the co-teaching strategies and on the paradigm shift from a traditional student teaching model to an apprenticeship model. Data collection included surveys and interviews to examine perceptions of CPCT of mentor teachers and teacher candidates. Additionally, East Carolina University collected data on which of the co-teaching strategies teacher candidates reported using during each two-week period of the internship and conducted classroom observations using a co-teaching observation protocol to document what strategies were evident in the classroom.

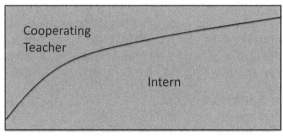

Shared Responsibility

FIGURE 10.2. Co-Teaching Model. Changing responsibilities throughout an internship (adapted from Cayton & Grady, 2016).

Lessons learned: The need for targeted training and patience. Initial implementation of CPCT in the two sites revealed that many mentor teachers, especially some of the most experienced teachers, were somewhat reluctant to embrace the new paradigm of internships and the co-teaching strategies. The teacher candidates, having little to no experience with other models, were much more open to trying co-teaching. In fact, mentor teachers often tried the strategies because they were proposed by the teacher candidates. At one school, two teacher candidates teaching similar classes worked together to create a set of stations for reviewing a unit on angles and triangles. They negotiated with their mentor teachers for each mentor/candidate pair to devote their planning period to implementing the activity in the other's classroom, so they had four teachers engaging in station teaching together. The results were so positive for the classroom students that both of these mentor teachers now use *station teaching* with their interns and encourage colleagues to try the strategy. As researchers and mathematics teacher educators, the members of the CPCT sub-RAC learned the importance of allowing time for change to happen. Experienced mentor teachers were not going to completely and immediately change their perspectives or practices just because they were trained in CPCT. However, with each successful experimentation, mentor teachers became less skeptical and more open to trying CPCT.

Another important lesson from the early implementation is that CPCT, as part of an apprenticeship model of teaching, can be transformative for the mentor teachers. This transformation was evident when mentor teachers reflected on the training that the cooperative pairs received on co-planning. As mentor teachers worked to help their teacher candidates focus on student learning, they reexamined their own teaching and planning processes. As one mentor teacher put it,

I think we have always focused on student learning, but not to the [same] level. CPCT made us reanalyze how we teach. I think it has exacerbated the idea of [targeting] what students are learning before we teach a lesson ... Now, day-to-day I am looking deeper at student learning than I probably did before. (Mentor Teacher, Spring 2015)

In implementing CPCT, the sub-RAC talked about its potential to improve the clinical experiences of teacher candidates but did not anticipate the powerful impact that it had on some of the mentor teachers.

During those first years, the sub-RAC also learned some important lessons about the forms of training and support that participants needed to fully engage in CPCT. One of the first realizations was that participants needed training on the co-teaching strategies that were more specific to high school mathematics. The available CPCT training materials contained examples only from elementary classrooms. Mentor teachers and teacher candidates struggled to see how strategies like parallel teaching (see Table 10.1) or station teaching might be used in high school mathematics classroom. This struggle was evident both in comments by mentor teachers during training and by the limited use of any of the strategies beyond *one teach, one assist* and *one teach, one observe.*

It also became clear that mentors and teacher candidates needed more guidance on how to engage in co-planning. While training made it clear that co-planning was essential and should focus on student learning, it did not provide the same types of specific ideas for planning together that the co-teaching strategies had provided for teaching together. Several mentor teacher teachers expressed frustration at not really knowing what else to do besides "giving the teacher candidate all of my materials" or "turning them loose to do it all on their own." Teacher candidates, having heard that they supposed to be co-planning, felt the lack of this with their mentor teachers. One candidate lamented,

> ... we do not do much co-planning. [My mentor] lets me write my own lesson plans and then we go over them together once they are done. She does give me tips on what she has done in the past and then modify them, as she wants them to be. I really like [my mentor] and the way she teaches but I wish she would work with me during planning rather than us working separately. (Teaching Candidate, Spring 2015)

Another teacher candidate expressed the concerns this way:

> I do not think that [my mentor] quite understands the co-teaching model. For our planning, I am still planning my Math 1 class alone, and she is planning her other two classes alone. She does not ask me my opinion on the other classes, and she does not offer much help with the planning of my class. She has provided me with ample amount of resources that have been very helpful. However, we have not done much co-planning at this point. (Teaching Candidate, Spring 2015)

These concerns made it evident that there was a need for more targeted training on how to co-plan. Sub-RAC action items to improve the implementation of CPCT in the future were: a) provide better training and support to both mentors and teacher candidates, and b) provide more guidance about how to co-plan.

First PDSA Cycle: A Focus on Promoting Student Learning

Overall experiences with CPCT to this point were sufficiently promising for the sub-RAC to undertake more intentional implementation and study CPCT across several different institutions using improvement science. The first cross-institution PDSA Cycle was created. During the first PDSA Cycle, the CPCT sub-RAC adopted the use of the apprenticeship model for learning (Brosnan, Jaede, Brownstein, & Stroot, 2010) as a means to suggest how the mentor teacher and the teacher candidate (collaborative pair) could use co-teaching during instruction. Member institutions agreed to promote the following dictum: "We will no longer teach teachers how to teach. From now on, we will teach teachers how to get students to learn" (Dr. Patricia Brosnan, personal communication, June 6, 2013). Collaborative pairs were encouraged to focus on student learning in their planning, instruction, and assessments. The collaborative pairs were also encouraged to attend to the *CCSS-M*'s Standards for Mathematical Practice (2010) and the mathematics teaching practices (NCTM, 2014), to develop students' mathematical knowledge and their ability to exhibit mathematical habits of mind (Cuoco, Goldenberg, & Mark, 1996).

The CPCT sub-RAC developed measures to document the nature of mentor teachers' and teacher candidates' knowledge and implementation of co-teaching strategies, mathematical content standards, the *CCSS-M*'s Standards for Mathematical Practice, the assessments utilized, and the lesson planning strategies used (Sears et al., 2017b). The sub-RAC also investigated how mentor teachers attended to the diverse needs of students within the classroom setting. Measures were developed for the beginning, middle, and end of the full-time clinical experience (Sears et al., 2017b). Sub-RAC members developed a pre-survey, a short survey to be given several times during the field experience (just-in-time survey), and an exit survey to collect data about the collaborative pair's knowledge of and experiences with co-planning and co-teaching and the *CCSS-M* across the span of their clinical experiences (Sears et al., 2017b). An observation protocol was created to document which co-teaching strategies were used during observations and to capture teacher candidates' reflections about their enacted lessons and co-planning efforts. The plan for this cycle was to a) implement CPCT at several sites with at least one collaborating pair at each site, b) provide more math-specific training for our participants, and c) provide more guidance on how to co-plan.

As part of the "Do" in the PDSA Cycle, the first action was to address the need for more guidance on co-planning. Sub-RAC members at East Carolina University, inspired by the co-teaching strategies, proposed a similar set of co-planning strategies. These strategies, discussed earlier in this chapter (see Table 10.2), describe six specific strategies for how two teachers might work together on planning. With these strategies in place, sub-RAC members developed training modules that attended to co-planning and co-teaching and incorporated mathematics-specific examples of the co-teaching strategies. These materials were piloted at the University of South Florida and portions of them were used at several other institutions in the partnership.

Data collection for this PDSA Cycle focused on studying the use of different CPCT strategies in the field experience and the perceptions that mentor teachers and teacher candidates had about the usefulness of CPCT in the field experience. The sub-RAC created and/or modified surveys to be given before the full-time field experience, at two points during the experience, and at the end. It was during the planning for this round of data collection that issues of the variability of field experiences at our member institutions first had to be addressed. Data was being collected from sites with semester-long internships and yearlong internships, along with fifth-year and graduate programs. This issue was addressed by having each institution implement the training and administer surveys as they saw fit.

Lessons Learned in Cycle 1: Effectiveness of CPCT and Challenges in Co-Planning. Given that earlier adjustments were made to allow time for mentor teachers to try out and to accept the value of CPCT, as well as the improved training provided to teacher candidates and mentor teachers in this PDSA Cycle, it was not surprising that mentor teachers were more positive about CPCT. They talked about benefits for themselves, their teacher candidate, and their classroom students. One mentor reported,

> Co-teaching was a good experience for me. I believe it was good for my intern as well. It allowed me to be a part of the teaching experience and helped me to scaffold the intern into teaching on her own." Another reflected on how co-teaching benefited both her students and her teacher candidate. She stated, "The benefits are that the students get two teachers rather than one and it allows the intern to be guided into teaching full time instead of just being thrown to the wolves. (Mentor Teacher, Spring 2016)

Teacher candidates also identified benefits to co-teaching. Several talked about the value of co-teaching in allowing the mentor to help in real time, during a lesson. One stated, "The benefits of co-teaching is when you do not have the confidence in teaching a certain topic, the [mentor teacher] is there to step in and help." Another put it this way:

> I think [co-teaching] helped to better prepare me to be a teacher because instead of just teaching on my own or watching my [mentor] teach, he could model the correct behavior while we were both in front of the classroom. He was also able to make up for any areas where I was lacking, and I think that made it more beneficial for our students. (Teacher Candidate, Spring 2016)

Other teacher candidates also talked about the value of CPCT for the students, noting that students received more individualized attention and instruction that was better aligned to assessments.

Although both teacher candidates and mentor teachers saw clear benefits to CPCT, they also identified some challenges. Most of these challenges came from teacher candidates who felt that there was too little time for co-planning and that their mentor teachers did not interact effectively with them during planning. One intern stated, "Co-planning is a sore subject. I felt that she would just tell me the topics but didn't explain how to teach them. The times we did co-plan worked really well in terms of teaching together and doing activities in class." She went on to say, "There were many times where we would have conversations but she did not want to hear what I had to say … I felt she didn't have the significant time to sit with me and discuss the day and future days." Similar statements by other teacher candidates and observations by university supervisors helped the sub-RAC to highlight that cooperating pairs were using a limited set of the co-teaching strategies and might not be taking full advantage of the time-saving aspects of some of the other strategies. It was clear that better training was still needed on both the co-planning strategies and the communication skills necessary to implement them. In addition, the lack of data from mentor teachers about co-planning pointed to the need to fine-tune the pre- and post- surveys for the mentor teachers. Action items coming out of this Cycle were a) redesign co-planning training materials to incorporate interactive planning activities for the cooperating pairs and b) when possible, try to incorporate more training on CPCT into the methods courses that teacher candidates took before their full-time field experiences.

Second PDSA Cycle: Refining Instruments and Facilitating Professional Development

After considering the lessons learned in cycle 1, and to accommodate the inclusion of more institutions in our work, plans for PDSA Cycle 2 were to a) refine our data collection instruments, b) improve our training materials, especially those focused in co-planning, and c) make training materials more readily available for all of our member institutions.

The "Do" phase of this PDS Cycle began with modification of the pre- and post- surveys for the mentor teachers. The pre- and post- surveys for the teacher candidates already asked separately about their experiences with co-teaching and with co-planning, so the mentor teacher surveys were modified to match. The hope was that this would provide a more nuanced picture of what was and was not working well for mentor teachers in CPCT. CERAC institutions also agreed to use a common instrument across the three groups of the CERAC to measure knowledge and use of mathematics teaching practices (NCTM, 2014) and to utilize the Mathematics Classroom Observation Protocol for Practices (MCOP^2; Gleason, Livers, & Zelkowski, 2017).

To provide more support for co-planning, the CPCT training materials were modified to include activities to help cooperating pairs practice some of the co-planning strategies and the communication skills necessary to effectively co-plan. With the scale-up to more institutions participating in CPCT, the sub-RAC sought to make the training materials readily available to all member institutions. The original three training modules were revised to include the co-planning activities and used in face-to-face trainings at the University of South Florida, East Carolina University, and California State University, Chico. An attempt was made to create video-recordings of these trainings to be used at other sites, but the materials were not of sufficient quality to be shared. Instead, slide presentations for the three modules were made available to other members of the sub-RAC for them implement at their sites. Across the sub-RAC, several institutions were also embedding training about CPCT into the methods courses that teacher candidates took before their full-time field experiences.

Lessons Learned in Cycle 2: The Benefits of CPCT and High-Quality Training. One lesson learned from the data in PDSA Cycle 2 was that both mentor teachers and interns were fully embracing CPCT and had an overall very positive experience with it. Mentor teachers talked about the effect of co-teaching on students' perceptions of the teacher candidate: "The benefits are having the intern and teacher to have somewhat of an equal role in the classroom and the students see them as a team and not one having more authority than the other." Mentor teachers also noted the value of co-teaching for students and for the lead teacher. One noted that "students have access to two teachers at a time. Also, the one observing/assisting can monitor the students while the one teaching focuses more on the lesson," while another observed that "co-teaching is a great strategy in a large classroom. Both teachers can help meet the individual needs of the students. It is also helpful for the students to see multiply ways of presentation."

Teacher candidates shared similar observations about the role of co-teaching in shaping students' perceptions of the two teachers. One intern noted, "When implemented correctly, students see both of you (the Cooperating Teacher and the intern) as the teachers and are comfortable with whoever is teaching." Teacher candidates also appreciated the sense of continual support that co-teaching gave them, especially when teaching mathematical content with which they were not especially comfortable. One candidate observed that the "biggest benefit is that we are not just thrown to the wolves. We have someone there to guide us in real time in case something goes awry, we can play it off as co-teaching."

The data from this PDSA Cycle provided evidence that the increased emphasis on interactive co-planning activities and communication during training had a positive influence in how mentor teachers and teacher candidates implemented CPCT. Mentor teachers reported that they felt that they could better help scaffold their interns as they learned to plan. As one mentor teacher put it, "Co-planning allowed me to be directly involved in my intern's lessons even if he was teaching them. It allowed me to offer advice on new ideas and strategies he wanted to try in his lessons and made a great time for him to feel comfortable asking questions." Teacher candidates talked about how co-planning helped their mentors provide valuable checks on their plans to ensure quality. One candidate noted, "The benefits are that co-planning provides support to the intern and ensures that students will receive the best instruction." Another observed that, "Co-planning helps the intern see and experience faults in their thinking ... the benefit is that clinical teachers are able to help us plan as opposed to us planning an entire lesson and then it not being what the teacher expects." Most of the challenges to co-planning noted by both mentor teachers and teacher candidates were those inherent to any type of co-planning, negotiating how to put together different ideas into a cohesive plan.

In examining the co-planning strategies being used by the collaborative teams, the *one reflects, one plans* and *one plans, one reacts* strategies were still the dominant strategies with few participants reporting using other co-planning strategies. However, as mentor teachers and teacher candidates began to tell stories about their experiences, it became evident that they were experimenting with some other strategies, often with positive results. For example, one mentor teacher, despite being pleased overall with the work of her teacher candidate, was experiencing frustration trying to get the candidate on the same page with lesson planning. The mentor had tried the multiple strategies that she had developed over her years working with teacher candidates but was still not seeing her candidate produce the kinds of lesson plans she expected. This mentor teacher had, like many other mentor teachers, seen little potential in the parallel planning strategy (see Table 10.2) since it seemed to be a wasteful duplication of effort. However, in desperation, the mentor teacher selected an upcoming lesson for which she and the teacher candidate would each write separate, complete lesson plans. The mentor teacher described the result as "like breaking a log jam." Looking together at these parallel lesson plans, the teacher candidates saw for the first time exactly how her plans differed from those expected by the mentor. The mentor, also for the first time, understood exactly what the teacher candidate had been trying to convey in her plans. This experience helped this cooperating pair communicate more clearly, making the rest of this field experience far more productive for both of them. The mentor teacher has been telling this story ever since, encouraging her colleagues to try out some of the co-planning and co-teaching strategies about which they are skeptical.

In addition to being more willing to engage in the use of CPCT, mentor teachers were able to better engage in individual and collaborative conversations with their interns because of their involvement in the apprenticeship for learning model. Employing the skills of CPCT increased the capacity of the mentors to focus on intern thinking and intern learning. Because of this strategy, they were better mentors. James, a middle-school mathematics teacher said, "CPCT has given me clear and simple tools for guiding discussions with my intern in planning, reflecting, and problem solving. It is easy to talk and share, but CPCT helps keep the focus on the intern's thinking and leads to a clear goal." Janice, a high-school teacher had a similar comment: "CPCT has given me specific strategies, actual phrases and wording, that can be used to promote deeper thinking in my intern teacher. I really appreciate the coaching that enables me to be a more effective mentor."

The overall lesson of PDSA Cycle 2 was that, with high-quality training, CPCT was highly valued by both mentor teachers and teacher candidates. The challenge was to continue this success by providing consistently high-quality training across sites. Face-to-face trainings are expensive to implement, and even some of the original sites were having difficulty continuing to provide thorough training. Including CPCT training in methods courses was helpful for teacher candidates but did not address the need for training new mentor teachers.

Third PDSA Cycle: Scaling Up Implementation and Data Collection Efforts

The third PDSA Cycle, developed in September 2017, focused on increasing the dissemination of professional development training modules to a broader range of member institutions, and collecting more data from across sites. Thus, the CPCT sub-RAC planned to facilitate professional development training that would be professionally video-recorded and shared online. During the professional development workshop, the collaborative pairs worked together to reflect on how they could implement the co-planning and co-teaching strategies within their practices. In addition, the developers of the MCOP^2 and other CERAC members with experience using the instrument provided training to select members of the MTE-Partnership, including members of the CERAC, to increase the fidelity of implementation of the instrument. The sub-RAC also created a data collection calendar template to increase the consistency of data collection across institutions while still allowing data collection schedules to be adapted for variety of programs represented in the sub-RAC.

Survey data from this PDSA Cycle provided strong evidence that CPCT had become normative in many of the sites. Mentor teachers and teacher candidates routinely used the language of CPCT to describe their lessons and planning and had difficulty even thinking about going back to earlier models. A university supervisor at one site noted that long-time mentor teachers at her institution would likely quit if they were told that they had to go back to a traditional student teaching model.

CPCT and the apprenticeship for learning framework also have influenced the mentors' perceptions about their role. They now view themselves as teacher educators, mediators of intern learning, and are focused on the day-to-day development of their interns as opposed to evaluators, hosts, or bystanders giving up control of their classrooms to those less qualified. The dynamics of the relationship shifted to a collaborative partnership with the shared goal of identifying evidence of student learning. Darrell, a high school mathematics teacher stated, "This experience has certainly changed my perspective on mentoring student interns. Initially, I was very apprehensive, mainly due to my colleagues discussing negative past experiences. This co-planning/co-teaching model has exceeded all of my expectations." Another mentor teacher, Catia, stated, "We work better together than in previous years that were more hands off." Stephanie, a middle-school mathematics teacher expanded on the improvements from the past as well, stating, "The idea of mentoring an apprentice and co-planning/co-teaching as a method of 'bringing someone along' are very different from my personal student teaching experiences. I feel the mentoring methods encouraged by [this project] are far more effective.

Mentor teachers noted a shift in the relationship between mentor and candidate from the traditional student teaching model. First, Stephanie, a middle school mathematics teacher noted, "I view the interns as equals now more than ever." Janelle, a high school teacher added, "I think this project made [clinical preservice] more of a partnership than a sit back and watch what happens environment." And, when talking about her relationship with her intern, Catia, a high school mathematics teacher added, "We are a team, not just teachers who share space."

As CPCT becomes more normative for the participants, members of the sub-RAC are finding that their roles as researchers are sometimes at odds with their roles as teacher educators. As teacher educators, they celebrate that mentor teachers and teacher candidates are doing better at working together and communicating. However, as CPCT becomes more comfortable for mentors and candidates, it sometimes becomes more difficult for researchers to recognize individual components of the model in their practice, such as different co-teaching or co-planning strategies. In many cases, cooperating pairs are going beyond the initial strategies in which they were trained and creating new ways to work smoothly together. When observing a classroom, it is not unusual now to see a mentor teacher and teacher candidate shift smoothly through three or four different co-teaching strategies during a single lesson or combine multiple co-teaching strategies into a model that defies easy categorization. Likewise, with co-planning strategies, study participants will report that they are using a particular co-planning strategy but then go on to describe their interactions in ways far different from the original vision of the named co-planning strategies. As teacher educators, their obvious comfort with CPCT and the positive effect it is having on field experiences is gratifying. As researchers, it is initiating a reexamination of some definitions and a reconsider of the focus of research on CPCT.

One topic that warranted more attention is the potential for CPCT to help support more equitable teaching practices. Neither the $MCOP^2$ nor current surveys provide adequate data to better understand what is happening in this area. One action item for the next PDSA Cycle is to attend more explicitly to equity by including additional items on our surveys and in our observation protocols.

Fourth PDSA Cycle: Attention to Equity

The fourth PDSA Cycle, currently in progress, focuses more intentionally on access and equity, one of the primary drivers from the CERAC driver diagram (see Figure 10.1). The plan is to collect and analyze data focused on access and equity in relation to CPCT. The CPCT sub-RAC seeks to explicate, rather than infer, how the collaborating pairs were attending to equity during their planning and enacted lessons. Thus, the exit survey was modified to ask explicitly how equity was attended to when co-planning and co-teaching was employed during clinical experiences for secondary mathematics. For this PDSA Cycle, more attention is being given to investigating how the co-planning and co-teaching strategies are being used. Rather than using the just-in-time survey, collaborating pairs are charting their use of the co-planning and co-teaching strategies, and reporting on how they explicitly attended to equity in their planning and enacted lessons. Additionally, the sub-RAC participants have agreed to use an online calendar to track the data collection across institutions and to foster accountability and increase the consistency of data gathered from collaborating pairs. Data collection is almost complete for this cycle; analysis of the data is in the early stages.

LESSONS LEARNED: EFFECTIVELY IMPLEMENTING CPCT

The experiences and research of the CPCT sub-RAC provide evidence that the use of co-planning and co-teaching during clinical experiences has potential to positively influence secondary mathematics clinical experiences across a variety of clinical settings. Sub-RAC members have successfully implemented CPCT in yearlong internships, semester-long field experiences, fifth-year programs, and others. CPCT has shown promise in helping mentor teachers and teacher candidates focus on student learning and better attend to the needs of their students. Mentor teachers and teacher candidates involved in this research have reported that engaging in CPCT during field experiences in secondary mathematics has helped participants develop better relationships, helped teacher candidates more easily move from their role as college student to the role of classroom teacher, and supported candidates in improving their collaborations, teaching, and planning skills.

This research also provides some possible insights for other institutions seeking to implement CPCT in field experiences. Professional development is essential to the adoption and implementation of co-planning and co-teaching strategies. Providing collaborative pairs with resources and practical examples of co-planning and co-teaching strategies will support their ability to enact the concepts. Professional development can take the form of online training, face-to-face training, professional learning communities, or book clubs that focus on co-planning and co-teaching. Training can and should be embedded in methods classes, mentor teacher training, and university supervisor training. Because change takes time, professional development training needs to be ongoing and needs to attend to technical, logistical, organizational, affective, and conceptual concerns that may arise. It can be advantageous to have a resident expert on the implementation team who can provide additional insights as needed. It is important to recognize that CPCT is more than a set of strategies for co-teaching and co-planning. Mentor teachers, teacher candidates, and university supervisors need to understand that CPCT entails a fundamental shift in the roles of teacher candidates and mentors in the classroom and in planning.

Also critical to successful implementation of CPCT is a shared vision of goals and expectations. In the research of the CPCT sub-RAC, the apprenticeship model for learning provided a shared focus on student learning for all collaborative pairs. In addition to training and a shared vision, it is important that institutions seeking to implement CPCT recognize that it will require time and continued support for CPCT to become normative in a new setting. Some elements of CPCT are more readily accepted and utilized by participants than others. Especially for experienced mentor teachers who already have set expectations for field experiences, some resistance is to be expected. This resistance can be overcome by providing more subject- and grade-specific training and providing ongoing encouragement and support for mentor teachers to try different aspects of CPCT. Once mentor teachers experience success with CPCT strategies or ideas, they are usually more open to trying other ideas and to sharing their success stories with their peers. The CPCT sub-RAC has found that by providing quality training, helping collaborative pairs build a common vision, and allowing for incremental acceptance of new ideas, CPCT can be successfully implemented in a variety of clinical settings.

ENDNOTES

1. Funded by the National Science Foundation Directorate for Education & Human Resources, Division of Undergraduate Education (DUE)—Improving Undergraduate STEM Education (IUSE), Development & Implement I & II: Engage Student Learning—Grant ID#s: 1726998, 1726362, and 1726853. Any opinions, findings, and conclusions or recommendations expressed in this material are those of the author(s) and do not necessarily reflect the views of the National Science Foundation.

REFERENCES

Association of Mathematics Teacher Educators. (2017). *Standards for preparing teachers of mathematics.* Raleigh, NC: Author. Retrieved from http://amte.net/standards

Bacharach, N. (2007). *Utilizing co-teaching during the student teaching experience.* Retrieved from http://www.aascu.org/ programs/teacher/pdf/07_st_cloud.pdf

Bacharach, N., Heck, T. W., & Dahlberg, K. (2010). Changing the face of student teaching through coteaching. *Action in Teacher Education, 32*(1), 3–14.

Brosnan, P., Jaede, M., Brownstein, E., & Stroot, S. A. (2014). *Co-planning and co-teaching in an urban context.* Paper presented at the Annual Meeting of the American Educational Research Association, Philadelphia, PA.

Bryant, M., & Land, S. (1998). Co-planning is the key to successful co-teaching. *Middle School Journal, 29*(5), 28–34.

Bryk, A. S., Gomez, L. M., Grunow, A., & LeMahieu, P. G. (2015). *Learning to improve: How America's schools can get better at getting better.* Cambridge, MA: Harvard Education Press.

Cayton, C., & Grady, M. (2016). *Co-planning strategies to support intern development.* Research report presented at the Fifth Annual Mathematics Teacher Education Partnership Conference, Atlanta, GA.

Cuoco, A., Goldenberg, E. P., & Mark, J. (1996). Habits of mind: An organizing principle for mathematics curricula. *The Journal of Mathematical Behavior, 15*(4), 375–402.

Darling-Hammond, L. (2010). Teacher education and the American future. *Journal of Teacher Education, 61*(1–2), 35–47.

Darragh, J. J., Picanco, K. E., Tully, D., & Henning, A. S. (2011). When teachers collaborate, good things happen: Teacher candidate perspectives of the co-teach model for the student teaching internship. *The Journal of the Association of Independent Liberal Arts Colleges of Teacher Education, 8*(1), 83–109.

Diana Jr., T. J. (2014). Co-teaching: Enhancing the student teaching experience. *Kappa Delta Pi Record, 50*(2), 76–80.

Dieker, L. A. (2001). What are the characteristics of "effective" middle and high school co-taught teams for students with disabilities? *Preventing School Failure, 46*, 14–23.

Friend, M., Cook, L., Hurley-Chamberlain, D., & Shamberger, C. (2010). Co-teaching: An illustration of the complexity of collaboration in special education. *Journal of Educational and Psychological Consultation, 20*(1), 9–27.

Gleason, J., Livers, S., & Zelkowski, J. (2017). Mathematics Classroom Observation Protocol for Practices (MCOP²): A validation study. *Investigations in Mathematics Learning, 9*(3), 111–129.

Grady, M., Cayton, C., Preston, R. V., & Middleton, C. (2018). *Co-planning with interns: Envisioning new ways to support intern development of effective lesson planning.* Brief research report presented at the 40th Annual Meeting of the North American Chapter of the International Group for the Psychology of Mathematics Education. Greenville, SC.

Guise, M., & Thiessen, K. (2017). From pre-service to employed teacher: Examining one year later the benefits and challenges of a co-teaching clinical experience. *Educational Renaissance, 5.*

Heck, T. W., Bacharach, N., & Dahlberg, K. (2008). Co-teaching: Enhancing the student teaching experience. In the *Eighth Annual International Business and Economics Research & College Teaching and Learning Conference Proceedings*, Las Vegas, NV. Retrieved from: https://www.stcloudstate.edu/soe/accreditation/_files/documents/standard3/Co-Teaching%20article%20-%20Enhancing%20the%20Student%20Experience%20-%20Exhibit%203.4.h.2.pdf

Howard, L., & Potts, E. A. (2009). Using co-planning time: Strategies for a successful co-teaching marriage. *TEACHING Exceptional Children Plus, 5*(4), Article 2. Retrieved from http://escholarship.bc.edu/education/tecplus/vol5/iss4/art2

Magiera, K., Smith, C., Zigmond, N., & Gebauer, K. (2005). Benefits of co-teaching in secondary mathematics classes. *TEACHING Exceptional Children, 37*(3), 20–24.

Martin, W. G., & Gobstein, H. (2015). Generating a networked improvement community to improve secondary mathematics teacher preparation: Network leadership, organization, and operation. *Journal of Teacher Education, 66*(5), 482–493.

Morton, B. M., & Birky, G. D. (2015). Innovative university-school partnerships co-teaching in secondary settings. *Issues in Teacher Education, 24*(2), 119–132.

Murawski, W. W. (2012). 10 tips for using co-planning time more efficiently. *Teaching Exceptional Children, 44*(4), 8–15.

Murawski, W. W., & Lochner, W. W. (2011). Observing co-teaching: What to ask for, look for, and listen for. *Intervention in School and Clinic, 46*(3), 174–183.

Murawski, W. W., & Swanson, H. L. (2001). A meta-analysis of co-teaching research: Where are the data? *Remedial and Special Education, 22*(5), 258–267.

National Council of Teachers of Mathematics. (2014). *Principles to actions: Ensuring mathematical success for all.* Reston, VA: Author.

National Governors Association Center for Best Practices, Council of Chief State School Officers. (2010). *Common core state standards: Mathematics.* Washington, DC: Author.

Scruggs, T. E., Mastropieri, M. A., & McDuffie, K. A. (2007). Co-teaching in inclusive classrooms: A metasynthesis of qualitative research. *Exceptional Children, 73*(4), 392–416.

Sears, R., Brosnan, P., Oloff-Lewis, J., Gainsburg, J., Stone, J., Biagetti, S., . . . & Junor Clarke, P. (2017a). *Co-teaching mathematics: A shift in paradigm to promote student success.* Paper presented at the Hawaii International Conference on Education, Honolulu, HI.

Sears, R., Brosnan, P., Oloff-Lewis, J., Gainsburg, J., Stone, J., Spencer, C., . . . & Andreason, J. (2017b). Using improvement science to transform clinical experiences with co-teaching strategies. In M. Boston & L. West (Eds.), *Annual perspectives of mathematics education (APME) 2017: Reflective and collaborative processes to improve mathematics teaching* (pp. 265–273). Reston, VA.: National Council of Teachers of Mathematics.

Strutchens, M., Iiams, M., & Sears, R. (2016). Clinical experiences. In B. R. Lawler, R. N. Ronau, & M. J. Mohr-Schroeder (Eds.), *Proceedings of the fifth annual Mathematics Teacher Education Partnership conference.* Washington, DC: Association of Public and Land-grant Universities. Retrieved from http://www.aplu.org/projects-and-initiatives/stem-education/SMTI_Library/clinical-experiences.pdf

Wassell, B., & LaVan, S. K. (2009). Tough transitions? Mediating beginning urban teachers' practices through coteaching. *Cultural Studies of Science Education, 4*(2), 409–432.

Yopp, R. H., & Young, B. L. (1999). A model for beginning teacher support and assessment. *Action in Teacher Education, 21*(1), 24–36.

Zeichner, K. (2010). Rethinking the connections between campus courses and field experiences in college-and university-based teacher education. *Journal of Teacher Education, 61*(1–2), 89–99.

CHAPTER 11

FOSTERING COLLABORATIVE AND REFLECTIVE TEACHER CANDIDATES THROUGH PAIRED PLACEMENT STUDENT TEACHING EXPERIENCES[1]

Marilyn E. Strutchens[2,] Jennifer Whitfield,
David Erickson, and Basil Conway

As discussed in Chapter 5 of this book, which provides an overview of the Clinical Experience Research Action Cluster (CERAC), one goal of the CERAC is to increase the quality of the student teaching experience and address the inadequate supply of quality mentor teachers to oversee the experience. As a result of a thorough review of the literature related to clinical or field experiences in education, the *paired placement model* for student teaching emerged as one way of addressing this goal. In this model, a pair of teacher candidates work daily with an experienced mathematics mentor/coach who is devoted full time to helping the teacher candidates address the craft of teaching, plan lessons jointly, and teach those same lessons while actively observing, reflecting, and revising (Leatham & Peterson, 2010b). According to Leatham and Peterson (2010b), within this setting teacher candidates are slated to quickly realize that not everyone learns the way they

The Mathematics Teacher Education Partnership: The Power of a Networked Improvement
Community to Transform Secondary Mathematics Teacher Preparation, pages 257–280.
Copyright © 2020 by Information Age Publishing
257

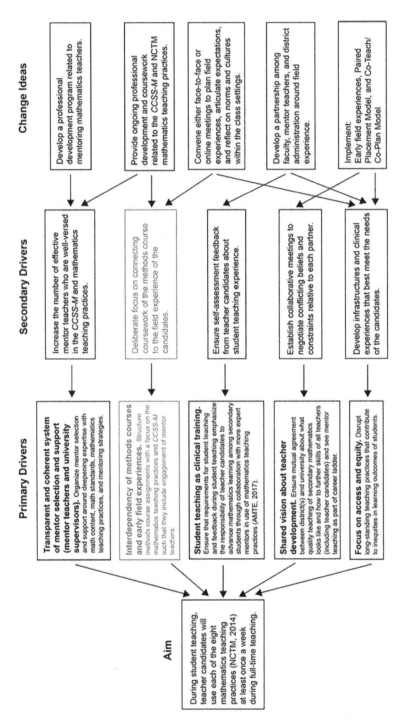

FIGURE 11.1. CERAC Driver Diagram. The Paired Placement sub-RAC focuses its work on all of these drivers, with the exceptions of the boxes that are faded as those are a main focus of the Methods sub-RAC.

learn, and their focus shifts to the learning of their students rather than their own learning. Furthermore, fewer mentor teachers are needed, and teacher candidates can be placed with mentor teachers who teach in alignment with the *Common Core State Standards for Mathematics* (*CCSS-M*; National Governors Association Center for Best Practices, Council of Chief State School Officers, 2010) and the National Council of Teachers of Mathematics' mathematics teaching practices (National Council of Teachers of Mathematics, NCTM, 2014). Moreover, the paired placement model is

> a clinical experience model designed to support more than just the teacher candidate or to provide extra classroom support for a teacher. The experience can become a system of simultaneous growth and renewal for the teacher candidate-mentor teacher-university supervisor team when they collaborate; all participants learn and lead while they work on behalf of students. (Association of Mathematics Teacher Educators, AMTE, 2017, p. 37)

Consequentially, implementation of the paired placement model has the potential of addressing the following AMTE Standards (2017): Standard P.4. Opportunities to Learn in Clinical Settings (p. 37), Standard C.4. Social Contexts of Mathematics Teaching and Learning (p. 12) and Standard C.2. Pedagogical Knowledge and Practices for Teaching Mathematics (p. 21). The paired placement model affords the teacher candidates the opportunity to focus on equity issues within classroom settings due to the collaboration and reflection that occur among the teacher candidates and mentor teacher focused on student participation and growth. More specifically, the Paired Placement sub-RAC of the CERAC focuses on the following components of the CERAC's driver diagram (see Figure 11.1).

LITERATURE REVIEW

Many teacher preparation programs adopt the traditional model for the student teaching experience whereby one teacher candidate is placed with one cooperating teacher. Generally, it is this one cooperating teacher who plays the primary role in developing the teacher candidate's teaching practice during the placement, and this arrangement alone may limit the full development of the teacher candidate's craft of teaching (Leatham & Peterson, 2010b; Peterson & Leatham, 2018). The traditional one-to-one model for student teaching has some components that can limit the impact the student teaching experience has on a teacher candidate's teaching practice. Peterson and Leatham (Leatham & Peterson, 2010a; Peterson & Leatham, 2018) cite lackluster outcomes, feeling a sense of survival over developing a proper skill set, a focus on self, isolation, and lack of leadership as some of the limitations of the traditional model for student teaching. Furthermore, teacher candidates who are placed in the traditional one-to-one model for student teaching often lack clear delineation of learning goals, practice inward reflection rather than outward reflection on items like student learning and are discouraged to collaborate with other prospective or in-service teachers (Peterson

& Leatham, 2018). These types of constraints limit the modes of support for the teacher candidate. The traditional one-to-one model also tends to emphasize the teacher candidate's execution of classroom management rather than planning, enacting, and reflecting on his or her craft of teaching (Leatham & Peterson, 2010b; Peterson & Leatham, 2018). To address some of the limitations of the traditional one-to-one model for student teaching, some teacher preparation programs have started implementing innovative and less traditional models for the student teaching experience.

The paired placement model is a nontraditional model that places two teacher candidates with one cooperating teacher and has been implemented across contexts at a number of universities across the nation. In this model, the three teachers work as a trio throughout the student teaching experience, and all three teachers develop and strengthen their teaching practices. The paired-placement model has been shown to have a variety of positive effects on teacher candidates during their student teaching experience (Bullough et al., 2003; Conway et al., 2017; Goodnough et al., 2009; Mau, 2013; Martin & Strutchens, 2018; Peterson & Leatham, 2018; Strutchens et al., 2019). These positive effects include increased collaboration, stronger sense of community, higher levels of self-confidence, willingness to engage in pedagogical risk-taking, and a broadening of reflective teaching practices.

Across the various contexts in which the paired placement model has been implemented, researchers reported the model developed a strong sense of community and helped to increase collaboration (Bullough et al., 2003; Conway et al., 2017; Gardiner & Robinson, 2009; Goodnough et al., 2009; Leatham & Peterson, 2010a; Mau, 2013; Peterson & Leatham, 2018; Martin & Strutchens, 2018; Strutchens et al., 2019). For example, increased sense of collaboration could be due to the fact that the two teacher candidates are always together throughout the school day, share similar experiences in the classroom, and are working toward similar goals and objectives. Increasing opportunity for collaboration provides professional support and more detailed feedback about the art of teaching (Goodnough et al., 2009), promotes exposure to multiple perspectives, more dialogue about teaching and student learning, offers more support during implementation of student-centered pedagogies (Gardiner & Robinson, 2009), and helps to engage the teacher candidates in more frequent and varied communication (Mau, 2013). Additionally, Peterson and Leatham (2018) reported that the teacher candidates saw the collaboration as a means that improved their lessons. Increasing the time that teacher candidates spend together both as a pair and as a trio with their mentor teacher seems to foster team building opportunities and consequently helps to develop the positive aspects of teamwork.

The benefits of the paired placement directly relate to the mathematics teaching practices that promote equitable teaching (Aguirre, Mayfield-Ingrams, & Martin, 2013; Boston, Dillon, Smith, & Miller, 2017; NCTM, 2014, 2018). Specifically, the model increased the teacher candidate's ability to establish mathematical goals that provide focus for student learning. The teacher candidates in the paired place-ment model exhibited stronger reflections geared at promoting student thinking about mathematics that builds procedural fluency from conceptual understanding (Peterson & Leatham, 2018). Through the reflections, the teacher candidates have situated goals within learning progressions that ensure all students have the op-portunity to think and grow mathematically. The decrease in student misbehavior and increase in effective pedagogical practices have boosted meaningful math-ematical discourse and purposeful question posing in the classroom and shifted the teacher candidate's focus to student learning (Conway et al., 2017; Peterson & Leatham, 2018). Providing these types of experiences during the student teaching semester helps to develop equitable teaching practices from the beginning instead of having to change practices as an in-service teacher.

Though there are many benefits of the paired placement model, it does not come without its challenges. Nokes et al. (2008) conducted a study to test the ef-fectiveness of the paired placement of teacher candidates in secondary school set-tings to see if such placements fostered the learning and development of teacher candidates and the learning of their students. In addition, in this study most of the pairs and their mentor teachers were not given advice on how to work together. After 15 weeks of student teaching, the participants (teacher candidate pairs, men-tor teachers, and students) were interviewed about (a) the perceived strengths and weakness of paired placements, and (b) the relationships that developed between the teacher candidates and with their mentor teacher. Results indicated that paired placement teacher candidates enjoyed a rich learning experience because of the tensions, dialogue, and reflections that grew out of being placed with a peer. The secondary settings allowed for a combination of individual and team teaching. Results also suggested that student learning was enhanced by having two teacher candidates in the same setting.

However, Nokes et al. (2008) also shared some weaknesses that were ex-pressed by participants in the study, including the voices of the teacher candidate pairs, mentor teachers, and students. Some weaknesses shared are (a) needed to have more opportunities for individual teaching of the teacher candidates; (b) students had to adjust to two teachers; (c) sometimes planning was difficult for some pairs and appeared to take more time than individual planning would have; and (d) students sometimes played the teacher candidates against each other. As a result of their findings, Nokes et al. (2008) provided several suggestions for how teacher educators can maximize the benefits of paired placements while minimiz-ing costs:

1.   Place teacher candidates with mentor teachers who genuinely value col-laboration and collaborative learning communities;
2.   Encourage teacher candidates and mentors to arrange schedules that al-low for opportunities for both team and individual teaching;

3. Encourage teacher candidates to devise schedules that support doing much of their planning together, regularly developing, revising, and refining teaching ideas and related practices;

4. Forewarn teacher candidates and mentor teachers that there may be disagreements between the teacher candidates regarding effective pedagogy and appropriate policies, and that this is expected and valued; and

5. Encourage paired student teachers to talk openly about their differences and make a real effort to learn from the tensions that emerge; let them know that learning often occurs as conflicts are resolved. (p. 2175)

In addition, Goodnough et al. (2009) described concerns of dependency, confusion about classroom management during the experience, loss of individuality, and competition between the prospective teachers. As suggested by Nokes et al. (2008), many of these challenges can be addressed by the teacher preparation program, but it requires program personnel to be vigilant and to stay in touch with the daily happenings within the model. Moreover, Conway et al. (2017) found challenges such as: (a) personal issues between paired teacher candidates, (b) perceived classroom management ability of the teacher candidates, (c) number of days teacher candidates teach in isolation, (d) preparing university supervisors and mentors for a non-traditional experience, and (e) support for teacher candidate collaboration afterward.

A sufficient amount of evidence in the current research pool describes the paired placement model, along with its benefits and limitations. A variety of teacher preparation programs have implemented the model at different levels and in various contexts. However, little literature has described how this model may be orchestrated effectively in different secondary mathematics classrooms and how preparation programs can collaborate to strengthen the model and make it flexible enough, so it is effective across contexts. The intent of this chapter is to describe the different contexts in which the paired placement model was implemented at four different institutions across the United States and further describe how Plan-Do-Study-Act (PDSA) Cycles were used to collaborate and strengthen the model at each institution.

PAIRED PLACEMENT SUB-RAC

For the past four years, the Paired Placement sub-RAC has been developing syllabi and workshop guides to help mathematics teacher education programs effectively implement the paired placement model across multiple contexts. Even though the sub-RAC has undergone some changes in partner teams, the sub-RAC has seen much success with the model. Each partnership team is composed of at least a mathematics teacher educator, a mathematician, and a school partner. However, during face-to-face meetings of the sub-RAC and Mathematics Teacher Education Partnership (MTE-Partnership) annual conferences, usually the mathematics teacher educators represent their teams. The teams' representatives bring the concerns of their other team members to the group. Initially, Auburn University, the University of Montana, New Mexico State University, and Texas A&M University were involved with the sub-RAC. The sub-RAC members reviewed one another's syllabi, program and state assessments, and student teaching assignments, and called on experts (e.g., Keith Leatham and Blake Peterson) for resources that they used in implementing the paired placement model. The sub-RAC also created surveys and PDSA Cycles to help make improvements on the implementation of the model.

Furthermore, the sub-RAC focuses on the aim of the CERAC that "during student teaching, teacher candidates will use each of the eight mathematics teaching practices (NCTM, 2014) at least once a week during full time teaching" and positive implementation of the paired placement model. During the implementation of the paired placement model, team members address the driver diagram (see Figure 11.1). Members of the Paired Placement sub-RAC interact through conference calls, Dropbox, communication platforms, and face-to-face meetings sponsored by the MTE-Partnership and the National Science Foundation. Currently, the sub-RAC consists of Auburn University, the University of Montana, Texas A&M University, Columbus State University, and the University of Hawai'i at Manoa. Next, we highlight how the paired placement model has been implemented at four sites.

University of Montana

The MTE-Partnership at Montana is composed of two school districts and the University of Montana-Missoula. As a Research II university in a rural state, the program values partnering with districts because more than 13% of the credits earned (715 hours in middle or high school classrooms) for a degree are clinical experiences, and the hours students spend in the field are reported as the most valuable part of their secondary teacher preparation program. Typically, the University of Montana graduates 10 students with degrees in secondary mathematics education each year, and 20% of these are post-baccalaureate or graduate students, as prospective teachers return from other careers with prior degrees to become teachers of students in secondary mathematics.

At the University of Montana, all university students seeking to teach secondary mathematics major in mathematics (41–42 credits), which includes two semesters of Calculus, one semester each of Linear Algebra, Abstract Mathematics (proof), Abstract Algebra, Number Theory, Probability and Statistics, History of Mathematics, Math Modeling/Technology, Geometry, and any advanced mathematics/statistics course elective, along with middle/secondary mathematics methods. Additionally, at least 12 credits of at most two sciences are required. Methods are taught in the College of Education and Human Sciences, but faculty work collaboratively with the Department of Mathematical Sciences on a weekly basis, reading and discussing current issues in mathematics education via seminars. In addition to the methods course housed in education, there are 16 clinical experience credits spread over three semesters and 15 credits of other education courses (educational psychology and measurement, exceptionalities and classroom management, ethics and policy issues, content area literacy, promoting well-being P–12, and reflective practice and applied research).

The Council of Accreditation for Educator Preparation accredits the secondary mathematics teacher education program. The Missoula program has paired secondary mathematics teacher candidates during the methods course semester for the past 24 years, and since Fall 2013, the program has paired secondary teacher candidates during student teaching when possible. With the methods course clinical experience, the program faculty schedules the time into students' schedules, so all are available for the 15 weeks, 45–hour experience at the same time and placed within a 15–minute commute of the campus. However, with the student teaching internship semester, that is not an option because of grade levels desired and locations selected across the state—and around the world—for the semester-long student teaching internship. The methods course prepares teacher candidates for the student teaching experience via the three-week unit, in which they teach in pairs in a high school mathematics classroom while the reminder of the cohort assists and provides immediate feedback after the lesson, looking for strengths of the lesson and necessary changes for the next day's lesson.

Moreover, during methods, the teacher candidates learn about the eight standards for mathematical practice (National Governors Association Center for Best Practices, Council of Chief State School Officers, 2010) and the eight mathematics teaching practices (NCTM, 2014), and the need for addressing each of these weekly. Prospective teachers read about the standards and teaching practices, integrate the standards and practices into their lesson plans the semester before student teaching, and reflect on the implementation and struggles in helping their own students reach mastery.

Because of the close partnership with the two school districts and the long-established pairing of clinical experience students during the methods course, Missoula has established rapport with many secondary teachers of mathematics willing to accept pairs of teacher candidates especially grateful for the additional help in the classroom in preparing students for the state assessments. Recently, retired secondary teachers of mathematics, who typically know the cooperating teachers, supervise teacher candidates during student teaching. This stable relationship is helpful during the 16–week semester experience. Paired prospective teachers have time to teach individually and observe multiple teachers during the last weeks of the semester, but start day one in co-planning and co-teaching lessons, ideally with all three working as a team.

One cooperating teacher commented, when "three teachers are cooperating for the betterment of the students, collaborating on lessons (ideas/feedback), [it almost] forces a professional learning community within the classroom." One intern reflected on the purpose of this new paired-placement model: "I think the goal of the internship was to prepare my teaching partner and I to teach our own classes and collaborate with future colleagues. I also think the goal was to provide a better experience for the students, as they had access to multiple teachers and were able to get more one-on-one help." Rather than directly telling each prospective teacher what they should do, we encourage an approach to grow into the role needed by providing time to listen to questions and concerns. One such concern is communication among the team as shared by an intern from the Fall 2017 semester: "Better communication among participants would help. While we didn't step on each other's toes, we didn't always share on how the other could best help out or help adjust the lesson on the fly." Co-planning and co-teaching are essential for optimum success of this model, and changes to address this concern were implemented in Fall 2018. The Missoula group learned across partnerships that their experiences were similar, and their students' frustrations and struggles in communicating were common. Productive struggle is encouraged throughout the teacher preparation program, encouraging all to think, reason, and realize that confusion, errors, and struggle are natural components of learning.

Auburn University

Auburn University serves as the lead and the public land-grant institution member of the Central Alabama Mathematics Teacher Education Partnership (CAMTEP) of the MTE-Partnership. Other institutional partners of CAMTEP include Tuskegee University; Alabama State University; the Alabama Mathematics, Science, and Technology Initiative; Auburn City Schools; Tallassee City Schools; and Dadeville City Schools. The secondary mathematics program at Auburn University is housed in the Department of Curriculum and Teaching in the College of Education and has 12 to 20 bachelor's and/or alternative master's program completers each year. Each of these programs are initial certification programs and are Specialized Professional Association (NCTM, 2014) accredited with the Council for the Accreditation of Educator Preparation (CAEP). Thus, the programs have to meet specific standards for CAEP and the Alabama State Department of Education. Auburn University also grants a doctoral degree, an educational specialist degree, and a master of education.

The programs are run by two nationally known mathematics teacher educators who strive to ensure that their teacher candidates receive the highest quality mathematics education that they can, based on research and innovative practices in the U.S. and the world. In addition, teacher candidates are provided with field experiences that are in alignment with their program goals and state and national standards.

Teacher candidates seeking initial certification at Auburn University take 42 hours of mathematics courses, which enable them to have a double major in mathematics and mathematics education. Other courses outside of mathematics education courses include foundation and core curriculum courses. The teacher candidates are required to take three methods courses:

- CTSE 5040: Technology and Applications in Secondary Mathematics, which focuses on use of technology that supports mathematics teaching and learning;
- CTMD 4010: Teaching Mathematics in the Middle School, which focuses on teaching middle school mathematics and the middle school mathematics curriculum; and
- CTSE 4030: Curriculum and Teaching in Secondary Mathematics, which focuses on teaching high school mathematics and the high school mathematics curriculum.

Throughout the methods classes, students are exposed to the importance of mathematical problem solving as the core of school mathematics by careful analysis of the *Standards for Mathematical Practices* in the *CCSS-M* (National Governors Association Center for Best Practices, Council of Chief State School Officers, 2010), which are presented as required content that they will need to address as mathematics teachers. The importance of teaching via problem solving as a way to both meet the practice standards and make sense of mathematics is also emphasized. A variety of readings (NCTM, 2009, 2014, 2018; Schroeder & Lester, 1989; Skemp, 1976; Van de Walle, Bay-Williams, Lovin, & Karp, 2018), vignettes, videos, and in-class "laboratory experiences" are used to illustrate these points.

Teacher candidates at Auburn University encounter the mathematics teaching practices (NCTM, 2014) in each of the methods courses aforementioned. They gain experience implementing the mathematics teaching practices (NCTM, 2014) during their practicum clinical experiences and their student teaching semester. Teacher candidates learn about the mathematics teaching practices through reading *Principles to Actions: Ensuring Mathematical Success for All* (NCTM, 2014). They also develop a deep understanding of the practices through examining teachers' instructional moves in videos, vignettes, and field placements. Teacher candidates write lesson plans in which they are required to use the mathematics teaching practices and implement them in their field placements. Moreover, the teacher candidates reflect on their experiences.

Teacher candidates at Auburn University discuss equity issues all throughout their program. They learn about micromessaging, stereotypes, the impact of tracking, barriers to students' success in mathematics classrooms, teacher expectations, the impact of identity, and equitable teaching strategies. These issues are discussed in class, and teacher candidates also reflect on what they experience in their field placements with university faculty. Below is a quote from a teacher candidate during student teaching reflecting on her struggle to maintain an equitable classroom that meets the needs of each and every student:

One of my strengths is having a growth mindset about each and every student. I genuinely believe each student is capable of doing mathematics, and I communicate that with my students. I sometimes struggle to give each student an opportunity to share their ideas in class. When I am not careful, I find myself showing the same students' work and/or calling on the same students. It is also challenging to ensure I am making the appropriate accommodations for every student with IEPs, 504s, DSI, etc. Sometimes, it is hard to find the line between offering equitable supports and downgrading the challenge of the math. (Teacher Candidate, Spring 2018)

Paired Placement Implementation. The team at Auburn University initially implemented the paired placement model in Spring 2014. Because this was the first implementation of the model, the team tried to select students who had a strong work ethic and who had compatible personalities. Mentor teachers who were selected to host paired teacher candidates either held advanced degrees from the program or had attended over 200 hours of professional development provided by projects directed by the secondary mathematics teacher education faculty at Auburn. The program coordinator wanted to ensure that the mentor teachers could assist the teacher candidates in developing the eight mathematics teaching practices. Team members provided two separate orientation workshops and sessions for teacher candidates using each model. During the paired placement workshop and orientation sessions, emphases were placed on the *Standards for Mathematical Practices* and the NCTM (2014) mathematics teaching practices, the craft of teaching as discussed by Leatham and Peterson (2010a), high leverage practices (Ball & Forzani, 2011), scheduling within the paired placement model, and pros and cons related to the paired placement model. Emphasis was also placed on the trio being cognizant of who will serve as the lead teacher for a given day. University-required assessments and the assessments that are used across the CERAC are also discussed. More recently, we have been orienting the teacher candidates and the mentor teachers together about the paired placement model due to the acquirement of a grant[3] from the National Science Foundation.

Student teaching at Auburn lasts for 15 weeks, in which the teacher candidate gradually takes on full responsibility for all of the mentor teacher's classes for 20 days, with 10 of those days being consecutive days. The full teaching load is a state requirement. However, recently there has been more flexibility with the 10 consecutive days being replaced with two sets of five consecutive days. Most of the pairs of student teachers have stated that planning their lead teaching schedule is the most difficult part of the experience due to the state regulations and the district holidays. Once the schedule for lead teaching and transitioning from teacher to each teacher candidate and from teacher candidate to teacher candidate is complete, the trio working with each other has gone well. Furthermore, coordinating the scheduling was one of the first improvements made as a result of implementing a PDSA Cycle. The Auburn team learned that it needed to place the pair of teacher candidates with the mentor teacher during their methods course in order to acclimate the pair to the teacher and the school so that during student teaching the teacher candidates are ready to take on teaching responsibilities quickly. See Table 11.1 for a sample schedule of rotation provided by one of the mentor teachers.

TABLE 11.1. Auburn University Rotation Teaching Schedule for a Pair of Teacher Candidates and Their Mentor Teacher.

Week(s)	Responsibilities
Week 1	Teacher candidates observe, help teach with homework, help with groups, etc.
Week 2	Teacher candidates co-teach with mentor teacher using mentor teacher's lessons
Week 3	Teacher candidates co-teach with co-planned lessons (transition week)
Week 4– Week 8	Teacher Candidate 1 takes main responsibility with Teacher Candidate 2 helping when needed or observing. Teacher Candidate 1's professional work sample (PWS) is completed during this time.
Week 9	Teacher Candidates co-teach with co-planned lessons (transition week)
Week 10– Week 14	Teacher Candidate 2 takes main responsibility with Teacher Candidate 1 helping when needed or observing. Teacher Candidate 2's PWS is completed during this time.
Week 15	Teacher Candidates co-teach with co-planned lessons (transition week)
Week 16	Teacher Candidates wrap up, do more observations of other teachers, hand classes back to the mentor teacher. (The pairs usually begin one week earlier than the traditional model interns.)

The following quotes from teacher candidates participating in the model indicate that the paired placement model is affording the predicted outcomes of the teacher candidates, such as confident, caring, collaborative, and reflective practitioners:

- "I am so glad that I got to have the experience of doing my internship alongside a peer. I truly believe that I learned more than I would have if I had been on my own. We constantly reflected with one another whether it was in the car to or from the school or in a more formal reflection time."
- "We were continually talking to one another about our experiences. ... After each class we talked about what worked and what didn't without realizing that we were reflecting on our teaching, which helped us improve."
- "It has been wonderful having two people to reflect with me about my lessons and to give me constructive criticism. Both Mrs. Brown and the other intern see different things in my teaching that helps me to become a better teacher."
- "I also think this experience helped me to become a much more collaborative teacher. Before this semester I would have tended to simply work alone and not work with my fellow teachers. However, this semester I saw the importance of working with peers. It was so great when the three of us worked together with students' best interest as our priority. I saw firsthand that lessons ran more smoothly, and students benefitted the most when we all put input into how best to approach and teach a lesson."

Additionally, there have been positive outcomes for the mentor teachers who have served as hosts for paired placements. As follows are quotes from a mentor teacher who captured the bidirectional benefits of the paired placement model to the teacher candidates and the mentor teacher:

- "Sought my colleagues' advice and tried to encourage collaboration in my department more, after being a part of the experience."
- "Felt more accountable for holding teacher candidates to the mathematical teaching practices and mathematical practice standards and using them myself."
- "Encouraged the implementation of a social justice lesson."
- "Deep discussions: We constantly focused on lesson goals and standards, and discussions around assessment and proper measurements were a necessity. Little discussion was on classroom management."
- "1–1 vs. 2–1 placement: There was a sense of collaboration during discussions rather than a sense of critiquing or judgment."

Overall, the implementation of the paired placement model has been successful at Auburn University. Following are insights about the model from a university supervisor who was a Ph.D. candidate in mathematics education and had previous experience teaching secondary mathematics. (The supervisor also recently completed a university supervisor apprenticeship with a mathematics teacher educator, which included supervising a teacher candidate individually and alongside the mathematics teacher educator.) Although the apprenticeship was completed with a traditional internship model, the collaboration between the supervisor, mentor teacher, mathematics teacher educator, and teacher candidate provided the supervisor a perspective into collaboration and multiple perspectives also present within the paired placement model. The following semester, the university supervisor was assigned one pair of teacher candidates; the assigned mentor teacher also had previous experience implementing the paired placement model. In order to meet the state requirement of 20 full-time teaching days, a second mentor teacher was occasionally utilized. For example, while one teacher candidate completed her full-time teaching requirements within the primary mentor teacher's classroom, the other teacher candidate was able to teach two or three class periods in the other mentor teacher's classroom. This arrangement provided the teacher candidate not completing her full-time requirements opportunities to teach mathematics and to still maintain a presence in the primary mentor teacher's classroom. Observations were planned in a manner that allowed the university supervisor to observe one of the candidates teach, while sitting alongside the second candidate. During the observation, the university supervisor and second teacher candidate often reflected and discussed possible improvements and strengths as related to the other teacher candidate's instruction. The university supervisor identified these observation opportunities as providing powerful learning opportunities. Following each observation, lesson debriefing included both interns, the mentor teacher, and the university supervisor. Given that all were provided equal opportunities to speak and reflect on the enacted lesson, a professional learning community of multiple perspectives developed. Lastly, the university supervisor observed an unspoken accountability between the interns to teaching using the mathematics teaching practices. The university supervisor hypothesized that because both interns understood the teaching expectations, neither intern was willing to deviate from best practices due to the accountability of having their paired intern present.

Since Spring 2014, Auburn University has had one pair in the Fall 2014, four pairs Spring 2015, two pairs Spring 2016, one pair Spring 2017, one pair Fall 2017, and two pairs Spring 2018. During the 2017–2018 academic year, the workshop for mentor teachers and the orientation for interns completing the paired placement were combined so that the teacher candidates and the mentor teachers could learn about the paired placement model together. This was a major improvement for implementing the model based on previous PDSA Cycles. The mentor teachers, university supervisors, and teacher candidates believed that the model allowed them to grow in a number of ways.

As follows are quotes from some of the teacher candidates and mentor teachers that highlight the reflection, collaboration, and support for all participants in the model, including the mentor teacher, teacher candidates, and students:

- "The benefits of the paired placement model include collaboration and peer mentoring. We make a great team! The interns are collaborating with one another on a regular basis and the three of us work together to plan curriculum mapping and specific lessons." (Mentor Teacher, Fall 2017)
- "Working together to better support the students has been great. One of the classes in particular needs more attention, and we are able to give it. It is also helpful to have two pairs of eyes watching me teach lessons during my first 10 days (from both a peer and a pro), as I feel like I am receiving genuine and helpful feedback from both." (Teacher Candidate, Spring 2018)
- "The biggest positive is moral support. Having a peer to encourage me that I can do it, the lesson was not as bad as I thought, etc., has been huge. Also, having three teachers has allowed opportunities for Dr. Green to collaborate with each of us one-on-one during the school day while the other one was teaching. Also, I can learn from Kate's teaching, and she can learn from mine. Having a peer is great because I can be real with her in a way that it is hard to be with someone in a role above me." (Teacher Candidate, Spring 2018)
- "I have more time to go through the lessons thoroughly with interns while the other is teaching. It allows us to do all such work within school hours after the initial few weeks of trying to squeeze everything into a planning period or after school. I am able to spend more time with each intern individually without taking away instruction or help time from students. Sometimes with interns, the instruction and help that students get is less than

what they would get with just the cooperating teacher due to inexperience and the cooperating teacher being pulled in different directions. With three of us in the room, that is not the case. Students get more out of this type of placement." (Mentor Teacher, Spring 2018)

The Mathematics Teaching Practices survey was completed by teacher candidates throughout their student teaching, and the Mathematics Classroom Protocol for Practices (MCOP²) was used as an observation tool for each teacher candidate participating in the pairs. Teacher candidates felt that both of these tools provided them with insights into their teaching practices and aided in their growth.

Texas A&M University

Texas A&M University is the lead university of the East Central Texas Mathematics Teacher Education Partnership. In addition, Texas A&M University is a large research university in College Station (approximately 100,000 in population) that has five high schools to place teacher candidates for all methods courses and student teaching experiences. This is generally about 60 secondary methods students across all content areas (not including student teachers) needing placements in the high schools. The middle grades math/science certification program (housed in the College of Education) is a large certification program that prevents placement of secondary students in the middle schools within College Station. This college town is situated about two hours from Houston and Austin and three hours from Dallas. Even though there are more high schools in urban areas in which teacher candidates can be placed, traveling to high schools while taking college courses is not reasonable. Paired placement teacher candidates are members of AggieTEACH, a secondary (Grades 7–12) math and science teacher preparation program for undergraduates. AggieTEACH is housed in the College of Science and is accredited by the Southern Association of Colleges and Schools Commission on Colleges (for degrees) and the Texas State Board for Educator Certification (SBEC). About 20 teacher candidates complete the program each year. Teacher candidates take six 3-credit-hour courses and a 1-hour course (see Table 11.2).

Teacher candidates who receive math certification are math majors and take Calculus I, Calculus II, Calculus III, Differential Equations, Foundations of Math (Intro to Proof), Geometry, Linear Algebra, Advanced Calculus I (Analysis), Math with Technology, Abstract Algebra, two upper-level (400) math elective courses, Statistics, Physics, and Computer Programming. Texas A&M also has a lighter program called University Studies—Math for Teaching. In this program students take Calculus I, Calculus II, Foundations of Math (Intro to Proof), Geometry, Linear Algebra, Advance Calculus I (Analysis), Math with Technology, Abstract Algebra, Statistics, and Physics.

Experienced teachers who have a master's degree in educational administration and are usually retired, or who have gotten out of the profession for some other

TABLE 11.2. AggieTEACH Program Coursework

Course Name	Credit Hours	Description
SCEN 201	1	Exploratory course taken in freshman or sophomore year to help students decide whether they want to pursue the teaching profession. Students are placed in a middle school and high school classroom for one class period once per week for a total of eight weeks.
INST 222	3	Course that focuses on cultural aspects of education.
INST 210	3	Course that focuses on special populations (special education, gifted-talented, disabilities, etc.).
RDNG 465	3	Course that helps teach the teacher candidates how to help develop literacy in the content area.
TEFB 322	3	Teaching field-based course first-semester junior year with 30 field hours.
TEFB 324	3	Teaching field-based course second-semester junior year with 30 field hours.
TEFB 407	3	Teaching field-based course first semester senior year with 30 field hours.

reason, usually supervise the students. University supervisors are selected based on years of teaching experience, discipline in which they are certified (though Texas A&M does not require mathematics experience), and degrees earned. Texas A&M prefers that the university supervisors have experience in the local schools and that the program has some history with the supervisors (i.e., they were teachers who hosted teacher candidates), but this is not always the case. Usually the supervisors do not have secondary mathematics teaching experience. Most of the supervisors are retired middle school principals or retired science teachers. For initial certification in the classroom teacher certification class, Texas A&M must provide at least one of the following:

- Clinical teaching for a minimum of 14 weeks (no less than 65 full days), with a full day being 100% of the school day; or
- Clinical teaching for a minimum of 28 weeks (no less than 130 half days, with a half day being 50% of the school day; or
- Internship for a minimum of one full school year for the classroom teacher assignment or assignments that match the certification category or categories for which the candidate is prepared by the EPP.

Paired placement students complete the 14–week experience. Teacher candidates can student teach at any school district that is on the list of partnering districts in the state. Teacher candidates provide a list of choices where they would like to be placed. This information is used by the placement office in consultation with the preparation program and school district personnel to place the teacher candidates in schools. This same process is used for early field experiences.

Teacher candidates learn about NCTM's (2014) mathematics teaching practices in their senior methods course (TEFB 407). Teacher candidates read and discuss NCTM (2014). In these assignments, students are required to (a) cite quotes from the readings and state why/how the quotes impacted them, (b) give their thoughts regarding the author's perspective of the purpose of assessment, (c) open discussions on the reading, and (d) explore the idea of conceptual understanding and outline their perception on conceptual understanding. Texas A&M is trying to put more of an emphasis on these practices.

Teacher candidates at Texas A&M develop a lens on equity in a 3-credit-hour class (INST 222) that emphasizes cultural diversity and equity; it is not specific to mathematics teaching and learning. In their senior methods classes, teacher candidates are required to read "The Pedagogy of Poverty Versus Good Teaching," an article written by Martin Haberman (1991). After reading the article, teacher candidates state some ideas with which they agree and disagree, compare their thoughts with their own experiences, and state how Haberman's article may impact their teaching practice.

Implementation of Paired Placement. The option to participate in the paired-placement model is only made available to teacher candidates who are placed in one of the five local high schools. Program faculty pair interested teacher candidates together based on individual personalities and teaching strengths/weaknesses. After the pairs of candidates are formed, program faculty find mentor teachers from the same school (and in the same department) to host the students. Having all pairs of students at the same school and in the same department has helped develop strong relationships among the teacher candidates in the model. For the first three weeks of the placement, the trio of teachers work together co-planning and co-teaching. In the fourth week, the pair of student teachers co-plan and co-teach with the mentor teacher overseeing the pair. After the fourth week, the pair continues to co-plan, but only one of the teacher candidates is in front of students, teaching.

The PDSA Cycles help preparation program personnel to stay in tune with the dynamics between the (a) teacher candidates, (b) teacher candidates and mentor teachers, (c) teacher candidates and the students they are teaching, and (d) supervisor and the trio. In some cases, the PDSA Cycles helped program faculty to determine if the pair of teacher candidates were working as a team (and not as two individuals), communicating well, and whether any problem areas existed in the communication or relationships between teacher candidates, mentor teachers, or supervisors. Sometimes the PDSA Cycle helped program faculty to make small changes, like write the name of the teacher candidate who was "in charge" that day on the board. Other times program faculty were able to make larger adjustments like making sure that the trio were co-planning, so each were teaching concepts in the same manner and that one teacher candidate was able to seamlessly pick up where the other one left off regarding instruction.

Viewing other sub-RAC member's syllabi and implementation plans (ideas for activities for teacher candidates who were not teaching) helped program faculty to structure the model. The faculty members were able to edit other sub-RAC members' syllabi to meet their needs. Other sub-RACS members also had very good ideas and resources regarding mentor and university supervisor training. Discussing and viewing the training resources helped Texas A&M faculty members to develop better training sessions, which led to smoother implementation for the pairs of teacher candidates. Also, discussing other activities, like shadowing a student for a day and book study, that the other sub-RAC members did was helpful. The discussions the sub-RAC had regarding helping teacher candidates focus on students' mathematical thinking were beneficial and led to an assignment where teacher candidates interviewed students about some of their mathematical thinking on assignments that they gave them. Conversations in the sub-RAC helped to keep this concept at the forefront and thus made purposeful activities for the teacher candidates to focus student learning and student mathematical thinking.

Jacksonville State University

Jacksonville State University was a member of the Central Alabama Mathematics Teacher Education Partnership and is a four-year university that offers bachelor's and graduate degrees in mathematics education through the School of Education. The university is accredited by CAEP. From 2015 to 2017, Jacksonville State University had an average of approximately eight undergraduate or alternative master teacher candidates in mathematics education each semester. Because the university was a regional university, many teacher candidates traveled from their hometowns, often as far as an hour away, to attend the university. The university often placed teacher candidates near their hometown and within a 60-mile radius of the institution to create an equitable situation for both the supervisor and the teacher candidate. Many of these placements were in rural areas; however, the institution ensured placement of teacher candidates in diverse urban, suburban, and rural areas as labeled by the state and based on proximity to the institution. Most placements for teacher candidates were guided by the two preceding criteria; however, other criteria also were given for mentor teachers such as having at least three years of experience teaching, a master's degree, and approval by their principal.

This wide range of placements made it difficult to pair teacher candidates as well as find mentor teachers who shared the same vision for mathematics education advocated by NCTM (2014). To mediate this difficulty, the field placements director worked with the mathematics educator to find strong mentor teachers with which to place teacher candidates. The mathematics educator consulted with a teacher professional development provider organization, Alabama Math and Science Initiative (AMSTI) to help identify strong mentor teachers. AMSTI was sponsored by the state to provide professional development to mathematics teachers in the field. This partnership between the university placement services, the state sponsored teacher professional development organization, and the mathematics educator significantly improved the placement of teacher candidates. In addition, the mathematics educator worked with the teacher development organization to help certify teacher candidates in AMSTI during their senior methods course with a concentration on middle grades curriculum that promoted the mathematics teaching practices (NCTM, 2014).

Mathematics education teacher candidates at Jacksonville State University who completed a traditional bachelor's degree had a minimum of 18 hours in humanities, 12 hours in social studies, 8 hours in science, 6 hours in computing, 42 hours in professional studies in education, and 39 hours in mathematics. A 3-credit-hour course in the professional studies program was specifically designed to focus on diversity and multicultural education. One 3–hour course was dedicated to the teaching of mathematics, while simultaneously completing a secondary senior practicum in which teacher candidates were observed by a mathematics teacher educator. These two courses were taken by teacher candidates before the final internship or clinical experience and is where teacher candidates were first introduced to the mathematics teaching practices. The instructor of the course used NCTM (2014) as one of the required texts for the course, infusing activities throughout the semester that highlighted the mathematics teaching practices. Due to the often-late assignment of placements for teacher candidates into practicum settings for methods courses, teacher candidates were stressed to fulfill practicum hours. In addition, their placements did not always represent effective teaching practices (NCTM, 2014). Thus, videos from NCTM's Principals to Actions Professional Learning Toolkit (https://www.nctm.org/PtAToolkit/) were used to help teacher candidates connect teachers' actions to the mathematics teaching practices. Modules were created in an online platform that required teacher candidates to watch videos and respond to focused questions related to the mathematics teaching practices. These modules were required to be completed early in the semester to replace practicum observation hours.

Nine of the hours in professional studies were dedicated to the internship or culminating clinical experience. Supervisors of the student teaching experiences at Jacksonville State University were often part-time employees who were retired administrators or science teachers. This often provided difficulty for focused discussions related to mathematical pedagogical knowledge, shared beliefs in best practices, and teaching practices (NCTM, 2014), from clinical experiences to student teaching. These differences in beliefs and beliefs of success of the traditional internship model also provided a challenge to implementation of the paired placement. The supervisor and department chair were initially reluctant to allow teacher candidates to participate in the paired placement because they believed they already had a working system. To mediate this and encourage the model, individual meetings were held with the department chair, supervisor, and mentor teacher to discuss the benefits of this model to provide mathematics and mathematical pedagogical support that were not available from the supervisor, in addition to other positive benefits from the research. Challenges of the model also were discussed, and potential remedies for these challenges were given. Research articles related to the paired placement were provided by email before the meeting to help support the discussion. Each meeting with these stakeholders led to support and implementation of the model.

Alabama state policies required teacher candidates to be the teacher of record for 10 consecutive days and 20 total days. For this reason, the teacher candidates who volunteered to participate in the paired placement started teaching early in their semester of internship. This was mediated by placing the teacher candidate with a strong mentor teacher in which they completed clinical experiences for their mathematical teaching methods course the semester before. This arrangement allowed for a faster ramp up of teaching by the teacher candidates. In addition, allowing teacher candidates to volunteer for the paired placement helped mediate personality conflicts between teacher candidates found previously in the literature.

After implementing the paired placement, the mentor teacher, teacher candidates, and the supervisor spoke highly of the model. Many of the benefits suggested by Strutchens et al. (2019) and Mau (2013) were evident, such as teaming, cooperation, and increased reflection. A retired administrator and science teacher serving as a supervisor stated,

> First this was a new experience for me, and I wondered how it would actually work in the classroom. I was pleasantly surprised when I made my first observation and post conference. This process allows the interns to plan together and to plan with their CT. It also permits them to critique each other, share ideas, team-teach, and explore different strategies and critique the process. This is an ongoing collaboration between teachers, and to me it is valuable for interns. (Spring, 2017)

A cooperating teacher also shared her positive experience with the paired placement. Her response showed not only an increase in pedagogical risk-taking on her own part, but also accountability for strong instructional practice as suggested by Conway et al. (2017) and Strutchens et al. (2019):

> I think this was the best thing that could have ever happened to me. I had to step up my game. I turned loose of the tight control I had over my classroom to the interns. I got to see a different side of my students. I learned I didn't have to give as much [direct] instruction as I was used to giving the students. I learned that kids at [this] age still love games, group activities, and stickers. (Cooperating Teacher, Spring, 2017)

A teacher candidate in the paired placement showed and discussed evidence toward an increase in equitable teaching practices as defined by NCTM (2014, 2018). The teacher candidate discussed the effectiveness of the model to encourage growth in each and every student toward the building of procedural fluency from conceptual understanding:

> One of the best parts about the paired placement was ... the relationship ..[that].. helped with procedural fluency and conceptual understanding. My cooperating teacher was able to effectively help my co-teaching partner and [me] recognize where the students in her classes are at in their understanding. This was beneficial

in helping my co-teaching partner and I plan to teach lessons that built upon pre-existing procedural fluency and conceptual understanding. This was accomplished during our instruction with more ease than a traditional placement. My co-teaching partner and I were able to utilize the different co-teaching models to help all students within the classroom. Whenever my partner was teaching, I was able to walk around and help the struggling students in the class keep up with the classroom pace, and my partner was able to do the same when I was lead teaching. We were able to work as a team to make sure all students were able to grow their learning throughout a lesson. The main students that were able to benefit from the paired placement model were the advanced and the struggling learners. We were able to focus on different learners and increase their understanding because we were able to have more one-on-one instruction than typically happens within a classroom. (Teacher Candidate, Spring, 2017)

Jacksonville State University used information from other universities' PDSA Cycles to help implement paired placements successfully in 2016. During the teacher candidates' paired placement, the researcher visited and checked on the paired placement on two different occasions to help support the teacher candidates and mentor teacher with the model. The researcher used questions from the PDSA Cycles to help ensure that the model was running smoothly, and challenges were being headed off before they surfaced. Jacksonville State University found strong evidence for the paired placement model to appropriately prepare teacher candidates for equitable mathematics teaching practices (NCTM, 2014, 2018).

Summary of the Settings

The sub-RAC members at the universities implemented the paired placement model for student teaching in accordance with the policies and other constraints that impact their contexts. The members of the sub-RAC hold conference calls and face-to-face meetings to develop PDSA Cycles that help them to improve the implementation of the model. Placing the pairs with their mentor teacher in an early field experience prior to the student teaching clinical experience was one result of a PDSA Cycle that has helped the teacher candidates to ramp up to taking over classes more quickly during student teaching. Also, the use of a PDSA Cycle early during the student teaching experience, in order to examine how the pair, the mentor teacher, and university supervisors are working together, has been fruitful in helping the sub-RAC members to head off any major problems that could occur due to a lack of communication among the participants. Moreover, as the sub-RAC members continue to monitor and refine the implementation protocols, they will be able to develop a manual that could lead others to successful implementation of the model. In the next section, we provide some lessons learned from implementing the model.

LESSONS LEARNED

Members of the Paired Placement sub-RAC have learned much from working together and implementing PDSA Cycles to increase the efficiency of the model. During each implementation of the paired placement, members of the sub-RAC utilize PDSA Cycles at various points throughout the student teaching semester to determine what aspect of the model needs improving. We have also held focus groups with university supervisors, teacher candidates, and mentor teachers to find out what worked well and what did not. What follows are lessons that we have learned that have helped us and others to implement the model more effectively.

First, it is important to help the teacher candidates and the mentor teachers become acquainted with all of the major components of the model. Providing an orientation session that details how the pair of candidates and the teacher should work together is key. Participants need to be aware of the importance of co-planning and co-teaching lessons, transitioning periods between teachers, the identification of the lead teacher for the day or week, and coordination of classroom roles and responsibilities. Providing the trios with this information helps the team to work more collaboratively and cohesively.

Second, scheduling shadowing, lead teacher roles, and full-time teaching for each teacher candidate—and transition periods between the candidates—are important to do at the beginning of the student teaching experience. We found that if possible, placing the pair with their mentor teacher the semester before student teaching helps with the scheduling process in that it allows the teacher candidates to become familiar with the teacher and students and enables the teacher candidates to take on teaching responsibilities more rapidly during student teaching. Furthermore, scheduling is one of the most difficult components of the approach. See Table 11.1 and the Auburn University section of this chapter for an outline of a possible schedule.

Third, during the orientation workshop, it is important to discuss possible pitfalls of the model. Discussing the possible weaknesses of the models, such as dependency, confusion about classroom management during the experience, loss of individuality, and competition between the teacher candidates, and others as described in the literature (Nokes et al., 2008; Goodnough et al., 2009), helps to deter the pitfalls from happening. In fact, it has been noted at many sub-RAC meetings that the mentor teachers and candidates benefit from knowing the possibility of the pitfalls happening and working to avoid them.

Fourth, we have found assigning particular foci for the teacher candidates to observe when they are not the lead teachers adds to their growth in pedagogical content knowledge, knowledge of their students, and effectiveness as a collaborator. The observations afford the teacher candidates the opportunity to stay engaged and committed to the learning process when they are not in charge of leading the class, which enables them to develop different lenses (equity, student learning, and effective teaching practices) for instruction throughout the internship experience and also to feel like a part of the class.

Finally, it important that the university supervisor facilitates the debriefing session between the lead teacher candidate, the mentor teacher, and the other teacher candidate. These debriefing sessions should benefit all of the teachers. Strengths and weaknesses of the lesson, student learning, and obstacles to learning taking place should be discussed. Also, if the teacher candidate who was not the lead teacher is given particular focus points to observe during a lesson, then his or her observations should be discussed during the debriefing of the lesson. If done well, these debriefing sessions can lead to the development of a professional learning community among the three teachers. The aforementioned lessons learned enable the Paired Placement sub-RAC to continue improving upon the model and to facilitate the growth of teacher candidates involved in the model.

ENDNOTES

1. Work on this chapter was supported in part by National Science Foundation grant #1726998, *Collaborative Research: Attaining Excellence in Secondary Mathematics Clinical Experiences with a Lens on Equity*. All findings and opinions are those of the authors, and not necessarily those of the funding agency.
2. Other contributors to this chapter include: Ruby Ellis of Auburn University and Charmaine Mangram of the University of Hawai'i at Manoa.
3. National Science Foundation grant #1726998, *Collaborative Research: Attaining Excellence in Secondary Mathematics Clinical Experiences with a Lens on Equity.*

REFERENCES

Aguire, J., Mayfield-Ingram, K., & Martin, D. B. (2013). *The impact of identity in K–8 mathematics: Rethinking equity-based practices.* Reston, VA: National Council of Teachers of Mathematics.

Association of Mathematics Teacher Educators. (2017). *Standards for preparing teachers of mathematics.* Raleigh, NC: Author. Retrieved from http://amte.net/standards

Ball, D. L., & Forzani, F. M. (2011). Building a common core for learning to teach: And connecting professional learning to practice. *American Educator, 35*(2), 17.

Boston, M., Dillon, F., Smith, M., & Miller, S. (2017). *Taking action: Implementing effective mathematics teaching practices in Grades 9–12.* Reston, VA: National Council of Teachers of Mathematics.

Bullough, R. V., Young, J., Birrell, J. R., Clark, D. C., Egan, M. W., Erickson, L., ... & Welling, M. (2003). Teaching with a peer: A comparison of two models of student teaching. *Teaching and Teacher Education, 19*(1), 57–73.

Conway, B., Erickson, D., Parish, C., Strutchens, S., & Whitfield, J. (2017, October). *An alternative approach to the traditional internship model.* Paper presented at the Georgia Association of Mathematics Teacher Educators, Eagle Rock, GA. Retrieved from http://digitalcommons.georgiasouthern.edu/gamte/

Gardiner, W., & Robinson, K. S. (2009). Paired field placements: A means for collaboration. *The New Educator, 5*(1), 81–94.

Goodnough, K., Osmond, P., Dibbon, D., Glassman, M., & Stevens, K. (2009). Exploring a triad model of student teaching: Pre-service teacher and cooperating teacher perceptions. *Teaching and Teacher Education, 25,* 285–296.

Haberman, M. (1991). The pedagogy of poverty versus good teaching. *Phi Delta Kappan, 73,* 290–294.

Leatham, K. R., & Peterson, B. E. (2010a). Secondary mathematics mentor teachers' perceptions of the purpose of student teaching. *Journal of Mathematics Teacher Education, 13*(2), 99–119.

Leatham, K. R., & Peterson, B. E. (2010b). Purposefully designing student teaching to focus on students' mathematical thinking. In J. W. Lott & J. Luebeck (Eds.), *Mathematics teaching: Putting research into practice at all levels. Association of Teachers of Mathematics Monograph* (No. 7, pp. 225–239). San Diego, CA: AMTE.

Martin, W. G., & Strutchens, M. E. (2018). Improving secondary mathematics teacher preparation via a networked improvement community: Focus on clinical experiences. In M. E. Strutchens, R. Huang, L. Losano, & D. Potari (Eds.), *Educating prospective secondary mathematics teachers.* Monograph Series Edited by Kaiser, G. (pp. 27–46). Cham, Switzerland: Springer.

Mau, S. (2013). Letter from the editor: Better together? Considering paired-placements for student teaching. *School Science and Mathematics, 113*(2), 53–55.

National Council of Teachers of Mathematics. (2009). *Focus in high school mathematics: Reasoning and sense making.* Reston, VA: Author.

National Council of Teachers of Mathematics. (2014). *Principles to actions: Ensuring mathematical success for all.* Reston, VA: Author.

National Council of Teachers of Mathematics. (2018). *Catalyzing change in high school mathematics: Initiating critical conversations.* Reston, VA: Author.

National Governors Association Center for Best Practices, Council of Chief State School Officers. (2010). *Common core state standards: Mathematics.* Washington, DC: Author.

Nokes, J. D., Bullough, R. V., Egan, W. M., Birrell, J. R., & Merrell Hansen, J. (2008). The paired-placement of student teachers: An alternative to traditional placements in secondary schools. *Teaching and Teacher Education, 24*(8), 2168–2177.

Peterson, B. E., & Leatham, K. R. (2018). The structure of student teaching can change the focus to students' mathematical thinking. In M. E. Strutchens, R. Huang, D. Potari, & L. Losano (Eds.), *Educating prospective secondary mathematics teachers.* ICME-13 Monographs, 9–26. Cham, Switzerland: Springer International.

Schroeder, T. L., & Lester, F. K. (1989). Understanding mathematics via problem solving. In P. Trafton (Ed.), *New directions for elementary school mathematics* (pp. 31–42). Reston, VA: National Council of Teachers of Mathematics.

Skemp, R. R. (1976). Relational understanding and instrumental understanding. *Mathematics Teaching, 77,* 20–26.

Strutchens, M. E., Sears, R., Whitfield, J., Biagetti, S., Brosnan, P., Oloff-Lewis, J., ... & Ellis, R. L. (2019). Implementation of paired placement and co-planning/co-teaching field experience models across multiple contexts. In T. Hodges & A. Baum (Eds.), *Handbook of research on field-based teacher education* (pp. 32–63). Hershey, PA: IGI Global.

Van de Walle, J. A., Bay-Williams, J. M., Lovin, L. H., & Karp, K. S. (2018). *Teaching student-centered mathematics: Developmentally appropriate instruction for grades 6–8* (Volume III, 3rd ed.). New York, NY: Pearson.

CHAPTER 12

FOCUS ON IMPROVING CLINICAL EXPERIENCES IN SECONDARY MATHEMATICS TEACHER PREPARATION[1]

Charmaine Mangram, Pier A. Junor Clarke,
Patrice Waller, Ruby L. Ellis, and Cynthia Castro-Minnehan

Throughout this section, the Clinical Experiences Research Action Cluster (CERAC) shared reviews of literature and current research endeavors focused on clinical experiences provided to secondary mathematics teacher candidates. Chapter 7 began with a review of literature and then provided an initial framing of research in this area by the Mathematics Teacher Education Partnership (MTE-Partnership). Chapter 8 described the general organization of the CERAC, which was formed to conduct research on improving clinical experiences in secondary mathematics teacher preparation. The following three chapters provided summaries of the three sub-Research Action Clusters (sub-RACs) who are working on particular aspects of clinical experiences.

Chapter 9 focused on early clinical experiences during methods classes. The authors reported on their work related to designing modules that provide mentor teachers and their teacher candidates with shared experiences around the teaching and learning of secondary mathematics. Zelkowski et al. (Chapter 9 of this book)

The Mathematics Teacher Education Partnership: The Power of a Networked Improvement Community to Transform Secondary Mathematics Teacher Preparation, pages 281–292.
Copyright © 2020 by Information Age Publishing

presented three modules, which are at different stages of development, and the impacts—and predicted impacts—of the modules on the development of well-prepared beginning teachers.

Chapter 10 focused on the use of co-planning and co-teaching strategies in clinical experiences, particularly student teaching. Grady et al. (Chapter 10 of this book) focused on their development of written materials and online videos to help mentor teachers and teacher candidates to implement various co-planning and co-teaching strategies together within secondary mathematics classrooms. The authors shared the importance of the mentor teacher and the teacher candidate pairs working together to meet the needs of each and every student.

Chapter 11 focused on the paired placement model for student teaching, in which two teacher candidates are placed with one mentor teacher. Strutchens et al. (Chapter 11 of this book) shared how they implemented and used Plan-Do-Study-Act (PDSA) Cycles to improve the implementation of the paired placement model across multiple contexts. The Paired Placement sub-RAC illuminated the major benefits of the model, such as bidirectional pedagogical growth between the mentor teacher and the pair of teacher candidates, collaborative and reflective teacher candidates, and a student-centered learning environment.

Finally, in this chapter, we outline suggestions for secondary mathematics teacher preparation programs who are planning on improving their clinical experiences. We share recommendations based on lessons that we have learned and are learning throughout our work across the CERAC.

Key elements must be in place in order for clinical experiences to be effective and meaningful for all stakeholders, including university faculty, university supervisors, administrators, mentor teachers, teacher candidates, and students. In this chapter we make several recommendations for secondary mathematics teacher education preparation programs for improving clinical experiences based on the following key elements: (a) strong relationships between university personnel and district partners, (b) well-articulated goals for the teacher candidates, (c) mentor teachers who are able to facilitate the pedagogical and content growth of teacher candidates, and (d) open communications between the mentor teacher and teacher candidate.

First, we recommend that faculty members who are interested in transforming their programs start by focusing on part two of the RAC's stated problem (see Chapter 8):

> Bidirectional relationships between the teacher preparation programs and school partners in which clinical experiences take place are rare. Such relationships that reflect a common vision and shared commitment to the vision of the *CCSS-M* [*Common Core State Standards for Mathematics*] and other issues related to mathematics teaching and learning are critical to the development and mentoring of new teachers.

According to Zeichner (2010), the disconnect between campus and school-based components of a program is a central problem that has plagued teacher education for years. Zeichner (2010) contended that a "'third space' in teacher education where academic and practitioner knowledge and knowledge that exists in communities come together in new less hierarchical ways in the service of teacher learning" (p. 89) is needed. Throughout the work of the CERAC, we are paying attention to this "third space." Each of our teams are composed of at least a mathematics teacher educator, a school partner, and a mathematician. Each of these voices is important in helping to shape the products that we are developing to support effective clinical experiences across the continuum from practicums in methods classes to student teaching.

Teams within the CERAC have made progress toward strengthening connections between school partners and campus work through meetings designed to help the aforementioned stakeholders think about what their needs are and how together we can negotiate ways of meeting the expressed needs. The modules and protocols that we have been developing around particular approaches to clinical experiences and the professional learning opportunities that we provide to mentor teachers and teacher candidates have been developed jointly and are benefitting all constituents. It is a win-win situation when school partners feel that their teachers are developing as strong mentors to teacher candidates and are strengthening their classroom skills at the same time. Also, the schools benefit from being able to hire well-prepared beginning teachers who have the knowledge and dispositions that are needed to teach well.

Second, we recommend that much of the discussion in the "third space" be centered on teacher candidates' needed knowledge in order to become well-prepared beginning teachers. The Association of Mathematics Teacher Educators' (2017) *Standards for the Preparation of Teachers of Mathematics* (AMTE Standards) is a useful document to share and discuss with school partners as it can serve as a guide to specific outcomes that should be attained by teacher candidates. After establishing the goals for the teacher candidate, it is important to think about when, where, and how these different proficiencies will be addressed. Through our grant work, we have developed a data collection tool that goes across the continuum of clinical experiences to help teams think about when they are going to collect data on the teacher candidate's development of the mathematics teaching practices and other equitable teaching strategies (see Appendix A).

Third, we recommend supporting the mentor teachers (i.e., cooperating teacher) early in the process of mentoring teacher candidates for the successful implementation of any of the models that we advocate. This finding is in alignment with existing research identifying the mentor teacher as one of the major factors in clinical experience success (Weasmer & Woods, 2003). Mentor teachers find value in knowing what teacher candidates are learning and doing in their university methods course(s), but they do not always have the time to adjust their own teaching to serve as examples of specific practices. Thus, mentor teachers are able to gain new teaching and facilitation strategies when teacher candidates bring fresh new ideas into mentor teachers' educational spaces. It is important that we try to continue to build a strong connection between mentor teachers' classroom, teacher candidates' learning experiences in methods course, and methods instructors' expectations of teacher candidates entering the teaching profession.

The RAC has also found that it is important to discuss potential co-planning strategies and co-teaching strategies (described in Chapter 10 of this book) with the teacher candidate and the mentor teacher together, if possible. Exposing teacher candidates and their mentor teachers to these strategies allows the pairs (or triads, depending on the model being implemented) to deliberately plan ahead how they will collaboratively plan lessons and also helps them identify each member's role during lesson implementation. This discussion helps avoid misunderstandings about roles and responsibilities between the co-teaching and co-planning team and their students.

To address this need, the co-planning and co-teaching sub-RAC developed modules, facilitated face-to-face professional development for the collaborating pairs, and created videos to be used to visualize various co-teaching approaches (Sears et al., 2017). Being cognizant that institutions can vary in their budget allocations for professional development, having access to online resources reduces cost and increases the likelihood that individuals are adequately prepared to enact the strategies effectively. Thus, there are a variety of ways that support for co-planning and co-teaching can be accomplished. Some institutions may choose to offer a one-day professional development session for teacher candidates and mentor teachers prior to the assignment of placement to learn about and enact strategies for co-planning and co-teaching. In other institutions, faculty members may offer online support through the use of online modules using voiced-over slide presentations and online video clips of some of the co-planning and co-teaching strategies in action (Sears et al., 2017). Finally, in other institutions, the faculty members may provide a one-hour meeting to discuss co-teaching and co-planning strategies with the mentor teacher and teacher candidates once the placements have been assigned.

Furthermore, data from the Co-Planning and Co-Teaching sub-RAC showed that station teaching, team teaching, and parallel teaching more readily increased equitable learning opportunities when compared to the one-teach one-observe co-teaching strategy (Friend et al., 2010; Murawski & Spencer, 2011). It is important to note, however, that team teaching and parallel teaching were seldom used within the mentor teacher and teacher candidate environment, despite the fact they were potentially more effective strategies. Whenever the one-teach-one-observe co-teaching strategy was observed, it typically reflected the more traditional modes of instruction in the United States that were documented in the Trends in International Mathematics and Science Study (TIMSS; Mullis, Martin, Foy, & Arora, 2012; Stigler & Hiebert, 2009). Therefore, if co-teaching strategies are to be utilized, mentor teacher and teacher candidate pairs ought to be encouraged to use a variety of the strategies, rather than relying solely on the one-teach one-observe approach.

Along with providing support for mentor teachers and teacher candidates around co-planning and co-teaching strategies, the Paired Placement sub-RAC discovered through PDSA Cycles across the various institutions the importance of placing teacher candidates with the mentor teacher during the semester prior to the student teaching (internship) semester, if possible. Early placement allows the mentor teacher and teacher candidate to establish rapport and also allows the mentor teacher to identify strengths and areas of growth related to each university teacher candidate's professional dispositions and their teaching practices before the teacher candidate takes full responsibility of the mentor teacher's course load. Early detection facilitates the faculty's ability to begin helping both the university teacher candidates and mentor teachers in addressing challenges early, which is also a key to success. If it is not possible to place teacher candidates with the mentor teacher the semester prior, it is recommended that, at minimum, the mentor teacher is introduced then to the pair of teacher candidates.

In addition, the CERAC recommends that emphasis on reflection is important for the implementation of all of these clinical experiences. For example, the Co-Planning and Co-Teaching sub-RAC noticed a need to reflect on means to build professional relationships that portray the professional standards for the discipline (NCTM, 1991). Particularly, the sub-RAC had to emphasize that during planning and personal correspondence there should be a purposeful focus on learning and on discourse that orchestrates mathematics teaching and learning (Stein, Engle, Smith, & Hughes, 2008). Hence, as mathematics teacher educators, ensuring that the AMTE Standards (2017) are attended to is important in preparing preservice teachers, relative to their knowledge, skills, and dispositions via the co-planning and co-teaching strategies.

Relatedly, the Methods sub-RAC found that there is value in allowing teacher candidates to make mistakes or identify errors in practice and then explore strategies that can help them to make sense of new knowledge and enhance teaching practices. Activity 1 across all of the Methods sub-RAC's modules uses this as a philosophy. In Module 1, the Standards for Mathematical Practice module, teacher candidates brainstorm student habits and think about how students might develop those habits in relation to what happens or is seen in mathematics classrooms. In Module 2, the lesson planning module, teacher candidates create a lesson plan based on an assessment that will be given without extensive opportunities to plan. Teacher candidates use the Mathematics Classroom Observation Protocol for Practice (MCOP^2) to evaluate this initial lesson plan to see the flaws in planning simply around procedures and how that may relate to their own learning experiences as a student. In this module, teacher candidates realize that there are major aspects of student engagement and teacher facilitation that they need to attend to during planning to improve learning experiences for students. In Module 3, the feedback module, teacher candidates bring feedback that they have received in previous courses to think about how feedback makes them feel. This reflection opens the conversation for characteristics of providing meaningful feedback.

These methods allow teacher candidates to connect on a personal level either through previous or current experiences and allows teacher candidates to be more open toward learning and enacting new strategies and practices. Ultimately, the feedback module will result in teacher candidates' understanding the power of how quality feedback of various forms also provides improved student learning and achievement in addition to instruction.

Furthermore, the Paired Placement sub-RAC provides the mentor teacher with different debriefing tools that enable the mentor teacher and teacher candidates to reflect on lessons together and how to facilitate students' growth. The teacher candidates often continue to reflect on their lessons at the end of the day as they share rides back to the university after their time in the schools. Overall, the teacher candidates and their mentor teachers learn and struggle through effective collaborative models, assignments, and activities that bridge the gap between clinical experiences and methods courses, which ultimately result in the mentor teachers becoming mathematics teacher educators.

Finally, we recommend discussing any known pitfalls to particular models for field experiences upfront with mentor teacher and teacher candidates to avoid problems. For example, the Paired Placement sub-RAC recommends discussing possible pitfalls (Goodnough, Osmond, Dibbon, Glassman, & Stevens, 2009) of the paired placement model early with the mentors and teacher candidates. An open and honest conversation about possible pitfalls allows the trio working together during paired placements to begin thinking about strategies for addressing and possibly avoiding them. It also shifts the triads' thinking toward collective problem solving. Once again, this conversation can be accomplished in many ways: one way that has worked for our group is to share with the triad an article or summary of the literature prior to meeting with the team and having a focused discussion to brainstorm strategies for preventing and overcoming the identified pitfalls.

The recommendations that we have made are based on the work that we have done and are doing currently. We see this journey as one in which we are constantly learning and making progress toward the goals that we, as higher education faculty and school partners, have for our teacher candidates, mentor teachers, university supervisors, students, and ourselves.

Reading List

For those interested in learning more about how to improve their clinical experiences, the CERAC provides the following list of readings and short synopses.

1. Association of Mathematics Teacher Educators. (2017). *Standards for preparing teachers of mathematics*. Raleigh, NC: Author. Retrieved from: http://amte.net/standards
   - The Association of Mathematics Teacher Educators presents these standards as a guide to improve teacher education programs in the

U.S. It serves as a vision for the initial preparation of K–12 teachers of mathematics, and it advocates for practices that support candidates in their preparation in becoming effective mathematics teachers who guide student learning.

2. Friend, M., Cook, L., Hurley-Chamberlain, D., & Shamberger, C. (2010). Co-teaching: An illustration of the complexity of collaboration in special education. *Journal of Educational and Psychological Consultation, 20*(1), 9–27.

   - The authors present a comprehensive view of co-teaching, including its origins in special education, a definition of co-teaching, and its differences from team teaching. Although there is much enthusiasm for co-teaching, it is a complex undertaking. This document illustrates the complexity of co-teaching and includes the relationships and roles of co-teachers, co-teaching's impact on student achievement, and program logistics. It also raises issues such as inconsistencies in definitions, professional preparation, and school culture.

3. Goodnough, K., Osmond, P., Dibbon, D., Glassman, M., & Stevens, K. (2009). Exploring a triad model of student teaching: Pre-service teacher and cooperating teacher perceptions. *Teaching and Teacher Education, 25*, 285–296.

   - Goodnough, Osmond, Dibbon, Glassman, and Stevens' article provides insight into the benefits and possible challenges mentor teachers and teacher candidates may experience with the triad model. Additionally, Goodnough et al. provide a description of co-teaching models that may emerge during this type of student teaching experience.

4. Leatham, K., & Peterson, B. (2010). Secondary mathematics cooperating teachers' perceptions of the purpose of student teaching. *Journal of Mathematics Teacher Education, 13*, 99–119.

   - Leatham and Peterson's article highlights the importance of the mentor teacher viewing the student teaching experience as an opportunity for teacher candidates to strengthen their teaching practices. Through the use of a metaphor of a shoe store apprentice, Leatham and Peterson describe the role of the mentor teacher in facilitating the teacher candidate's growth. Similar to a shoe store apprentice's need to make shoes and run a shoe store, teacher candidates need to learn how to facilitate student learning and manage a classroom, with the greatest emphasis placed on facilitating student learning.

5. Mullis, I. V., Martin, M. O., Foy, P., & Arora, A. (2012). *TIMSS 2011 international results in mathematics*. Amsterdam, The Netherlands: International Association for the Evaluation of Educational Achievement.

   - The TIMSS (2011) provides international results for mathematics achievement for fourth- and eighth-grade students in 63 countries.

It includes achievement over time for participants in the previous TIMSS assessments in 1995, 1999, 2003, and 2007. This document also provides a mathematics report that describes the educational contexts for mathematics, including home environment support, students' backgrounds and attitudes toward mathematics, the mathematics curriculum, teachers' education and training, classroom characteristics and activities, and school contexts for mathematics learning and instruction.

6. Murawski, W. W., & Spencer, S. (2011). *Collaborate, communicate, and differentiate: How to increase student learning in today's diverse schools*. Thousand Oaks, CA: Corwin Press.
   – This guide provides a detailed, focused treatment of collaboration. It is presented in a practical and easy-to-access format for K–12 educators and administrators, as well as other stakeholders such as parents and policy makers.

7. National Council of Teachers of Mathematics. Commission on Teaching Standards for School Mathematics. (1991). *Professional standards for teaching mathematics*. Reston, VA: Author.
   – This document presents professional standards, developed by the National Council of Teachers of Mathematics (NCTM). It details what mathematics teachers need to know to teach mathematics in the new framework detailed in an earlier document, "Curriculum and Evaluation Standards for School Mathematics." It also details how to evaluate mathematics teaching and presents standards for professional development of mathematics teachers. This document provides next steps that will need to be addressed to continue to implement the standards.

8. Peterson, B., & Leatham, K. (2018). The structure of student teaching can change the focus of students' mathematical thinking. In M. E. Strutchens, R. Huang, L. Losano, & D. Potari (Eds.), *Educating prospective secondary mathematics teachers*. Monograph Series edited by Kaiser, G. (pp. 9–26). Cham, Switzerland: Springer.
   – Peterson and Leatham's article provides insight into the impact of changing the traditional structure of the student teaching experience. Peterson and Leatham's model placed two teacher candidates with one mentor teacher. They found that this model resulted in increased collaboration and shifted teacher candidates' focus from classroom management and behavioral issues to a focus on students' mathematical thinking.

9. Sears, R., Brosnan, P., Oloff-Lewis, J., Gainsburg, J., Stone, J., Spencer, C., ... & Andreason, J. (2017). Using improvement science to transform clinical experiences with co-teaching strategies. *Annual perspectives of mathematics education (APME) 2017: Reflective and collaborative*

processes to improve mathematics teaching (pp. 265–273). Reston, VA: National Council of Teachers of Mathematics.
- Sears et al. discuss how improvement science influenced their work with mentor teachers and co-teaching strategies. By utilizing PDSA) Cycles, Sears et al. explored factors that impacted the sustainability and effectiveness of the co-planning and co-teaching model.

10. Stein, M. K., Engle, R. A., Smith, M. S., & Hughes, E. K. (2008). Orchestrating productive mathematical discussions: Five practices for helping teachers move beyond show and tell. *Mathematical Thinking and Learning, 10*(4), 313–340.
- In this article, the authors propose a pedagogical model for using student responses to inquiry-based tasks more effectively in whole-class discussions. They argue that this model can make student-centered approaches to mathematics instruction more attainable and more feasible for teachers. The model specifies five practices that teachers can use: anticipating, monitoring, selecting, sequencing, and making connections between student responses. The authors suggest that these five practices can help teachers gain confidence and efficacy over their inquiry-based instruction because they learn ways to reliably form student discussions.

11. Stigler, J. W., & Hiebert, J. (2009). *The teaching gap: Best ideas from the world's teachers for improving education in the classroom.* New York, NY: Simon and Schuster.
- In this book, the authors use the phrase "teaching gap" to illustrate the differences in mathematics teaching methods they observed across 63 countries studied in TIMSS. They emphasize that lessons in the U.S. lack rich problem solving and emphasize discrete and disconnected procedures. They argue that to improve student achievement in the U.S., the performance of teachers must be improved through greater opportunities to learn about teaching. They emphasize that U.S. schools need to commit to continuous improvement systems that make changes that are sustained over the long term.

For current information on the CERAC, please visit mtep.info/CERAC.

ENDNOTE

1. Work on this chapter was supported in part by National Science Foundation grant #1726998, Collaborative Research: Attaining Excellence in Secondary Mathematics Clinical Experiences with a Lens on Equity. All findings and opinions are those of the authors, and not necessarily those of the funding agency.

APPENDIX A: SAMPLE DATA COMMITMENT FORM

Description	Semester 1			Semester 2		
	Beginning of Semester	Changes		Beginning of Semester	Changes	
* Program enrolled in — middle/high/secondary — undergrad/grad/postbacc						
* Number of students						
* Demographics of cohort — math background — initial vs. career-changers — race/ethnicity, gender						
Kind of Field Experiences	Describe Plans	Implementation Notes		Describe Plans	Implementation Notes	
* Expected program completion — semester/year						
* Program component — methods course, internship, other						
* Duration and frequency — number of days/weeks — full days vs. length of visit						
* Context/placement — characteristics of school — grade levels — number of placements						

Interventions	Describe Plans	Implementation Notes	Describe Plans	Implementation Notes
* Methods subRAC – which modules – when in the semester				
* CPCT – specific activities or expectations – when in the semester				
* Paired Placement – specific activities or expectations – when in the semester				
Instruments	**Describe Plans**	**Implementation Notes**	**Describe Plans**	**Implementation Notes**
* MCOP[2] – how many times and when				
* MTP Survey – how many times and when				
* Completer Survey – when				

(At the end of the semester)
 – Reflections on lessons learned throughout the semester and adjustments to the plan needed.

Note: Add columns for additional semesters.

REFERENCES

Association of Mathematics Teacher Educators. (2017). *Standards for preparing teachers of mathematics*. Raleigh, NC: Author. Retrieved from http://amte.net/standards

Friend, M., Cook, L., Hurley-Chamberlain, D., & Shamberger, C. (2010). Co-teaching: An illustration of the complexity of collaboration in special education. *Journal of Educational and Psychological Consultation, 20*(1), 9–27.

Goodnough, K., Osmond, P., Dibbon, D., Glassman, M., & Stevens, K. (2009). Exploring a triad model of student teaching: Pre-service teacher and cooperating teacher perceptions. *Teaching and Teacher Education, 25*, 285–296.

Mullis, I. V., Martin, M. O., Foy, P., & Arora, A. (2012). *TIMSS 2011 international results in mathematics*. Amsterdam, The Netherlands: International Association for the Evaluation of Educational Achievement.

Murawski, W. W., & Spencer, S. (2011). *Collaborate, communicate, and differentiate: How to increase student learning in today's diverse schools*. Thousand Oaks, CA: Corwin Press.

National Council of Teachers of Mathematics. Commission on Teaching Standards for School Mathematics. (1991). *Professional standards for teaching mathematics*. Reston, VA: Author.

Sears, R., Brosnan, P., Oloff-Lewis, J., Gainsburg, J., Stone, J., Spencer, C., ... & Andreason, J. (2017). Using improvement science to transform clinical experiences with co-teaching strategies. *Annual perspectives of mathematics education (APME) 2017: Reflective and collaborative processes to improve mathematics teaching* (pp. 265–273). Reston, VA: National Council of Teachers of Mathematics.

Stein, M. K., Engle, R. A., Smith, M. S., & Hughes, E. K. (2008). Orchestrating productive mathematical discussions: Five practices for helping teachers move beyond show and tell. *Mathematical thinking and learning, 10*(4), 313–340.

Stigler, J. W., & Hiebert, J. (2009). *The teaching gap: Best ideas from the world's teachers for improving education in the classroom*. New York, NY: Simon and Schuster.

Weasmer, J., & Woods, A. M. (2003). The role of the host teacher in the student teaching experience. *The clearing house, 76*(4), 174–177.

Zeichner, K. (2010). Rethinking the connections between campus courses and field experiences in college- and university-based teacher education. *Journal of Teacher Education, 61*, 89–99.

SECTION IV

OPPORTUNITIES FOR RECRUITMENT AND RETENTION

This section reviews the literature on recruitment and retention—in the credentialing program and into the profession—and describes what members of the MTE-Partnership community are learning about improvement. Chapter 13 summarizes research on teacher recruitment and retention; reports on relevant practices at MTE-Partnership institutions; and describes the MTE-Partnership's leverage of the Networked Improvement Community research model for improving recruitment and retention at its institutions. Chapter 14 describes early work in the Partnership to provide a resource to members to guide the design of a marketing strategy to attract potential teachers to the career. Having achieved this product, the research team refocused its efforts more broadly on recruitment and retention in a credential program. This work is described in Chapter 15. The final chapter of this section, Chapter 16, presents the emerging work of a research team learning about the retention of secondary mathematics teachers early in their careers. The research reported in this section shows that effective teacher recruitment and retention, more than any other improvement concern, involves multiple stakeholders, not only across the teacher preparation program, but also in partnerships with school districts, schools, and state agencies.

CHAPTER 13

RECRUITMENT AND RETENTION IN SECONDARY MATHEMATICS TEACHER PREPARATION[1]

Ed Dickey, Dana Pomykal Franz, Maria L. Fernandez, and Beth Oliver

The process of teacher recruitment and retention was identified as a primary driver identified by the Mathematics Teacher Education Partnership (MTE-Partnership) as part of the theory of improvement described at the beginning of this book. This chapter summarizes research on teacher recruitment and retention; gives an overview of relevant practices at MTE-Partnership institutions; and describes the MTE-Partnership's strategy for improving recruitment and retention at its institutions. Although the chapter briefly discusses factors such as salary and strategies, its focus is on factors that can be addressed by partnerships between school districts and universities.

The content of this chapter served as the genesis of two Research Action Clusters (RACs) within the MTE-Partnership: one focusing on the recruitment of secondary mathematics teachers and the other focusing on retention of teachers once in the field. The recruitment cluster later expanded its work to include both the diversity of candidates and the retention of the recruited teacher candidates through program completion. The work of these RACs is thoroughly discussed in the following chapters. Each chapter highlights the specific literature that guided their work. The following literature review is the initial review that guided the

The Mathematics Teacher Education Partnership: The Power of a Networked Improvement Community to Transform Secondary Mathematics Teacher Preparation, pages 295–317.

development of the original driver diagram and laid the groundwork for all subsequent work on recruitment and retention.

REVIEW OF LITERATURE

This section reviews research findings and reports on strategies used to recruit or retain teachers and teacher candidates. It works backward from what is known about retention and recruitment of mathematics teachers in schools, to retention of secondary mathematics teacher candidates within preparation programs, to selection and recruitment of students into secondary mathematics teacher preparation programs.

Retention and Recruitment of Mathematics Teachers

Consider what is known about retention of mathematics teachers in schools: A frequently cited statistic is, "Almost a third of America's teachers leave the field sometime during their first three years of teaching, and almost half leave after five years" (National Commission on Teaching and America's Future, 2002). The cost of teacher attrition is considerable. A conservative estimate from the Alliance for Excellent Education (2005) found that every year in the United States, schools spend $2.2 billion to recruit, hire, and train teachers to replace those who left their positions.

This cost estimate for all teachers suggests that the cost of replacing mathematics teachers has been a significant burden for at least a decade. In 1999, 54% of secondary schools had job openings for mathematics teachers and a sizable proportion (22%) reported serious difficulties filling the openings. Although there have since been substantial increases in the number of qualified mathematics teachers, hiring problems remain. Often these problems vary by school rather than by district. Analyses of nationally representative samples find that attrition differences between schools may occur even within the same district (Ingersoll & Perda, 2010), and the "largest variations in overall teacher turnover by location are not between regions, states, or districts but those between different schools, even within the same district" (Ingersoll & May, 2012, p. 456). Attrition and turnover are often most acute for schools serving high-minority and high-poverty populations. Studies of nationally representative samples find that "high-poverty, high-minority, urban and rural public schools have among the highest rates of both attrition and migration of math and science teachers" (Ingersoll & Perda, 2010, p. 588). This attrition is not, however, because these teachers tended to move to districts that were more affluent or had fewer minorities (Ingersoll & May, 2012).

Factors Affecting Attrition. Ingersoll and May's (2012) analysis of data collected in 2004–2005 from a nationally representative sample of teachers (the National Center for Education Statistics Teacher Follow-up Survey) finds that of those who left classroom teaching, one-third retired and one-third took jobs in the educator sector (e.g., publishing, curriculum development, or administration).

Far fewer obtained employment outside of education, enrolled in college and university programs, or left teaching to care for family members. Mathematics and science teachers were not more likely than other teachers to obtain employment outside of education (e.g., to take jobs related to technology as commonly conjectured). Three hypotheses for this finding are: jobs in industry are not plentiful for mathematics and science baccalaureates; jobs in industry are not plentiful for mathematics and science teachers due to their lesser ability; mathematics and science teachers are extremely committed to education (Ingersoll & May, 2012). Another hypothesis is mobility. Although interstate agreements exist,[2] certification and licensure requirements often vary by state and are not invariably reciprocal. A teacher licensed in one state may be required to fulfill additional requirements for licensure in another state.

Factors Affecting Turnover. In analyzing teacher turnover, Ingersoll and his colleagues identify three types of factors: organizational conditions (e.g., principal leadership); school characteristics (e.g., school demographics); and characteristics of individual teachers (e.g., years of experience or type of preparation) (Ingersoll & May, 2012; Ingersoll, Merrill, & May, 2012).

Organizational Conditions. Among the strongest organizational predictors of retention for mathematics teachers was the provision of useful, content-focused professional development, but not, perhaps surprisingly, salary (Ingersoll & May, 2012). For mathematics and science teachers, turnover was lower at schools with better principal leadership.

School characteristics. Minority teachers are overwhelmingly employed in public schools serving high-poverty, high-minority, and urban communities (Ingersoll & May, 2011) and are retained in high-minority school at higher rates than White teachers (Guarino, Santibanez, & Daley, 2006). In addition to increasing the pool of teachers, a diverse teacher workforce arguably has benefits for students. Students assigned to teachers of a similar race or ethnicity have had statistically significant academic achievement gains (Dee, 2005; Goldhaber & Hansen, 2010). Also, by increasing the diversity of teaching professionals, all students benefit by seeing individuals from diverse backgrounds in professional positions, possibly assisting to dispel negative beliefs they may have about minorities in professional careers (Clewell & Villegas, 1998). These professional role models give hope to students from racially and ethnically diverse backgrounds that they too can grow up to assume professional positions in the future (Ahmad & Boser, 2014; Goldhaber & Hansen, 2010).

Teacher Characteristics. Perhaps unexpectedly, type of bachelor's degree (e.g., mathematics or education), was not found to have an effect on attrition. Instead, characteristics that predicted retention involved pedagogical preparation; one was clinical experience (practice teaching as well as opportunities to observe other teachers and receive feedback on their own teaching). A second was what Ingersoll, Merrill, and May (2012) called pedagogy, which included whether a teacher had taken methods courses.

Ingersoll, Merrill, and May (2012) only examined attrition after a year of teaching. The common wisdom is that classroom management is the first hurdle for the beginning teacher, explaining the importance of pedagogical preparation. Consistent with this common wisdom and Ingersoll et al.'s finding, an earlier study found that attrition rates after three years were highest for teachers with the shortest preparation—summer training vs. one or more years (Darling-Hammond, 2000, pp. 19, 37–38).

Factors Affecting Recruitment. An approach that can impact recruitment of mathematics teachers to urban and rural schools, as well as any school, is to streamline the hiring process (National Comprehensive Center for Teacher Quality [NCCTQ], 2007). Highly qualified applicants for teaching positions can be discouraged by the bureaucracy of the hiring process and sometimes find other jobs by the time some schools and districts get around to calling them for an interview. Schools and school systems, particularly those that are hard to staff, cannot afford such an event.

Although states and districts have offered a variety of financial incentives including increased pay to recruit teachers in high-needs areas for the schools that need them, no single incentive has been found successful on a large scale (NCCTQ, 2007). Financial incentives suggested for increased recruitment of teachers in hard-to-staff fields such as mathematics include signing bonuses, higher base starting salary, and housing assistance (Business–Higher Education Forum, 2007; NCCTQ, 2007). Two examples suggest that they can have at least a short-term effect.

Guilford County Public Schools Mission Possible. To address the shortage of teachers in certain subjects (including mathematics) in Guilford County, North Carolina, a differentiated pay structure including staff development for teachers was implemented (Business–Higher Education Forum, 2007). This program was highly successful with 174 applicants applying to teach mathematics in the district the year it was approved as compared with only seven the year before, 87% of the "Mission Possible" participants returned the following year (Klein, 2007; Rowland, 2008; Thomasian, 2011).

Denver, Colorado ProComp. In 2005, the Denver County Public Schools implemented ProComp, a compensation system that includes bonuses for hard-to-serve populations (e.g., socioeconomic status of students) and hard-to-staff subjects (e.g., secondary mathematics). A 2010 evaluation found that schools with greater rates of ProComp participation had 11% higher retention rates (Thomasian, 2011).

Teachers often seek positions at schools close to their hometowns or colleges (Boyd, Lankford, Loeb, & Wyckoff, 2005). Reininger's (2012) analysis of a national dataset found that close to 60% of teachers took positions within 20 miles of the high school from which they graduated. Goldhaber and Cowan (2013) found that over the past 20 years, about one-third of teachers accepted positions within 20 miles of their preparation programs in the state of Washington.

Summary. Although financial considerations are frequently invoked to explain teacher attrition, they are not supported by the findings above. Instead, these findings suggest the importance of partnerships among local stakeholders in the preparation, hiring, and nurturing of teachers as well as the importance of location-specific strategies for preparation programs. Because graduates are likely to come from and return to local schools, colleges and universities that prepare teachers may often be able to place their graduates in local schools, maintain contact with them, and provide them useful, content-based professional development.

Selection and Retention of Teacher Candidates

Next, consider what is known about the retention of mathematics teacher candidates in mathematics teacher preparation programs. Preparation of mathematics teachers varies by state (due to differences in certification and licensure requirements) and by institution (due to differences in program requirements). In some states, teacher candidates are required to obtain a degree in mathematics prior to, or concurrent with, their preparation in methods of teaching. Others require teacher candidates to obtain a degree in mathematics education. Still others add alternative pathways that allow virtually any bachelor's degree holder to become a mathematics teacher. However, there are many commonalities to consider. Research consistently has shown that how teachers are prepared is critical in their success in the classroom (American Association of Colleges for Teacher Education, 2013; Shutz, Crowder, & White, 2001; see also Chapter 1 of this book).

In all teacher preparation programs, the selection and retention of candidates is a critical endeavor. Research has shown that licensure and other exams (e.g., the SAT) may predict performance differently for different cultural groups (Goldhaber & Hansen, 2010), and, moreover, "there is relatively little empirical work linking teacher licensure test scores to student achievement" (p. 2). At the same time, research has also shown that student achievement is increased when the diversity of the classroom is reflected in the diversity of the teaching force (McKenzie, Skria, Scheurich, Rice, & Hawes, 2011). Obtaining a high-quality and diverse teaching force, then, demands that teacher preparation programs consider not only the ability to perform well on specific tests and the predictive power of those tests, but also key attributes which research indicates lead to effective teaching practice including professional dispositions (e.g., integrity, intellectual spirit, stewardship, commitment to justice). Field experiences throughout teacher preparation programs, then, are critical for identifying and retaining candidates who are of high quality, not only academically, but with respect to dispositions, for mathematics classrooms.

An additional and significant issue facing the preparation of high-quality mathematics teachers is the discrepancy between the number of teachers needed to meet the needs of schools nationwide and the number of program candidates entering the teaching profession for the first time each year (Ingersoll & Perda, 2010). Programs that have been successful in retaining and recruiting high-quality candidates to fill high-needs classrooms have often had the benefit of financial incentives including grant-funded stipends, scholarships, and forgivable loans. These incentives allow high-quality candidates to consider the teaching profession by allowing ease of access to education and preparation programs (National Council for Accreditation of Teacher Education, 2010).

Equally important as recruiting high-quality prospective mathematics teachers into teacher preparation programs is retaining individuals with the dispositions to be highly effective teachers. Minority students and those from low-income families often who might be the first in their families to attend college or those with lower quality pre-college education may struggle more in mathematics than their peers. To recruit, prepare, and retain a workforce of teachers who are representative of the diversity in classrooms and who possess the knowledge, skills, and dispositions necessary to be successful in these diverse classrooms, measures must be in place to support prospective teachers with the qualities of effective teachers, but who struggle in college mathematics. One example of a program that helped students who struggle in college mathematics is found in the work of Treisman (1992) and his colleagues, who investigated what caused difficulties for students, particularly Black students, in college calculus. Various hypotheses were investigated, including economics, prior preparation, and motivation. Differing study habits had the highest impact. African American students who studied alone had much higher failure rates than the Chinese American students who studied in groups, shared specifics about their courses, and checked one another's answers. Failure rates for African American students decreased after Treisman and his colleagues offered an intensive workshop program as an accompaniment to calculus courses, with "a challenging, yet emotionally supportive academic environment" (p. 368) that served students of all ethnicities. This example suggests the importance of course and program design in supporting undergraduates who have difficulty in navigating college coursework, as illustrated recently by the findings from the Mathematical Association of America's study Characteristics of Successful Programs in College Calculus[3].

Seymour and Hewitt (1997) offer insights into why undergraduates leave science, technology, engineering, and mathematics (STEM) programs. Their study, "Talking About Leaving," analyzed interviews with undergraduates at seven colleges and universities and found that what students perceived as poor teaching was the most significant influence on STEM majors' decisions to switch fields. A replication of this study began in 2012.

Summary. Selection criteria such as test scores should be carefully considered, particularly for students from underrepresented groups. Two factors that may play a role in retention of teacher candidates are financial incentives and program characteristics. Financial incentives include scholarships and forgivable loans. Program characteristics include the quality of undergraduate instruction, which, in the past, was mentioned by prospective STEM majors as a reason for switching majors. Having students attend an intensive workshop program coordinated with their calculus course is one strategy that has produced improved outcomes.

Selection and Recruitment of Teacher Candidates

Finally, consider what is known about the selection and recruitment of mathematics teacher candidates in mathematics teacher preparation programs.

Recruitment Practices. Target populations and recruitment practices can be classified according to how much time each takes to bring about the preparation and hire of a certified teacher (Clewell, Darke, David-Googe, Forcier, & Manes, 2000).

Marketing may be thought of as a comprehensive process that promotes becoming a teacher and provides information and experiences with a product (e.g., programs for mathematics teacher preparation). Marketing processes may address any of the target populations shown in Table 13.1. One aspect of marketing involves advertising through varied channels such as newspapers and alumni magazines, educational and/or career advisors, college-recruitment specialists, knowledgeable faculty advisors, career fairs, brochures sent to mathematics teachers as ambassadors to share with their current or former students, highway billboards, movie theatre ads, and websites (Reys & Reys, 2004).

Slow Track. Marketing, beyond just advertising, may also involve a slow-track recruitment process of information sharing and "product" trialing. NCCTQ (2007) suggests "grow-your-own" programs including promotional and recruitment efforts beginning as early as middle school that encourage students who excel in mathematics to pursue a career in teaching. In middle and high schools, these types of recruitment programs include experiences such as mentoring by teachers, an introduction-to-teaching course, teaching and tutoring activities, student education clubs, and events or workshops for parents and other family members (U.S. Department of Education, 2012). Such programs were originally implemented to recruit students into teaching careers in any field, but have stimulated programs specific to mathematics. Examples are South Carolina ProTeam, South Carolina Teacher Cadets, and North Carolina Teaching Fellows. Although information exists on numbers of students in such programs, their effects on students' choice of major have not been studied. Other evidence, however, supports the design of these programs. Students choose to become teachers because they enjoy helping others, sharing information, and interacting socially (Barker & Reyes, 2001).

TABLE 13.1. Recruitment Classifications

Name	Target Population	Years to Certification
Slow track	Students in Grades 6–12	5–8
Moderate track	Undergraduates	3–4
Fast track	Baccalaureates, paraprofessionals, etc.	0–2

Positive experience with teaching, and teachers, peers, and/or family, are all influences that have been found to affect the choice to become a teacher (Shutz, Crowder, & White, 2001).

Moderate Track. Similar programs are directed to college students, encouraging them to pursue education programs and teacher certification. As with the slow-track approaches, these college-level programs include experiences that provide opportunities for non-education majors to become familiar with becoming a teacher through courses, mentoring, teaching activities, cohort-based opportunities, and informational events for undergraduates and (where appropriate) their parents or guardians. As part of these programs, undergraduates are also offered financial support through stipends and scholarships, as well as academic support for success in the program. These programs provide opportunities for undergraduates to participate in trial teaching through an introductory course on learning and teaching and related teaching experiences.

Examples of successful programs that incorporate some or all of these features include UTeach, Learning Assistant programs for STEM teachers, the Robert Noyce Teacher Scholarship Program funded through the National Science Foundation, and Call Me MISTER (Mentors Instructing Students Toward Effective Role Models) designed to recruit Black men as elementary teacher candidates.

UTeach. This program began in 1997 at the University of Texas at Austin. UTeach students receive both their undergraduate degree in a STEM field and their teaching certificate in four years. One enticement to STEM majors to consider this dual enrollment program is that the first two education courses are offered free of charge. Approximately 88% of UTeach graduates enter the teaching profession and 80% of UTeach graduates who begin teaching are still in schools five years later.

Learning Assistant (LA). This program began at the University of Colorado. Since then, the pool of well-qualified K–12 physics teachers has increased by a factor of 3 or more, and the program has engaged scientists significantly in the recruitment and preparation of future teachers (Otero, Pollock, & Finkelstein, 2010). In a study of how mathematics teacher candidates were recruited from academically strong mathematics undergraduates, an LA program was found to confirm and maintain interest in becoming a teacher when positive prior teaching experiences, family, and peer influences were part of a student's goal history, as well as when the program provided positive teaching experiences and peer influences that supported the recruitment into mathematics teaching of a student who was intrinsically motivated but unsure about becoming a teacher (Fineus & Fernandez, 2012).

Noyce Scholarship Program. This program has been underwritten by the National Science Foundation since 2002 with the goal to provide scholarships and stipends to students entering a STEM teaching field in high-needs areas. The American Association for the Advancement of Science is working with the Noyce program to disseminate successful programs and recruitment strategies. Many notable programs across the U.S. have been funded such as Eastern Michigan's *Developing Urban Education Teachers in STEM classrooms.*

Call Me MISTER. This program has increased the percentage of Black male teachers at elementary schools in South Carolina by 40% over the last 12 years (American Association of Colleges for Teacher Education, 2013).

Fast Track. Fast-track recruitment targets various populations such as paraprofessionals, military personnel, and professionals with STEM backgrounds. Teach for America (TfA) is a well-publicized example. Research examining TfA outcomes over the last decade suggests that TfA recruits had a positive effect on student achievement in mathematics when they had received training and certification beyond their typical two-year commitment (Heilig, Cole, & Springel, 2010). However, studies find that the attrition rate of TfA teachers is high: 80% or more by the fourth year of teaching (Boyd et al., 2012; Heilig, Cole, & Springel, 2010). Donaldson and Johnson (2011) found that only 27.8% of TfA teachers remain in teaching after five years and only 14.8% stay in the school where they were originally assigned.

Many of the institutions involved in the MTE-Partnership include fast-track options. In California, for example, teacher credential programs are generally post-baccalaureate programs.[4] Admittance to these programs requires that students complete a waiver program or pass exams to show subject matter competency. The survey described in the next section of this chapter found that preparation programs at Partnership institutions include options for *career-changers,* individuals who have earned a bachelor's degree and later decided to pursue teacher certification or licensure. Many of these programs are connected with state alternative certification options or national efforts such as Math for America or Woodrow Wilson Fellowships.

In general, traditionally prepared teachers are more likely to remain in the field than alternatively certified teachers (Alt & Henke, 2007; Heilig, Cole, & Springel, 2010). However, alternative certification pathways vary (National Research Council, 2010), and differences in alternative preparation pathways have been associated with differences in retention. As mentioned earlier, more teaching preparation is associated with higher rates of teacher retention (Darling-Hammond, 2000; Ingersoll, Merrill, & May, 2012).

Financial Incentives. Financial incentives, such as scholarships, loan-forgiveness programs, and grants, have a long history and are still the prevailing recruitment strategies. These incentives have been drawn on and used to market becoming a mathematics teacher and as part of comprehensive recruitment programs within the slow, moderate, and fast tracks. The NCCTQ (2007) recommends that formal recruitment programs targeted at high school students should have financial assistance toward certification in place along with other recruitment components that provide encouragement, mentoring, and training.

The Noyce Scholarship Program begun in 2002 is a familiar source of financial incentives in the forms of scholarships or stipends for students choosing to pursue mathematics or science teaching as a career. Noyce projects are able to target undergraduate and graduate students (both education and non-education majors), mid-career professionals, paraprofessionals and so on, thus contributing to moderate- and fast-track recruitment. Some Noyce projects target students with strong STEM backgrounds as potential mathematics or science teachers at high-needs schools (Liou & Lawrenz, 2011). Survey responses from principal investigators indicate that Noyce funding greatly increases their ability to recruit a wide variety of individuals (Liou, Desjardins, & Lawrenz, 2010). Also attractive in recruitment is that Noyce provides funds for mentoring of these teachers into the first years of their teaching. In recent years, Noyce has begun focused research into promising practices of recruitment and financial incentives. Noyce scholarships have been offered as part of programs such as LA and UTeach.

Although scholarships and loan forgiveness programs have been used as teacher recruitment strategies since the National Defense Education Act of 1958, research supporting their effectiveness is not extensive (Liou, Kirchoff, & Lawrenz, 2010). Surveys of Noyce scholarship recipients suggest that these financial incentives were, in general, not highly influential in decisions to become teachers. Only 3.5% indicated that becoming a teacher was contingent on funding (Liou & Lawrenz, 2011). However, it was perceived as more influential by non-Whites than Whites and by career-changers than those preparing for a first career (Liou & Lawrenz, 2011). For non-Whites and career-changers, the funding was perceived as more influential when it covered more of the tuition costs. Given that career changers and non-Whites traditionally have higher rates of retention as teachers and in high-needs schools (Guarino, Santibanez, & Daley, 2006; Liou & Lawrenz, 2011), these findings suggest strategies for future scholarship funding. Nonetheless, further study is needed to examine Noyce scholar retention, satisfaction with the Noyce program, and perceptions about teaching in high-needs schools over time as scholars complete their contractual obligations and decide whether to remain in teaching (Liou, Desjardins, & Lawrenz, 2010). Research is also needed to examine the effects of other scholarship incentives, such as the underutilized federal Teach Grants for individuals who want to teach in high-needs fields and schools and the Teacher Quality Partnership Grants in Title II of the Higher Education Act, which support one-year clinical preparation for candidates (National Council for Accreditation of Teacher Education, 2010).

Partnerships for Recruitment. Through collaboration, districts or schools and universities can increase the pipeline of mathematics teachers and assure that these teachers receive the education and experiences that prepare them for the content they will teach as well as the students they will serve. District or school partnerships with universities can invaluably benefit the recruitment approaches and strategies discussed previously. Clearly the relationship would only add to the marketing approaches for recruitment by reaching a broader base through the partnership. Alternative routes to certification are often based in local colleges or universities (NCCTQ, 2007). Moreover, early-college and dual enrollment systems, summer bridge programs, and collaborations that help align and maintain alignment of high school and college mathematics courses for students' success in college can help increase the success and recruitment of students into teacher education programs (NCCTQ, 2009).

Summary. Recruitment activities range from fast-track efforts that target professionals and career-changers to slow-track efforts focused on middle and high school students. They may include financial incentives such as scholarships and loan forgiveness for students and salary or signing bonuses for new or practicing teachers. The efficacy of the latter has not been well studied, aside from short-term local initiatives. However, responses from Noyce scholarship recipients suggest that scholarships exert greater weight for non-White students and career-changers.

Retention of undergraduates includes retention and support in mathematics courses. In the past, poor teaching in undergraduate courses has been a major factor for attrition of mathematics majors, including teaching majors (Seymour & Hewitt, 1997). Mathematics workshops for undergraduates have helped to retain students who are traditionally underrepresented (Treisman, 1992).

Entry into a teacher education program may involve selection criteria such as grades and test scores. Because these may predict future success—in college or in teaching—differently for different cultural groups, the use of these and other measures (including licensure test scores) needs to be monitored via appropriate data collection and analysis.

Retention of practicing teachers is affected by numerous factors. Those that can be affected by preparation programs include attention to pedagogical preparation and quality of field and internship experiences, as well as initial placement of newly qualified teachers. Opportunities for professional learning particularly during the first years of teaching are also important. These may be workshops, or school-based and online professional learning and growth opportunities.

FINDINGS FROM THE MTE-PARTNERSHIP SURVEY

In March 2013, the MTE-Partnership planning group conducted a survey of the institutional teams. A total of 39 Partnership teams were represented in the responses with the team leader typically responding for the team and describing the program at the lead institution. This section gives an overview of findings related to teacher candidate recruitment and retention. See Appendix A in this chapter for statistical details.

The findings below should be considered illustrative rather than representative of the MTE-Partnership as a whole. These data are intended to provide a picture of the landscape as described by MTE-Partnership institutions at its inception. It should be noted that not all MTE-Partnership institutions responded to the 2013 survey, and more institutions have since joined the MTE-Partnership.

- Program characteristics: Approximately 70% of respondents required a mathematics major. The remaining respondents described requirements that are equivalent to or nearly the same as a mathematics major. Some programs include options for middle grades certification as well as variations motivated by state policies or regulations. As noted earlier, most MTE-Partnership programs include post- baccalaureate/fast-track options for career-changers.
- Recruitment challenges: The most significant challenge related to recruitment by Partnership programs centered on recruiting mathematics majors, but other factors were frequently mentioned. Recruiting diverse candidates was also cited as a challenge.
- Tracking and support for graduates: Although most respondents tracked candidate progress toward degree completion, about one-third did not do this systematically. Rigorous mathematics requirements and disenchantment with teaching as a career were mentioned as the factors that contributed to candidates not completing programs. Other reasons cited were lack of interpersonal skills and complicated state certification or documentation requirements.

Few programs (30%) in 2013 tracked graduates into the first three years of teaching. A similar number provided assistance to graduates as they enter the teaching profession. Many respondents mentioned informal tracking and advisors who stayed in touch with graduates, but few described a formal process. Some states required that assistance be provided, and several programs provided assistance informally or through advanced degree programs. With the adoption of standards from both the Council for the Accreditation of Education Professionals (CAEP, 2013) and the Association of Mathematics Teacher Educators (AMTE, 2017), institutions are now challenged to gather and maintain data on how program completers perceive their preparation and the satisfaction of employers with program completers' preparation.

Of the programs that do track graduates into the first years of teaching, frustration with school related factors was cited as the most common reason for leaving the profession, followed closely by low salaries for teachers. One respondent noted that some new teachers leave teaching to start a family but later return to the profession.

The survey ended with six statements from the Mathematics Teacher Education Assessment used during the initiation of the Partnership, a request for respondents to rate the degree to which each was being addressed, and to comment. Approximately 70% of respondents disagreed that programs were "resourceful in recruiting talented mathematics teacher candidates from a variety of sources" while about 60% disagreed that their program "actively recruits diverse and underserved populations into mathematics teacher preparation." Most (70% to 90%) agreed that their programs:

- Have admissions criteria that focus on candidates with genuine interest in teaching mathematics.
- Screen "candidates who are not only academically accomplished, but also demonstrate interest, persistence, high expectations, and success in working effectively with learners."
- "Set standards for program admission that require candidates in mathematics education, without exception, to demonstrate strong performance in college course work, especially in mathematics content areas."
- Provide "academic support through faculty availablity, student tutors, peer support, online services, or other resources to ensure that teacher candidates possess the necessary background in mathematics content to improve student success in mathematics."

In summary, mathematical preparation equivalent to that of an undergraduate major is a requirement in programs at many lead MTE-Partnership institutions. Most say that recruitment of mathematics majors to become teachers is challenging. Perceptions about teaching as a career was also mentioned as a recruitment and program retention obstacle. Few programs track the success of their graduates into their teaching careers or get any systematic information about program strengths and weakness from graduates after they enter the teaching profession. Of the information that is received, most is secured informally or anecdotally.

THEORY OF IMPROVEMENT FOR RECRUITMENT AND RETENTION

This section of this chapter describes a theory of improvement developed to address the problem of recruitment and retention of secondary mathematics teacher candidates. First, factors posited to address the problem are presented in a driver diagram. Second, suggested measures and data useful in measuring progress relative to the problem are discussed. Lastly, potential interventions addressing the problem are presented.

Driver Diagram

Using the literature review and survey responses from MTE-Partnership institutions, a working group posed the recruitment and retention problem illustrated by Figure 13.1. The general improvement target is given in the left-most column; this target is a primary driver identified by the MTE-Partnership as a whole.

The primary drivers are elements that are likely to create movement toward the improvement target—attract and maintain an adequate supply of secondary school mathematics teacher candidates. The existence of a marketing plan that is both *purposeful* and *sustainable* was hypothesized as a secondary driver. Here, purposeful means addressing local needs, demographics, problems, and situations. Both the literature review and the MTE-Partnership's collective experience indicated that

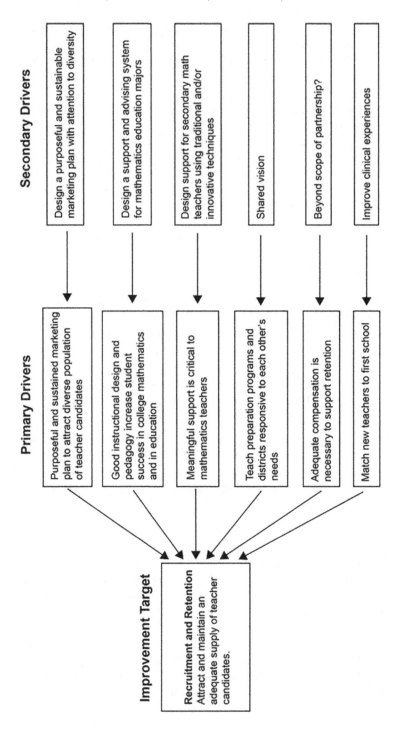

FIGURE 13.1. Driver Diagram for Recruitment and Retention. In This Driver Diagram for Recruitment and Retention, the General Improvement Target Is Given at Left and This Target Is a Primary Driver Identified by the MTE-Partnership as a Whole.

strategies that succeed at one institution will not necessarily be successful at others; therefore, consideration of local factors was hypothesized to be essential. The literature also pointed to a long history of teacher recruitment efforts that were initiated with funding but terminated when funding ended. A marketing plan with features that allow institutions to sustain the effort was hypothesized as equally important. Also important was an operational definition of diversity that was sensitive to local and institutional demographics. For example, goals for diversifying programs to include more candidates who are African American may not be realistic for an institution serving a state or region in which a small percentage of the population is African American. Similarly, Hispanic or Latino/a populations vary significantly across regions and also change dramatically over time, so programs are encouraged to examine demographic data for their state and region so as to set recruitment and retention goals that document who the program completers are, representative of the student population of the state or region served.

Retaining teacher candidates through program completion was considered another secondary driver. Here the group concluded that good instructional design and pedagogy, both in terms of courses, clinical experiences, and the overall structure of the teacher education program comprised the key drivers to program retention. Research on program retention suggests that this driver requires collaboration between mathematics and education faculty as well as cooperation and partnership with schools and districts so as to develop meaningful clinical experiences, building on a common vision (see Chapter 1 of this book).

Retention in the profession is a complicated and multifaceted problem with responsibilities and expectations that cross various stakeholder groups. The key driver for retention was hypothesized to be the support systems in place for practicing secondary mathematics teachers. Literature describing the fostering of professional learning communities describe both traditional and innovative techniques for supporting teachers in a manner that is likely to impact retention positively. The interplay among the teacher candidates themselves, the teacher education program developing the candidates, and the school and district employing the teachers, impacts the program and potential improvements. Items such as teacher salaries, administrative support, hiring practices, and personal preferences were suggested as requiring attention when addressing retention. Also important and related to clinical experiences is the possibility of matching teacher candidates' preferences, philosophical stances, and dispositions to that of the first school at which they are employed.

Suggested Data and Measures

This section lists types of data and measures specific to measuring improvement in the recruitment and retention problem depicted in the driver diagram. In order to design and implement effective recruitment and retention plans, schools, districts, and states must have reliable, well-disaggregated data. Descriptions of key resources for collecting and analyzing such data are included in a 2009 NCCTQ report.

CAEP (2013) recognizes its responsibility in working with states and the Council Chief State School Officers to assist providers in developing needed data-gathering and reporting capacities, while stating that program providers are responsible for maintaining a system of ongoing data collection and reporting.

Program Recruitment. Measures for program recruitment include numeric data on actual candidates: those who express an interest in the program, those who are admitted, and those who complete the program. Also important are measures of quality, particularly those related to the CAEP "high quality" expectation: scores on standardized tests (e.g., GRE, Praxis) as well as prior and current grade-point averages in course work in mathematics, education, and in general. Data on candidate ethnicity, gender, and race are also critical to meeting diversity goals.

Because needs for teachers vary by location, state or local data about secondary mathematics teacher shortages should be gathered and considered as part of assessing the effectiveness of recruitment efforts. Evidence of collaboration with school districts and other entities serve as indicators of awareness of employment needs or opportunities. Similarly, local data on financial aid and forgivable loans are a critical component to teacher recruitment and must also be gathered and examined.

The quality of recruitment efforts often depends on the financial and human resources required to implement recruitment strategies, so data on funds and staff time needed for recruitment are important. Lastly, evidence of marketing and recruitment directed at high schools and colleges that are racially and culturally diverse should be gathered.

Program Retention. Measures for retention within teacher preparation programs include numerical measures of those meeting progression requirements (e.g., grade-point average, test, key assessments) as well as graduation rates, i.e. program completers, and measures of disposition. Data on support structures, such as scholarships, internship, student loans, student organizations and cohorts, preparation assistance for licensing examinations, availability of tutoring or supplemental instruction, and quality of advisement, are also valuable for assessing program retention effectiveness.

The many characteristics of clinical experiences such as the timing, length, and diversity of the experience; the matching process of the candidate to a school; and the experience embedded within a professional development school network provide useful data for assessing program retention goals tied to the critically important field experiences. Other factors are related to the mentor teacher: mathematical knowledge, receptivity to teacher candidate, willingness to grant autonomy to a teacher candidate, provision of frequent feedback to the teacher candidate, effective classroom management, and creative lesson planning. Disposition data can include evidence of actions related to academic or professional misconduct but should also include outcomes of assessments tied to integrity, intellectual spirit, professionalism, and a sense of justice as demonstrated by actions and behaviors.

Retention Within Profession. Measures for retention within the teaching profession include graduation rates and rates at which graduates receive state licensure or certification. Equally important are data on the actual hiring as secondary school mathematics teachers, the salary received, the type of position held, and type school of employment (middle or high school), as well as the location. Details about whether the hiring school was the candidate's first choice, whether the accepted position was based on a sole offer, and how many offers the candidate received help assess the degree of job availability. Also useful are data on incentives that may have been provided during the first year of teaching (e.g., signing bonus, differential pay, reduced teaching load, aides, technology resources, loan forgiveness, scholarship payback requirements, housing assistance, professional development funding, supplies, etc.). Information on whether the candidate is involved with informal or formal mentoring programs or part of a teacher induction process should be gathered. If a mentor is assigned, noting whether the mentor is a mathematics teacher or a general mentor is also useful.

More specific to retention, school demographic (e.g., urban, rural, suburban, serving high needs, and ethnic/racial diversity, student achievement) and geographic data will help the program measure its impact. Data on the local economy including overall teacher vacancies provide more context information as do data on the school's effectiveness or the local community's support of teachers and education.

The candidate's satisfaction with the position, their sense of preparedness, career trajectory, and being welcomed by the new school all inform variables tied to teacher retention and serve as useful measures. Also useful to note are school structures or programs likely to impact retention positively. These might include teacher perception of administrative support including effectiveness with school discipline, effectiveness of professional development, teaching load (number of courses and different preparations), expectations for extracurricular activities (committees, leadership, clubs), presence of a faculty appreciation or social committee, and participation in team planning. Also, information on whether the teacher continued to be in contact with faculty or leaders in or peers graduating from the teacher preparation program can be useful.

Finally, teachers who leave the profession could be surveyed with respect to the impact (if any) of their preparation programs on their decision to leave teaching.

Next Steps: Research Action Clusters

Based on the literature reviewed and information from the MTE-Partnership survey, a working group developed three possible Research Action Clusters (RACs). Each RAC addresses one of the primary drivers shown in the driver diagram: recruitment, retention within teacher education programs, and retention within profession.

MATH: Marketing for Attracting Teacher Hopefuls. This RAC initially focused on developing a purposeful and sustained plan for recruitment rather than the frequently spotty, temporary, or haphazard efforts that are not based on research evidence. Interventions demonstrated an understanding of marketing and were created within a framework that ensured that the campaign was purposeful and sustainable. Teams tested and continue to test a range of interventions in different markets and targeting different groups to determine factors in their success. Beyond just advertising, the campaigns included, and continue to include, attention to known persuasive influences (peers, family, teachers, teaching experiences) and will include consultation with multiple stakeholders to understand the market and its reality. An important outcome of the MATH RAC was the creation and dissemination of a Secondary Mathematics Teacher Recruitment Campaign Implementation Guide (Ranta & Dickey, 2015) made available to AMTE and MTE-Partnership members online, http://bit.ly/MATHImplGuide.

Sustained commitment to local school districts is also seen as an important marker for recruitment including how to match teacher candidates to schools and districts. Measures for success included pre- and post-data on enrollments, as well as demographics, of mathematics teacher candidates. Interim mechanisms were and will be developed for collecting feedback from stakeholders and teacher candidates.

Keeping the Value: Retaining High-Quality Mathematics Teacher Candidates. This potential RAC was suggested to design a support and advisement system for mathematics education majors. The drivers tied to this RAC were combined with recruitment drivers into the Program Recruitment and Retention (PR²) RAC (see Chapter 15 of this book), established in June of 2017. In addition to focusing on the recruitment of diverse secondary mathematics teachers, the PR² RAC also seeks to support candidate retention in, and completion of, credential programs, which is especially important when working with candidates of color, as they may have had negative experiences in the public education system themselves and may receive pressure from family members to seek out careers with higher pay or status (Partelow, Spong, Brown, & Johnson, 2017).

Survivors to Thrivers: Meaningful Support to Retain Secondary Mathematics teachers. This potential RAC was suggested to provide support for first-year secondary mathematics teachers through induction programs, mathematics learning communities, partnerships, job-embedded professional development, online communities, systems for reward and compensation, and systems of feedback and growth. In June 2015 the MTE-Partnership established the Secondary Teacher Retention and Induction in Diverse Education Settings (STRIDES) RAC using some of the drivers suggested within the Survivors to Thrivers concept. STRIDES focuses specifically on new teacher induction, both through mentoring and helping administrators to better support new teachers.

Measuring success will involve collecting data on teacher retention for one, three, and five years after graduation; teacher leadership and professional growth; and teacher and employer satisfaction. Relationships of interest are the impact of professional learning on student outcomes and classroom practices and the effects of traditional or innovative methods for support initiatives.

APPENDIX A. STATISTICS FROM THE PARTNERSHIP SURVEY

Current and Projected Secondary Mathematics Teacher Candidate Graduates

Year	Total	Range	Mean	Median	10 or Fewer
2011–2012 (actual)	432	0–50	13.5	10	20
2012–2013 (projected)	521	0–48	15.8	14	13

Which of the following best describes your secondary mathematics teacher certification program?

	Percent	Number
Requires:		
bachelor's degree in mathematics	72.4%	21
bachelor's degree in education	37.9%	11
master's degree in mathematics	17.2%	5
master's degree in education		0
other (please specify)		15
Number of institutions responding		29

Which of the following challenges does your program face related to recruiting secondary mathematics teacher candidates? (check all that apply)

recruiting math majors to pursue teacher certification	80.0%	24
offering financial assistance to those applying	60.0%	18
finding applicants with required academic qualifications	60.0%	18
finding applicant with dispositions appropriate for teaching	56.7%	17
Other (please specify)		8
Number of institutions responding		30

Does your program systematically track teacher candidates' progress after admission and to graduation?

Yes	65.7%	23
No	34.3%	12
Number of institutions responding		35

Which of the following challenges does your program face related to retaining secondary mathematics teacher candidates through program completion? (check all that apply)

Lose candidates because of:		
rigorous mathematics requirements	75.0%	24
rigorous education requirements	18.8%	6
rigorous internship or student teaching requirements	21.9%	7
financial burden of program participation	28.1%	9
disenchantment with teaching as a career	6.0%	20
Other (please specify)		11
Number of institutions responding		32

	Percent	Number
Does your program track program graduates into the first 3 years of teaching?		
Yes	24.2%	10
No	69.7%	23
Number of institutions responding		33
Which of the following challenges do graduates of your program face as secondary mathematics teachers within the teaching profession over the first 3 years? (check all that apply)		
Our program does not track graduates over the first 3 years of their teaching	74.1%	20
We lose induction year teachers because they:		
do not perform to the level expected by employing districts	3.7%	1
find better paying jobs in other sectors	22.2%	6
find more satisfying jobs in other sectors	14.8%	4
become frustrated by school related factors	33.3%	9
do not find mathematics teaching fulfilling	7.4%	2
Other (please specify)		9
Number of institutions responding		35

ENDNOTES

1. This chapter is a revision of a white paper originally authored by Ed Dickey, Beth Oliver, and Maria L. Fernandez. Dana Pomykal Franz was a lead contributor to the revised paper, with contributions from Maria L. Fernandez, Julie McNamara, James Martinez, and Lisa Amick. Janet Andreasen and Stacy Reeder contributed to earlier drafts of the original white paper.

2. See the National Association of State Directors of Teacher Education and Certification listing of interstate agreements at https://www.nasdtec.net/page/Interstate.

3. For details, see http://launchings.blogspot.com/2014/01/maa-calculus-study-seven.html.

4. There are a few exceptions, but the majority of colleges and universities in California currently require a bachelor's degree prior to entering a credential program.

REFERENCES

Ahmad, F. Z., & Boser, U. (2014). *America's leaky pipeline for teachers of color: Getting more teachers of color into the classroom.* Washington, DC: Center for American Progress.

Alliance for Excellent Education. (2005). *Teacher attrition: A costly loss to the nation and to the states.* Washington, DC: Author.

Alt, M., & Henke, R. (2007). *To teach or not to teach? Teaching experience and preparation among 1992–93 bachelor's degree recipients 10 years after college* (NCES 2007-163). Washington, DC: National Center for Education Statistics.

American Association of Colleges for Teacher Education. (2013). *The changing teacher preparation profession: A report from AACTE's professional education data system.* Washington, DC: Author.

Association of Mathematics Teacher Educators. (2017). *Standards for preparing teachers of mathematics.* Retrieved from amte.net/standards.

Barker, S., & Reyes, R. (2001). Why be a science teacher? In N. Valanides (Ed.), *Science and technology education: Preparing future citizens, Proceedings of the IOSTE Symposium in Southern Europe* (pp. 57–68). Paralimni, Cyprus: University of Cyprus.

Boyd, D., Grossman, P., Hammerness, K., Lankford, H., Loeb, S., Ronfeldt, M., & Wyckoff, J. (2012). Recruiting effective math teachers: Evidence from New York City. *American Educational Research Journal, 49*(6), 1008–1047.

Boyd, D., Lankford, H., Loeb, S., & Wyckoff, J. (2005). The draw of home: How teachers' preferences for proximity disadvantage urban schools. *Journal of Policy Analysis and Management, 24*(1), 113–132.

Business–Higher Education Forum. (2007). *An American imperative: Transforming the recruitment, retention, and renewal of our nation's mathematics and science teaching workforce.* Washington, DC: Author.

Clewell, B. C., Darke, K., David-Googe, T., Forcier, L., & Manes, S. (2000). *Literature review on teacher recruitment programs.* Washington, DC: U.S. Department of Education.

Clewell, B. C., & Villegas, A. M. (1998). Diversifying the teaching force to improve urban schools: Meeting the challenge. *Education and Urban Society, 31*(1), 3–17.

Council for the Accreditation of Educator Preparation. (2013). *CAEP accreditation standards and evidence: Aspirations for educator preparation—Recommendations from the CAEP Commission on Standards and Performance Reporting to the CAEP Board of Directors.* Washington, DC: Author.

Darling-Hammond, L. (2000). Teacher quality and student achievement: A review of state policy evidence. *Educational Policy Analysis Archives, 8*(1), 1–44.

Dee, T. S. (2005). A teacher like me: Does race, ethnicity, or gender matter? *The American Economic Review, 95*(2), 158–165.

Donaldson, M. L., & Johnson, S. M. (2011). Teach for America teachers: How long do they teach? Why do they leave? *Phi Delta Kappan, 93*(2), 47–51.

Fineus, E., & Fernandez, M. L. (2012). An investigation of participants' perspectives about a Learning Assistant program and their thinking about becoming a mathematics teacher. In M. S. Plakhotnik, S. M. Nielsen, & D. M. Pane (Eds.), *Proceedings of the 11th Annual College of Education & GSN Research Conference* (pp. 54–61). Miami, FL: Florida International University.

Goldhaber, D., & Cowan, J. (2013). *Excavating the teacher pipeline: Teacher training programs and teacher attrition.* CEDR Working Paper 2013-5. Seattle, WA: University of Washington.

Goldhaber, D., & Hansen, M. (2010). Race, gender, and teacher testing: How informative a tool is teacher licensure testing? *American Educational Research Journal, 47*(1), 218–251.

Guarino, C., Santibanez, L., & Daley, G. (2006). Teacher recruitment and retention: A review of the recent empirical literature. *Review of Educational Research, 76*(2), 173–208.

Heilig, J. V., Cole, H. A., & Springel, M. A. (2010). Alternative certification and the Teach for America: The search for high quality teachers. *Kansas Journal of Law & Public Policy, 20*, 388–412.

Ingersoll, R., & May, H. (2012). The magnitude, destinations, and determinants of mathematics and science teacher turnover. *Educational Evaluation and Policy Analysis, 34*(4), 435–464.

Ingersoll, R., & May, H. (2011). *Recruitment, retention and the minority teacher shortage* (CPRE Research Report #RR-69). Philadelphia, PA: Consortium for Policy Research in Education.

Ingersoll, R., Merrill, L., & May, H. (2012). Retaining teachers: How preparation matters. *Education Leadership, 69*(8), 30–34.

Ingersoll, R., & Perda, D. (2010). Is the supply of mathematics and science teachers sufficient? *American Educational Research Journal, 47*(3), 563–594.

Klein, G. (2007, September 14). Congress proposes higher pay for urban teachers. *Media General News Service.* Retrieved from http://www1.gcsnc.com/good_news/pdf/wsls.pdf

Liou, P., Desjardins, C., & Lawrenz, F. (2010). Influence of scholarships on STEM teachers: Cluster analysis and characteristics. *School Science and Mathematics, 110*(3), 128–143.

Liou, P., Kirchoff, A., & Lawrenz, F. (2010). Perceived effects of scholarships on STEM majors' commitment to teaching in high need schools. *Journal of Science Teacher Education, 21*, 451–470.

Liou, P. Y., & Lawrenz, F. (2011). Optimizing teacher preparation loan forgiveness programs: Variables related to perceived influence. *Science Education, 95*(1), 121–144.

Mathematics Teacher Education Partnership. (2014). *Guiding principles for secondary mathematics teacher preparation.* Washington, DC: Association of Public and Land-grant Universities. Retrieved from mtep.info/guidingprinciples

McKenzie, K. B., Skria, L., Scheurich, J. J., Rice, D., & Hawes, D. P. (2011). Math and science academic success in three large, diverse, urban high schools: A teachers' story. *Journal of Education for Students Placed at Risk, 16*(2), 100–121.

National Commission on Teaching and America's Future. (2002). *Unraveling the "teacher shortage" problem: Teacher retention is the key.* Retrieved from http://www.ncsu.edu/mentorjunction/text_files/teacher_retentionsymposium.pdf

National Comprehensive Center for Teacher Quality. (2007). *Recruiting quality teachers in mathematics, science, and special education for urban and rural schools.* Washington, DC: Author.

National Comprehensive Center for Teacher Quality. (2009). *Key issue: Recruiting science, technology, engineering, and mathematics (STEM) teachers.* Washington, DC: Author.

National Council for Accreditation of Teacher Education. (2010). *Transforming teacher education through clinical practice: A national strategy to prepare effective teachers. Report of the Blue Ribbon Panel on Clinical Preparation and Partnerships for Improved Student Learning.* Washington, DC: Author.

National Research Council. (2010). *Preparing teachers: Building evidence for sound policy.* Committee on the Study of Teacher Preparation Programs in the United States, Center for Education. Division of Behavioral and Social Sciences and Education. Washington, DC: National Academies Press.

Otero, V., Pollock, S., & Finkelstein, N. (2010). A physics department's role in preparing physics teachers: The Colorado Learning Assistant model. *American Journal of Physics, 17*(11), 299–308.

Partelow, L., Spong, A., Brown, C., & Johnson, S. (2017). *America needs more teachers of color and a more selective teaching profession.* Washington, DC: Center for American Progress. Retrieved from https://www.americanprogress.org/issues/education-k-12/reports/2017/09/14/437667/america-needs-teachers-color-selective-teaching-profession/

Ranta, J., & Dickey, E. (2015). *Secondary mathematics teacher recruitment campaign implementation guide.* Washington, DC: Association of Public and Land-grant Universities. Retrieved from http://bit.ly/MATHImplGuide

Reininger, M. (2012). Hometown disadvantage? It depends on where you're from. Teachers' location preferences and the implications for staffing schools. *Educational Evaluation and Policy Analysis, 34*(20), 127–145.

Reys, B. J., & Reys, R. E. (2004). Recruitment of future mathematics teachers—An action plan by one university. *Mathematics Teacher, 97*(2), 92–95.

Rowland, C. (2008). *Mission possible: A comprehensive teacher incentive program in Guilford County, North Carolina.* Washington, DC: Center for Educator Compensation Reform. Retrieved from http://cecr.ed.gov/guides/summaries/GuilfordCounty-CaseSummary.pdf

Seymour, E., & Hewitt, N. M. (1997). *Talking about leaving: Why undergraduates leave the sciences.* Boulder, CO: Westview Press.

Shutz, P. A., Crowder, K. C., & White, V. E. (2001). The development of a goal to become a teacher. *Journal of Educational Psychology, 93*(2), 229–308.

Thomasian, J. (2011). *Building a science, technology, engineering, and math education agenda: An update of state actions.* Washington, DC: National Governors Association Center for Best Practices.

Treisman, U. (1992). Studying students studying calculus: A look at the lives of minority mathematics students in college. *College Mathematics Journal, 23*(5), 362–372.

U.S. Department of Education, Office of Postsecondary Education. (2012). *Teacher shortage areas nationwide listing 1990–1991 through 2012–2013.* Washington, DC: Author.

CHAPTER 14

MARKETING STRATEGIES FOR ATTRACTING PROSPECTIVE SECONDARY MATHEMATICS TEACHERS

Maria L. Fernandez

The Marketing to Attract Teacher Hopefuls (MATH) Research Action Cluster (RAC) was born out of a disciplined process within the Mathematics Teacher Education Partnership (MTE-Partnership) Networked Improvement Community (NIC). The MTE-Partnership was formed in 2012 under the auspices of the Association of Public and Land-grant Universities to address major challenges facing secondary mathematics teacher preparation. Of particular importance was the significant shortage of well-prepared secondary mathematics teachers, both in terms of quantity (cf. Ingersoll, Merrill, & May, 2012; Ingersoll & Perda, 2010) and quality of instruction (cf. Banilower et al., 2013). Based on survey data collected from the MTE-Partnership and the *Guiding Principles for Secondary Teacher Preparation Programs* originally created in 2012 (MTE-Partnership, 2014), recruitment of secondary mathematics teacher candidates into teacher preparation programs, along with retention in the programs, was deemed of high importance. More recently, the Association of Mathematics Teacher Educators (AMTE) in the *Standards for Preparing Teachers of Mathematics* (2017) included recruitment

The Mathematics Teacher Education Partnership: The Power of a Networked Improvement Community to Transform Secondary Mathematics Teacher Preparation, pages 319–336.
Copyright © 2020 by Information Age Publishing
319

and retention of teacher candidates as one of five key standards for preparing teachers of mathematics: "An effective mathematics teacher preparation program attracts, nurtures, and graduates high-quality teachers of mathematics who are representative of diverse communities" (p. 26).

As the MTE-Partnership NIC structure was emerging, teams of institutions of higher education and K–12 school districts were invited to join one of four working groups to expand upon the areas deemed of highest importance for the partnership in relation to its common aim. The MTE-Partnership aim was to transform secondary mathematics teacher preparation to ensure an adequate supply of new teachers who can promote mathematical excellence in their future students, leading to college and career readiness. The working groups arose from the first annual conference of the MTE-Partnership and were formed around particular problem areas identified by the MTE-Partnership to be impeding progress toward the common aim. The working groups were the precursor to the establishment of the RACs. One of these groups and problem areas was recruitment and retention to attract and maintain an adequate supply of candidates.

UNDERSTANDING THE PROBLEM OF RECRUITMENT AND RETENTION

In parallel with the other MTE-Partnership working groups, the group studying recruitment and retention reviewed relevant literature to more deeply understand available resources and research previously conducted on recruitment and retention of teachers, particularly mathematics teachers. Through study of the problem area, members of the working group wrote a white paper, now Chapter 13 in this volume.

The literature suggested the importance of bringing about the recruitment, preparation, and hiring of individuals as certified teachers from target populations along three tracks: *slow track*, *moderate track*, and *fast track*. These tracks can be used to classify recruitment practices. For example, practices for recruiting middle or high school students into teaching are classified as slow track (5 to 8 years) recruitment practices, while those for recruiting college freshmen and sophomores are classified as moderate-track (3 to 4 years), and those for college juniors and seniors, as well as college graduates, are fast track (0 to 2 years; Clewell, Drake, Davis-Googe, Forcier, & Manes, 2000). Education policy analysts have suggested that recruitment should target specific populations at critical points in the recruitment pipeline through particularly tailored programs and strategies (Clewell et al., 2000).

Marketing to increase awareness, knowledge, and interest in secondary mathematics teacher preparation is of key importance in recruiting individuals into programs, targeting specific populations as appropriate. Marketing can be thought of as a comprehensive process of promoting becoming a mathematics teacher, as well as providing information about and experiences with the "product" (e.g., trial teaching, an early field experience in which potential future teachers "try out" teaching for a short time period). Marketing practices may be implemented across all target populations, involve varied strategies, and fall into all three classifications for recruitment practices: slow track, moderate track, and fast track. One aspect of marketing involves advertising through varied avenues. Reys and Reys (2004) shared a variety of marketing channels pursued at their institution to recruit mathematics teacher candidates. These channels included newspapers and alumni magazines, billboards on interstates, posters on campuses, movie theater ads, educational and/or career advisors, college-recruitment specialists, faculty advisors knowledgeable in becoming a mathematics teacher, career fairs, brochures sent to mathematics teachers as ambassadors to share with their students or to capable mathematics students possibly previously identified by teachers, and websites. More recent channels for advertising programs may involve email and social media. Evaluation of the success in reaching mathematics teacher candidates through varied channels was an area thought to be beneficial for the MTE-Partnership.

Marketing, beyond just advertising, may involve slow, moderate, and fast track recruitment practices that include information sharing and teaching related experiences such as trial teaching and tutoring. The National Comprehensive Center for Teacher Quality (NCCTQ, 2007) suggested "grow-your-own" type programs including promotional and recruitment efforts beginning as early as middle school that encourage students who excel in mathematics to pursue a career in teaching. These types of recruitment programs in middle and high schools, as well as college, include experiences such as an introduction-to-teaching course, teaching/ tutoring activities, student education clubs, and events or workshops for parents and other family members (U.S. Department of Education, 2012). Examples of these programs geared toward middle and high school students include the South Carolina ProTeam, South Carolina Teacher Cadets, and North Carolina Teaching Fellows (U.S. Department of Education, 2012). Similar programs have been directed toward college students to encourage them to pursue education programs and teacher certification. Examples include Learning Assistant (LA) programs (Otero, Pollock, & Finkelstein, 2010) and UTeach programs (Brainard, 2007) for recruiting and retaining STEM teacher candidates. These programs target non-education majors with interests in STEM and provide opportunities to learn more about and experience what it is to be a teacher (e.g., opportunities for students to trial teach through an introductory course on learning and teaching, related early teaching experiences). As part of these programs, students are often offered financial support through stipends and scholarships, as well as academic support for success in the program.

Financial incentives, such as scholarships, loan-forgiveness programs, and grants, have a long history and prevalence as strategies used for recruitment and retention of individuals to seek teaching as a career. Incentives have been used to market becoming a mathematics teacher and as part of comprehensive recruitment programs within the slow, moderate, and fast tracks. At the college level, the

National Science Foundation Robert Noyce Teacher Scholarship Program (Noyce Program) begun in 2002 as a source of financial incentives for students choosing to pursue mathematics or science teaching as a career. According to results from a Noyce Program Principal Investigator (PI) survey, PIs felt that their Noyce project greatly increased their ability to recruit into teaching a wide variety of individuals (Liou, Desjardins, & Lawrenz, 2010). However, findings related to Noyce Scholars suggest that the financial incentives from the scholarship were not viewed as highly influential in the scholars' decisions to become teachers; only 3.5% indicated their becoming a teacher was contingent on the funding (Liou & Lawrenz, 2011). Noyce funding was perceived as having a greater influence by non-Whites than Whites to become teachers and teach in high-need schools, and was also more influential for career-changers than for those preparing for their first careers (Liou & Lawrenz, 2011). Given that career-changers and non-Whites traditionally have higher rates of retention as teachers and for remaining in high-need schools (Guarino, Santibanez, & Daley, 2006; Liou & Lawrenz, 2011), these findings can be used to plan the future funding of scholarships for recruiting mathematics teachers. Other types of financial incentives to increase recruitment of teachers in hard-to-staff fields such as mathematics include signing bonuses, higher base starting salary, and housing assistance (NCCTQ, 2007). Research on these incentives for recruiting teachers is lacking; however, there is some initial information that suggests that these types of monetary incentives are "less effective than the fast-track alternative training portion of the reform in inducing individuals to enter the teaching profession" (Guarino, Santibanez, & Daley, 2006, p. 195).

ESTABLISHING THE MATH RAC

The MTE-Partnership undertook a two-year long process of understanding the problem space including the following: developing guiding principles that underlie secondary mathematics teacher preparation (Martin & Strutchens, 2014); identifying problem areas impeding progress toward the aim (Martin & Gobstein, 2015); developing literature reviews in those areas; and subsequently identifying possible improvements in those areas. MTE-Partnership teams were invited to join one of four working groups to elaborate on four primary drivers in areas deemed of highest importance by the partnership (see Chapter 1): creating a vision, clinical preparation, content knowledge, and recruitment and retention. This process led to the formation of five RACs to address the four primary drivers supporting the overall NIC aim of increasing the number and quality of secondary mathematics teachers. These RACs were Clinical Experiences; Active Learning Mathematics; Mathematics of Doing, Understanding, Learning, and Educating for Secondary Schools (MODULE(S)²); Marketing to Attract Teacher Hopefuls (MATH); and Secondary Teacher Retention and Induction in Diverse Educational Settings (STRIDES).

MTE-Partnership teams were asked to join a RAC. A RAC capacity-building meeting was held on improvement science and analytics, including building knowledge of NICs, RACs, driver diagrams, and Plan-Do-Study-Act (PDSA) cycles used to study change ideas. The MATH RAC was initially represented by four members from different institutions of higher education that offer secondary mathematics teacher preparation programs: University of South Carolina (RAC leader), Boise State University, Florida International University, and the University of Arizona. Other institutions involved in the MATH RAC over time included: California State University, East Bay, California State University, Bakersfield, California State University, Chico, East Carolina University, Fresno State University, Mississippi State University, San Diego State University, San Jose State University, and Texas A&M University.

The four MATH RAC members at the improvement science capacity-building meeting familiarized themselves with one another's programs and worked toward deepening their understanding of the work. They deliberated ideas related to three central questions of improvement science, later discussed among the broader MATH RAC membership:

1. What specifically are we trying to accomplish?
2. What change might we introduce and why?
3. How will we know that a change is actually an improvement?
   (Bryk et al., 2015, p. 114)

With respect to the first central question, a primary goal of the MATH RAC was to increase enrollments in the secondary mathematics teacher preparation programs at their institutions. The RAC members felt that increasing the enrollment necessitated increasing awareness, extent of inquiries about the programs, and interest in the programs; thus, marketing would be of high importance in achieving this goal. The MATH RAC leader revealed that through his institution he had received a limited amount of funding to develop a marketing campaign for his institution's secondary mathematics teacher education program. He suggested that RAC members could contribute to the creation of branding and marketing campaign modules, as members developed, implemented, and studied marketing strategies new to these institutions. These modules could be shared more broadly among MTE-Partnership NIC members. Discussion of what the RAC was trying to accomplish helped RAC members clarify their work, making it more user-centered and problem-specific, the first of six guiding principles of improvement science (Bryk et al., 2015). This first guiding principle proposes that all activity is conducted in order to address a major problem in a system, and that the problem be examined from the users' point-of-view.

In relation to the second central question of improvement science, focused on change, the MATH RAC members discussed their current systems, along with marketing changes and strategies they could introduce to increase awareness and number of inquiries about their programs. At the time, approaches to marketing their programs had been limited. Often, potential program applicants of their own volition needed to seek out the existence of secondary mathematics teacher preparation programs within RAC member institutions through exploration of institutional websites or communication with advisors. By engaging in understanding the system producing the current extent of program inquiries and enrollments, the MATH RAC was implementing the third principle guiding improvement science (Bryk et al., 2015). In order to improve a system, the users need to understand the system along with the processes that comprise it. The RAC members determined it would be valuable to introduce active marketing approaches into the system. Some of these approaches could be gleaned from the literature (e.g., NCCTQ, 2007; Reys & Reys, 2004). These approaches could target specific populations in ways that would inform individuals in the targeted populations about the programs and would persuade them to seek further information.

Active marketing approaches considered included strategies such as sending out post-cards or emails to targeted populations, handing out flyers at new student orientations, or making announcements about the programs in courses taken by individuals from targeted populations. Some of the targeted populations considered were STEM undergraduate students, STEM graduates, and high school students who participate in student organizations such as Future Educators of America or Mathematics Honor Society. While discussing the active approaches, the MATH RAC suggested the need to pay special consideration to the layouts and designs of the emails, postcards, or flyers, as well as the production of program videos to include in emails, use at orientations, and post on websites with links provided on postcards and flyers. These considerations were believed important for branding and marketing the programs.

In order to address the third central question of improvement science, how to know if a change is an improvement (Bryk et al., 2015), the MATH RAC considered how to determine if a change to the marketing strategies was actually an improvement. RAC members deliberated about possible ways to measure the changes in amounts of inquiries and enrollments. The determination of ways to measure the extent of program inquires was deemed substantially more complex than determining changes in enrollment. Inquiries may occur through multiple channels and involve various individuals in an institution; on the other hand, program enrollments are maintained as part of official institutional records. Surmounting this challenge is important to the success of improvement science as suggested by the fourth guiding principle, "We cannot improve at scale what we cannot measure" (Bryk et al., 2015, p. 14).

MATH RAC members discussed possible ways to capture and record inquiries that occur about a program. One approach suggested was to post a short program inquiry survey on the program website for interested individuals to request further information. Another suggestion was to create a Google document so that different individuals from an institution, who may be contacted about the program through varied channels, could record the inquiry on the same electronic document. Through these means, RAC members could also capture information about what marketing strategies spurred individuals' inquiries about the programs. This

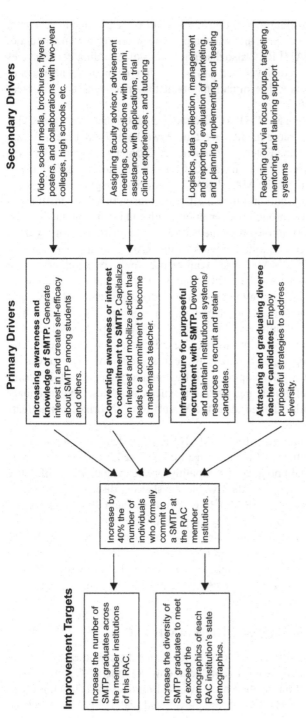

FIGURE 14.1.   MATH RAC Driver Diagram. Initial driver diagram for the past MATH RAC within the MTE-Partnership Networked Improvement Community. SMTP stands for secondary mathematics teacher preparation program.

information would contribute to meeting the second principle of improvement science, understanding variation, and what approaches work for whom and under what conditions. The MATH RAC also discussed ways of recording PDSA cycles as marketing strategies were developed and trialed, in conjunction with ways of using inquiry data to make determinations about the effectiveness of different marketing approaches studied.

The establishment of the MATH RAC also required development of the MATH RAC driver diagram. The recruitment and retention working group driver diagram found in the Recruitment and Retention chapter (see Chapter 13, Figure 13.1) was used as a starting-off point for the MATH RAC driver diagram. Figure 14.1 provides the driver diagram for the MATH RAC. This driver diagram was worked on as part of the capacity-building meeting by the four initial RAC institutions and was further explained through conversations with and input from the broader MATH RAC member institutions. It includes improvement targets, primary drivers, and secondary drivers to guide the work of the RAC. The improvement targets outline the aim of the MATH RAC with respect to the overarching aim of the MTE-Partnership to increase the number and diversity of mathematics teacher education graduates by increasing the number of individuals who formally commit to a secondary mathematics teacher preparation program. The primary drivers represent main areas of influence that the MATH RAC hypothesizes are necessary to advance the improvement targets. These include increasing secondary mathematics teacher preparation program awareness and knowledge, converting awareness to commitment, developing an infrastructure for purposeful recruitment, and attracting and graduating diverse teacher candidates. The secondary drivers consist of potential change ideas the MATH RAC theorizes can activate their related primary drivers.

Additionally, it was important for the MATH RAC to determine the logistics for collaboration among the RAC member institutions. The RAC members agreed that the RAC leader would organize the monthly online meetings or conference calls. These meetings would provide a venue for sharing, discussing, planning, and conducting the work of the RAC. RAC members could share and discuss information about what approaches they were trialing at their institutions and what had they learned from implementing these approaches, including sharing successes, failures, and plans for the future. They could also assist one another with the design, layout, and carrying out of varied approaches, as well as contribute ideas from their work toward the marketing and branding campaign modules to be developed.

THE WORK OF THE MATH RAC

Back at their home institutions, the MATH RAC members worked with other members of their partnership teams to conduct the work of the RAC at their institutions. A few of the institutions had been awarded funding for UTeach replication programs so their MATH RAC efforts dovetailed with their UTeach-related efforts. An important component of the UTeach program is the recruitment of students interested in STEM into an introductory 1-credit course to learn about their field and participate in trial teaching.

On a monthly basis, the MATH RAC leader called virtual meetings through the use of video-conference calling to share recruitment and marketing approaches, discuss the PDSA work at the various institutions, provide recommendations and feedback on the marketing and branding campaign models being developed, and plan for future collaborative efforts. The RAC members led the organization, orchestration, and implementation of the improvement science work at their home institutions. As indicated by AMTE (2017), effective mathematics teacher preparation programs implement a strategic process that tactically involves faculty and staff to recruit candidates. At their home institutions, MATH RAC members collaborated to varied extents with others such as mathematics education faculty, secondary education program directors, program master teachers, mathematicians, and district personnel in planning and carrying out the work of the RAC associated with marketing the programs and recruiting teacher candidates. Additionally, institutional marketing directors and staff, new student orientation directors and staff, institutional recruitment directors and staff, advising directors and advisors, mathematics teacher preparation students, mathematics teachers, and others with necessary expertise were also consulted or involved during the marketing and recruitment processes founding PDSA cycles at each institution. The involvement and orchestration of others with diverse expertise to address the given problem is aligned with the sixth principle guiding improvement science to accelerate improvements through leveraging the expertise of others (Bryk et al., 2015).

MATH RAC members leading the improvement science effort at their institutions in consultation and coordination with others worked on the development of a mode of data collection to gather information at their institutions about inquiries concerning the secondary mathematics teacher preparation programs. For example, at one institution, coordination with the secondary education program director and faculty in the program led to the development of a system involving a short web-based program-inquiry survey and paper-based sign-up forms to use at face-to-face events to capture inquiries about the program. All potential candidates with documented inquiries about the program from the online web-based survey or the sign-up forms were personally contacted by email to provide further information, respond to other questions they may have, and make arrangements to meet. The methods used for gathering inquiry information were of central importance in conducting PDSA cycles in order to accelerate learning about the extent that different marketing and recruitment approaches were stimulating interest in the secondary mathematics teacher education programs.

Mathematics Teacher Recruitment Guide

A substantive outcome of MATH RAC related efforts was the development of the *Secondary Mathematics Teacher Recruitment Campaign Implementation*

Guide (Ranta & Dickey, 2015). This guide consists of nine modules. The modules were written with limited funding from the MATH RAC leader's institution. As the modules were developed, they were shared and discussed among the MATH RAC members. Some ideas were trialed at RAC institutions and feedback was shared at monthly MATH RAC meetings and annual MTE-Partnership conferences. The modules contain documents and examples of recruitment and marketing-related resources that can be drawn on to develop and implement approaches for recruiting teacher candidates through various means, as well as create and implement a complete multifaceted recruitment campaign. An initial version of the complete guide was shared and feedback was discussed at the third annual conference in the summer of 2014 with the final version presented at the fourth annual conference in summer 2015. Table 14.1 contains information about the nine modules in the recruitment guide. All of the modules are couched in and refer to examples within the context of secondary mathematics teacher candidate recruitment.

MATH RAC PDSA-Related Work

The PDSA cycles at MATH RAC institutions provided disciplined inquiry related to the practices carried out for marketing the secondary mathematics teacher education programs. This process anchored practice in disciplined inquiry in order to learn and improve quickly, a fifth guideline of improvement science (Bryk et al., 2015). For example, one MATH RAC institution conducted PDSA cycles on marketing of the secondary mathematics teacher preparation program as part of new student orientations (see Table 14.2). A RAC institutional member in conjunction with the director of the secondary teacher education program and the director of new student orientations discussed available opportunities during the orientations to conduct the marketing of the secondary teacher preparation programs in mathematics and science at the institution. Three different venues for the marketing were determined: (1) a tabling event during lunch of a new student orientation that included STEM students, (2) a professional information session focused on STEM teaching during a new student orientation also including STEM students, and (3) a resource fair tabling event during a new student orientation that included STEM students. Plans were made to implement and study the amount of student interest during marketing of the program at each of these events during a two-day orientation. Informational flyers containing program website links shared with interested students during the events incorporated design and layout ideas based on recommendations provided by MATH RAC members from efforts at their institutions.

Table 14.2 contains pertinent information related to the PDSA cycles. Data were collected through student sign-in sheets at each of three marketing venues during a new student orientation that included new STEM majors, among others. During the first cycle, the tabling event during the one-hour lunch break was not successful. The marketing produced two students who signed in; however, their interest was on the attention-grabbing Oobleck (a cornstarch-based substance that

TABLE 14.1. Description of MTE-Partnership Recruitment Guide Modules

Modules	Brief Summary
1. Overview	Provides an overview of the recruitment campaign process, including 10 common steps of any communication and marketing campaign. Underscores the importance of communication with appropriate experts in your institution to create a unified voice and common messaging, as well as the value of evaluation from the start of the campaign and often.
2. Campaign Planning	Provides information and guidance to plan a successful recruitment campaign with a focus on recruiting individuals to become mathematics or STEM teachers. Discusses identifying the target audience, research to support appropriate strategies, and planning the launch, as well as introducing the Research, Planning, Implementation and Evaluation model.
3. Research	Discusses three components that campaign research should include: (1) initial background research including reviewing other campaigns; (2) primary research, either qualitative or quantitative, to discover facts and attitudes that impact messaging; (3) data acquisition and analysis.
4. Branding	Explains the processes associated with developing a brand identity to help university or college mathematics teacher education programs distinguish themselves from other pathways to a career in teaching. Provides suggestions for developing a strong, easily recognizable program logo in collaboration with your institution's media relations office, as well as recommendations for effective photography and information graphics.
5. Social Media	Discusses the power of social media, along with insights for managing and maximizing the impact of the message, particularly for recruitment of mathematics teachers. Provides descriptions of a variety of social media and ways of using them for teacher recruitment, including rules of usage.
6. Public Relations	Introduces different types of public relations initiatives, as well as steps in planning the implementation of an initiative to maximize impact. Shares information for use of different types of social media for public relations including the demographics of the typical audiences of the varied media.
7. Advertising through Paid Media	Presents different types of paid media including strengths and weaknesses of each and possible costs. Shares information about digital/internet and mobile advertising, along with production tips, reaching target audiences and integrating the use of varied media in a cohesive campaign.
8. Website Identity	Provides information on the varied purposes for websites, planning and mapping out the site to meet the desired purpose(s), and tips on copy, general design, graphics, and visual material. Discusses approaches for promoting the website and ways to maximize search engine optimization. Includes an example of an outline for a mathematics teacher education program website.
9. Lessons Learned/ Evaluation	Provides guidance for evaluating the success of a campaign. Discusses possible misconceptions of effectiveness, as well as sample quantitative and qualitative evaluation research, review of the budget, and providing recommendations for improvement.

TABLE 14.2. Plan-Do-Study-Act Cycle Marketing Program as Part of New Student Orientations

Approach: New student orientation that includes STEM
STEP: Share program information and flyer

	Plan	Do	Study	Act
Cycle: First try	Test number of inquiries at tabling event during lunch (1 hour); using sign-in sheets	Observed difficulty in capturing student attention during lunch; students had varied interests	Had two students sign-in; no interest in STEM-related field; curious about attention-getter Oobleck; process did not work well	Conclusion: need event with focused attention and students with interest in STEM or math
Cycle: Second try	Test number of inquiries at professional information session "STEM Teaching" (30 minutes); using sign-in sheets	Observed STEM students attending pre-health/ medical professional session; student flex-time so had other options	Had one math education major sign in out of four; others no interest in teaching math; process did not work well	Conclusion: need to target STEM and math students more broadly and more directly
Cycle: Third try	Test number of inquiries at resource fair (1 hour); announcement targeted STEM students during orientation overview; using sign-in sheets	Observed students with focused attention; interest from STEM students not familiar with the program	Had 15 STEM students sign in; five potential math education; process worked well	Conclusion: need to continue attending orientation resource fairs to target STEM and math students

can act as a solid and a liquid) and Möbius strips (a surface with one continuous side) being demonstrated to attract students to the program table, rather than being interested in the program. Many students, with and without parents, were walking through the tabling event headed to eateries within the Student Union; and, they were not particularly interested in learning about mathematics or science education-related majors.

For the second cycle, the professional information session, "STEM Teaching," which was also held during a new student orientation, produced the names of three students. One student was already majoring in mathematics education and the other two majoring in early childhood and elementary education and were not interested in teaching secondary school mathematics. This 30-minute session had been announced during the overview of the orientation; however, it was on the orientation schedule at the same time as medical school and other health career professional sessions that were attracting the broader group of STEM majors. Additionally, it was during a flex time while students were free to explore aspects of the university of their own choosing. The time of the session during orientation and the way it was posted in the orientation schedule attracted students who seemed to already know that they wanted to pursue a major in teaching although not necessarily in the STEM areas.

For the third cycle, the one-hour resource fair tabling event during a new student orientation produced 15 names of STEM students who demonstrated interest in exploring a major in mathematics or science education with at least five leaning toward mathematics teaching. All students that signed-in across the three venues (cycles) were given a flyer with information about the program and the initial 1-credit program course to trial teaching. For the resource fair event, a brief announcement about the program was provided during the orientation overview and interested students were instructed to visit the program table during the one-hour resource fair. Having the students' focused attention during the resource fair, which followed the brief announcement about the program targeted at STEM majors during the orientation overview, developed awareness and marketing of the secondary mathematics and science teacher preparation programs targeted at STEM-interested students and majors.

The PDSA cycles involving marketing the programs during the new student orientations, including highlighting program information during the orientation overview, demonstrated a gap in understanding about the new student orientations that ultimately contributed to an important change in the marketing of the mathematics and science teacher preparation programs at the institution. It was recognized that marketing the mathematics and science education programs during a time in the new student orientations when STEM students' attention was focused, such as with the program tabling event during the resource fair, and not competing at the same time with other events, such as when the professional session STEM Teaching event and the tabling event during lunch, produced greater numbers of students seeking program information. As indicated by Bryk et al. (2015), the PDSA inquiry cycles help to reveal gaps in understanding, particularly in the early stages of change, that were previously unrecognized but critical to successful change. The learning from this PDSA suggested to other MATH RAC institutions the value of working with new student orientation directors to identify best ways of marketing to and recruiting students, particularly, targeting STEM majors with strong mathematics backgrounds, during new student orientations. This PDSA also founded further PDSA cycles to study messaging during the orientation overview to persuade more STEM/math majors to inquire about the mathematics and science education programs during the resource fair tabling event.

Another example of a PDSA cycle studied different approaches to market, increase awareness, and generate interest in a secondary mathematics education program at an institution (i.e., primary drivers in the MATH RAC driver diagram in Figure 14.1). The marketing approaches included sending emails (using a mass email service) directly to 1,600 incoming students interested in STEM degrees, sending postcard flyers directly to the same 1,600 students, handing out program flyers at 12 (half) of the new student orientations during events where STEM-interested students could be targeted, and handing out program flyers during 10 (one-eighth) of the introductory lower-division STEM courses being offered during that semester. All of these avenues for marketing the programs directed the students to the secondary mathematics and science teacher preparation program website to complete a survey if they were interested in knowing more about the program and an initial free, 1-credit course to trial teaching mathematics or science. Data on how students first learned about the program were collected during the PDSA through the online survey on the program website and a similar survey given at the start of the semester during the 1-credit course for students to trial teaching mathematics or science. For students who completed both surveys, duplicate data points were removed.

The findings from this PDSA revealed the following results about how students (potential program applicants) learned about the program. Of the 94 students completing the surveys and demonstrating an interest in the mathematics teacher preparation program, the following results were found: 23% said they learned about the program through a new student orientation; 21% from a program email sent directly to them; 17% from the program website; 12% from program information heard and a program flyer given in an introductory mathematics or science class; 10% from a friend (responding to other-open text response); 9% from an advisor at the institution (responding to other-open text response); 5% from a postcard sent directly to them; and 5% reported not recalling (responding to other-open text response).

This PDSA helped the program administrators understand how to prioritize approaches to market the program to STEM-interested students using approaches that were more effective and efficient. Given that 9% of the students indicated hearing about the program from their advisor and the program had not yet begun to work closely with the advisors, this improvement cycle highlighted the need for the program to work more closely with the advisors, particularly the STEM advisors, to inform potential applicants about the program. It was also recognized that the program should expand (beyond 10) the number of introductory mathematics and sciences courses in which to share program information and flyers. These expansions would form a foundation for new PDSA cycles. Finally, the results of the PDSA triggered discontinuing the postal mailing of program information postcards individually to students, as the cost of this recruitment approach did not result in levels of student interest meriting the expense given the other more fruitful approaches. This information was shared at the monthly MATH RAC meetings where other institutions could benefit from the lessons learned through this PDSA inquiry. Other institutions with membership in the RAC also began to send out emails and visit STEM classes to provide program information and flyers. Additionally, at the annual MTE-Partnership conferences updates on the progress of the MATH RAC and findings related to recruitment have been shared across the NIC, as have updates from all of the RACs.

One challenge with which the MATH RAC grappled was determining how to measure best ways of marketing and recruiting program applicants from diverse populations. Some RAC members have felt that some individuals experience sensitivity when asked about their race/ethnicity or gender; thus, it has been challenging to determine effective ways to gather this information without potentially turning-off prospective applicants at the point where they are showing some interest by seeking further program information. One MATH RAC institution asked for the student institutional identification numbers of students seeking information about the programs; however, seeking race/ethnicity and gender using student numbers is time-intensive. Additionally, this may preclude the institution from gathering this information for individuals who have not yet attended the institution and are seeking program information. This challenge provides opportunities for investigating and solving this measurement-related problem in the future.

FOCUS ON IMPROVEMENT

The work of the MATH RAC as part of the MTE-Partnership NIC described in this chapter is driven by the shortage of well-prepared secondary mathematics teachers, particularly with respect to staffing schools with higher teacher turnover. At the end of this section, please see the reading list of recommended resources as an institution begins to examine its recruitment practices with aims to improve. The shortage of well-prepared secondary mathematics teachers is investigated, and findings are reported in the article by Ingersoll and Perda (2010). Given the problem space, this chapter provides insights that other institutions can draw on to establish and carry out the work of a RAC to investigate recruitment of secondary mathematics teacher candidates within their settings, including using PDSA cycles.

This work of the MATH RAC was supported by the six guiding principles for improvement science described by Bryk et. al. (2015), a useful resource for those interested in implementing improvement science methods. Additionally, this chapter provided descriptions and findings about approaches and strategies for recruiting secondary mathematics teacher candidates into mathematics education programs that were investigated by MATH RAC institutions and are aligned with the 2017 AMTE Standards related to recruitment of future teachers of mathematics, including highlighting the importance of collaboration among faculty and staff within an institution to more effectively recruit mathematics education teacher candidates.

Clewell et al. (2000) offers information about teacher-recruitment programs, as well as effective elements of state, local, and higher education recruitment efforts that can be implemented and studied within the reader's setting. Finally, this chapter describes the *Secondary Mathematics Teacher Recruitment Campaign Implementation Guide* developed as part of the MATH RAC. This resource is composed of nine modules that readers can draw on to plan, implement, and evaluate a recruitment campaign or use to plan and study varied approaches and strategies for promoting program awareness and interest to recruit students into mathematics teacher education programs. The recruitment guide is available through Open Canvas, an online platform adopted by the MTE-Partnership to conduct the work of the NIC, as well at http://bit.ly/MATHImplGuide.

Reading List

As follows is a list of resources to support improvement of mathematics teacher candidate recruitment.

1. Bryk, A., Gomez, L. M., Grunow, A., & LeMahieu, P. (2015). *Learning to improve: How America's schools can get better at getting better.* Cambridge, MA: Harvard Education Press.
   - This book draws on ideas from improvement science to describe a process of disciplined inquiry organized around six guiding principles and involving networked improvement communities to accelerate learning in important areas of education.

2. Ingersoll, R. M., & Perda, D. (2010). Is the supply of mathematics and science teachers sufficient? *American Educational Research Journal, 43*(3), 563–594.
   - Through analysis of nationally representative data from multiple sources, this study found that mathematics and science teaching are the most hard-to-staff fields, particularly in schools with higher turnover, due to complex factors behind these problems.

3. Clewell, B., Drake, K., Davis-Googe, T., Forcier, L., & Manes, S. (2000). *Literature review on teacher recruitment programs.* Washington, DC: U.S. Department of Education.
   - This document describes aspects of teacher-recruitment programs— including slow, moderate, and fast track programs—and highlights effective elements of state and local recruitment efforts, as well as describing ways to meet the need for teachers, particularly in hard-to-staff areas.

4. Ranta, J., & Dickey, E. (2015). *Secondary mathematics teacher recruitment campaign implementation guide.* Washington, DC: Association of Public and Land-grant Universities. Retrieved from http://bit.ly/MATHImplGuide
   - This material is composed of nine modules containing information, examples, and resources that can be drawn on to plan and implement a secondary mathematics teacher candidate multifaceted recruitment campaign, as well as planning and implementing select recruitment approaches and strategies to increase numbers of secondary mathematics teacher candidates.

CONCLUSION

This chapter describes the trajectory of the MATH RAC, which was created as part of the MTE-Partnership NIC, to address in part the shortage of mathematics teachers through disciplined inquiry and sharing of effective program recruitment approaches and strategies. As part of the trajectory, an implementation guide for a secondary mathematics teacher recruitment campaign was written, composed of nine modules (Ranta & Dickey, 2015). After the recruitment modules were created, the MATH RAC ended, and the Program Recruitment and Retention (PR²) RAC was established (see Chapter 15). This new RAC will continue to investigate recruitment; however, the focus will be more directly on recruiting diverse mathematics teacher candidates into secondary mathematics education programs. Additionally, PR² will focus on approaches and strategies for the retention of mathematics teacher candidates in their programs through to graduation.

REFERENCES

Association of Mathematics Teacher Educators. (2017). *Standards for preparing teachers of mathematics*. Raleigh, NC: Author. Retrieved from http://amte.net/standards

Banilower, E. R., Smith, P. S., Weiss, I. R., Malzahn, K. A., Campbell, K. M., & Weis, A. M. (2013). *Report of the 2012 National Survey of Science and Mathematics Education*. Chapel Hill, NC: Horizon Research.

Brainard, J. (2007, December 21). Texas offers a model for training math and science teachers. *Chronicle of Higher Education, 54*(17), A8–A10.

Bryk, A., Gomez, L. M., Grunow, A., & LeMahieu, P. (2015). *Learning to improve: How America's schools can get better at getting better*. Cambridge, MA: Harvard Education Press.

Clewell, B., Drake, K., Davis-Googe, T., Forcier, L., & Manes, S. (2000). *Literature review on teacher recruitment programs*. Washington, DC: U. S. Department of Education.

Guarino, C., Santibanez, L., & Daley, G. (2006). Teacher recruitment and retention: A review of the recent empirical literature. *Review of Educational Research, 76*(2), 173–208.

Ingersoll, R. M., Merrill, L., & May, H. (2012). Retaining teachers: How preparation matters. *Educational Leadership, 69*(8), 30–34.

Ingersoll, R. M., & Perda, D. (2010). Is the supply of mathematics and science teachers sufficient? *American Educational Research Journal. 43*(3), 563–594.

Liou, P. Y., Desjardins, C. D., & Lawrenz, F. (2010). Influence of scholarships on STEM teachers: Cluster analysis and characteristics. *School Science and Mathematics, 110*(3), 128–143.

Liou, P. Y., & Lawrenz, F. (2011). Optimizing teacher preparation loan forgiveness programs: Variables related to perceived influence. *Science Education, 95*(1), 121–144.

Martin, W. G., & Gobstein, H. (2015). Generating a networked improvement community to improve secondary mathematics teacher preparation: Network leadership, organization, and operation. *Journal of Teacher Education, 66*(5), 482–493.

Martin, W. G., & Strutchens, M. E. (2014, April). *Priorities for the improvement of secondary mathematics teacher preparation for the Common Core era*. Paper presented at the AERA Annual Meeting, Philadelphia, PA. Retrieved from www.aera.net/

Publications/OnlinePaperRepository/AERAOnlinePaperRepository/tabid/12720/Owner/314509

Mathematics Teacher Education Partnership. (2014). *Guiding principles for secondary mathematics teacher preparation.* Washington, DC: Association of Public and Land-grant Universities. Retrieved from mtep.info/guidingprinciples

National Comprehensive Center for Teacher Quality. (June 2007). *Recruiting quality teachers in mathematics, science, and special education for urban and rural schools.* Washington, DC: Author. Retrieved from http://www.gtlcenter.org/sites/default/files/docs/NCCTQRecruitQuality.pdf

Otero, V., Pollock, S., & Finkelstein, N. (2010). A physics department's role in preparing physics teachers: The Colorado learning assistant model. *American Journal of Physics, 17*(11), 299–308.

Ranta, J., & Dickey, E. (2015). *Secondary mathematics teacher recruitment campaign implementation guide.* Washington, DC: Association of Public and Land-grant Universities. Retrieved from http://bit.ly/MATHImplGuide

Reys, R., & Reys, B. (2004). Recruiting mathematics teachers: Strategies to consider. *Mathematics Teacher, 97*(2), 92–95.

U.S. Department of Education, Office of Postsecondary Education. (2012). *Teacher shortage areas nationwide listing 1990–1991 through 2012–2013.* Washington, DC: Author.

CHAPTER 15

SUPPORTING SECONDARY MATHEMATICS TEACHER PREPARATION PROGRAM RECRUITMENT AND RETENTION EFFORTS[1]

Julie McNamara, Dana Pomykal Franz, and Maria L. Fernandez

Recently released standards and guiding principles recommend the recruitment of diverse prospective teachers. For example, the Association of Mathematics Teacher Educators' (AMTE, 2017) *Standards for Preparing Teachers of Mathematics* state: "An effective mathematics teacher preparation program attracts, nurtures, and graduates high-quality teachers of mathematics who are representative of diverse communities" (Standard P.5, p. 26). Additionally, one of the Mathematics Teacher Education Partnership (MTE-Partnership)'s *Guiding Principles for Secondary Mathematics Teacher Preparation Programs* (2014) calls attention to the need for secondary mathematics teacher preparation programs to recruit and support diverse teacher candidates: "Guiding Principle 8. Student Recruitment, Selection, and Support—The teacher preparation program actively recruits high-quality and diverse teacher candidates into the program, and monitors and supports their success in completing the program" (p. 8). Recommendations call-

The Mathematics Teacher Education Partnership: The Power of a Networked Improvement Community to Transform Secondary Mathematics Teacher Preparation, pages 337–350.
Copyright © 2020 by Information Age Publishing

ing for the recruitment of diverse teacher candidates are not unique to mathematics education. The Council for the Accreditation of Educator Programs (Council for the Accreditation of Educator Preparation, CAEP, 2013), one of the primary accrediting bodies for colleges of education across the United States, has a standard directly related to recruitment noting that STEM classes are "hard to staff":

> The provider [of the education program] presents plans and goals to recruit and support completion of high-quality candidates from a broad range of backgrounds and diverse populations to accomplish their mission. The admitted pool of candidates reflects the diversity of America's P–12 students. The provider demonstrates efforts to know and address community, state, national, regional, or local needs for hard-to-staff schools and shortage fields, currently, STEM, English-language learning, and students with disabilities. (CAEP, 2013, Standard 3.1, p. 1)

The national call to increase the diversity of the prospective teacher pool stems from the continued disparities in educational outcomes across the United States. Serious discrepancies in the educational opportunities offered to K–12 students of color as compared to their White peers continues to persist in our schools (ACT & United Negro College Fund, 2016). This incongruity is particularly evident in STEM fields, in part due to the shortage of qualified teachers and also the practice of placing teachers with the least amount of experience in classrooms with the fewest resources and the most vulnerable students (ACT, 2018). These classrooms tend to have high numbers of students of color and students from lower socioeconomic backgrounds. This opportunity gap results in what is often referred to as the achievement gap, which in turn means that fewer students of color are prepared for college or career upon leaving high school.

In addition, the demographics of teachers currently in the classroom do not match those of the nation's children; yet, there is significant evidence that students benefit from having teachers of color (D'Amico, Pawlewicz, Earley, & Mcgeehan, 2017). By 2024, students of color are expected to make up 56% of the student population, while the teaching force will remain primarily White. This statistic has changed very little in the years since 2000 (U.S. Department of Education, 2016). Using data from the Education Longitudinal Study of 2002 of over 16,000 teachers, Gershenson, Holt, and Papageorge (2016) investigated teacher-student demographic mismatch and found that Black teachers held significantly higher expectations than did White teachers for Black students. These benefits extend to both academic and non-academic outcomes, especially for students of color (Gist, 2016). In addition, teachers and schools have been found to have lower expectations for Black and Latinx students' achievement as compared to their Asian and White peers, and thus may provide differential educational opportunities to their students of color (McKown & Weinstein, 2008; Muller, Riegle-Crumb, Schiller, Wilkinson, & Frank, 2010; Pringle, Lyons, & Booker, 2010).

Secondly, due to challenges that often face prospective teachers as they navigate secondary teacher education programs, an additional focus on retaining prospective teachers in secondary teacher preparation programs is critical. While much research exists on teacher retention in the K–12 classroom (see Chapter 13 of this book), there does not exist similar research on retaining students in secondary teacher education programs. Multiple pathways lead to teacher certification. Bachelor's degrees, master's degrees, post-baccalaureate programs, and other certificate programs all credential teachers. These pathways may take anywhere from less than a year to more than five years to complete. Additionally, entrance requirements vary greatly and have many additional "gates" to full admission to secondary teacher programs and eventually licensure to teach (National Research Council, 2010). Based on informal interviews with admissions officers, there is evidence suggesting that only 50% of prospective teachers who begin the application to licensure programs actually end up completing the application process. Thus, students need to have clear guidance from academic advisors and program administrators, as well as support from faculty and peers, for navigating through the application process as well as meeting all of the program requirements once admitted.

Retaining all prospective teachers is needed to increase the number of well-prepared secondary mathematics teachers. The needs of prospective teachers of color may be somewhat greater than those of their White classmates due to the feelings of isolation that are commonly expressed, especially in programs that are primarily made up of White faculty and White prospective teachers (Gist, 2016). In addition, the "residue of institutional racism" (p. 933) may result in prospective teachers of color needing more financial and academic support than other students. Continuing to find the best methods to retain across all prospective teachers is vital.

The issues of recruitment and retention in secondary teacher education programs are pervasive across the U.S. The initial Research Action Cluster (RAC) that emerged from an analysis of the issues related to recruitment and retention agreed to focus on basic methods for recruitment (see Chapter 13 of this book for a brief discussion of the Marketing to Attract Teacher Hopefuls (MATH) RAC). However, as the work expanded, members of the MATH RAC recognized that while the initial mission of the RAC had been achieved through the creation of the *Secondary Mathematics Teacher Recruitment Campaign Guide* (Ranta & Dickey, 2015), explicit attention to diversifying the prospective teacher pool still needed to be addressed, along with retention of teacher candidates to graduation. Following the 2016 MTE-Partnership Conference, the MATH RAC was re-envisioned to address this new focus. The new Program Recruitment and Retention (PR^2) RAC began work immediately.

IMPROVEMENT TARGET AND PRIMARY DRIVERS

In January of 2017 a subgroup of the newly formed PR^2 RAC met to develop a new improvement target (or aim) and subsequent primary drivers (Bryk, Gomez, Grunow, & LeMahieu, 2015). Initial discussions included brainstorming of the critical issues of recruitment and retention, development of statements to reflect these issues, discussion to prioritize these statements, and refinement of the statements. The RAC identified the following improvement target: *To increase the number of well-prepared secondary mathematics teachers entering the mathematics teaching workforce by at least 40% from each participating program by Summer 2022, reflecting the diversity targets of each program.* A target date of Summer 2022 was agreed upon to allow a five-year period beginning with the summer 2017 MTE-Partnership Conference to conduct multiple Plan-Do-Study-Act (PDSA) cycles and begin to make definitive changes in the recruitment and retention practices across institutions. The 40% increase was proposed by the RAC leadership team and reviewed and agreed to by RAC members. The intention was to ensure that each secondary teacher preparation program within the RAC works toward achievement of a 40% increase in program completers, and the target also addresses an average increase of 40% across all RAC participant institutions. This group acknowledges that a 40% target might be an ambitious target for some programs with recent growth or substantial changes but is a realistic and significant target for RAC programs and the MTE-Partnership as a whole.

The institutions that make up the PR² RAC are diverse, ranging from large public research universities and university systems to small regional colleges, and spread across the U.S.; thus, the target of reflecting the diversity of each program is critical. The PR² RAC acknowledges that racial, cultural, gender, and many other demographic factors vary significantly by locales, but each program will set diversity targets that mirror the demographics of the state or local community to be served by their graduating secondary mathematics teacher candidates as well as that of the institution. For example, a program serving a state in which the African American or Latinx citizens make up 30% of the population will seek to ensure 30% of its secondary mathematics teacher candidates are African American. If the institution's student body is 15% African American, that percentage might serve as an intermediate goal. The degree to which the program is below the goal will inform the priority and adoption of change ideas or interventions implemented to make progress toward reaching the target.

Primary Drivers

As shown in Figure 15.1, the PR² driver diagram has five primary drivers, three of which are consistent with the work of the prior MATH RAC. The first primary driver, attending to issues of diversity, equity, and social justice, was a target for and an implied concern of the MATH RAC but is now positioned more overtly. Moving this driver to the first position of the primary drivers showcases its importance to the work of the PR² RAC and as a focus for all subsequent drivers. Two additional primary drivers, building program awareness and interest and improving perception of teaching, both are closely tied to program marketing and new teacher recruitment efforts. The secondary drivers serve as resources for RAC members who seek to enlist new applicants, explore new strategies to increase the outreach to and admission of new prospective teachers to their teacher preparation programs, and to use the *Secondary Mathematics Teacher Recruitment Campaign Implementation Guide* (Ranta & Dickey, 2015) to aid their recruitment efforts.

The PR^2 RAC has expanded its attention with two additional drivers: advocating with policy makers and advising and supporting teacher candidates. The latter focuses on program retention issues included in the original *MTE-Partnership Working Group 4 White Paper on Recruitment and Retention* (a revision of this White Paper is found in Chapter 13). In keeping with the needs identified by research, the PR^2 RAC is investigating strategies to retain and nurture teacher candidates through program completion. This issue is defined as an important component of increasing the production of secondary mathematics teachers. The RAC also acknowledges the important role that policy makers have in the process of secondary mathematics teacher recruitment and include that driver in the hopes of impacting policies that research shows influence on teacher recruitment.

Secondary Drivers

The PR^2 RAC acknowledges that direct and significant efforts to address each of the five drivers of its improvement target are beyond the current capacity of the RAC membership. As a networked improvement community, the PR^2 RAC has systematically identified change ideas or initiatives from within each of the secondary drivers that we individually or collectively will work to implement as a means of reaching the improvement target. These ideas included below were first proposed when the driver diagram was established. The action to be taken on different ideas is determined by priorities set by members and the RAC leadership. As ideas are tried and analyzed, new change ideas or possible interventions are likely to be generated. The RAC members are committed to implementing actions, measuring and analyzing results, and revising plans in a manner that moves the RAC forward, toward the target of increasing the number of secondary mathematics teachers.

As previously discussed, the PR^2 member institutions are diverse, which adds to the complexity of the work that is undertaken. While each institution is committed to the five primary drivers, given the range of types, sizes, and locations of the member institutions, each institution needs to determine how best to address the primary and secondary drivers. In addition (with a few exceptions) the work of candidate recruitment and retention is not included in the responsibilities of faculty. As is the case with such diverse institutions, some of the faculty are at large teaching institutions and thus have large teaching loads each term, and others are at research institutions, with a high expectation for research and publication. Still others are at institutions that have both a high teaching load and the expectation of research and publication. For these reasons, the work of the RAC is somewhat more complex than the work of other RACs in the partnership. Still this RAC strives to create a framework to study recruitment and retention across the U.S.

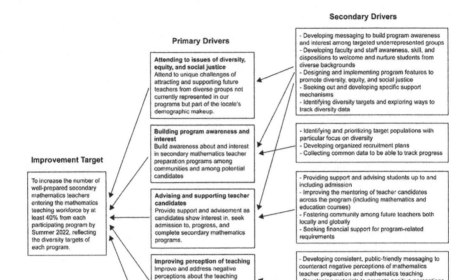

FIGURE 15.1. Program Recruitment and Retention Driver Diagram. The Program Recruitment and Retention RAC driver diagram has five primary drivers, three of which are consistent with the work of the prior MATH RAC. The secondary drivers serve as resources for RAC members.

PDSA CYCLES: PROGRESS AND CHALLENGES

Following the 2018 MTE-Partnership Conference, the immediate goal of the RAC was to systematically gather data to help member institutions better understand the impetus behind prospective teachers choosing their programs, as well as the reasons given by prospective teachers for leaving the programs. The PR² RAC conjectures that this information will help refine recruitment efforts and better meet the needs of prospective teachers once they commit to their programs. As previously discussed, member institutions are diverse both in their mission and the students they serve. The secondary teacher education programs range from four-year undergraduate degree and credential programs to two-year post-baccalaureate master's and credential programs, and are situated in small rural settings as well as large urban ones, so gathering this data for each individual program is essential. The data will also be aggregated to understand common themes and guide work going forward.

Individual Institutions

One of the challenges faced by the members of the PR^2 RAC is the diversity across the different programs in the RAC. Examples of the recruitment efforts from four of the programs are below. Given that each program is unique, the work of the PR^2 RAC is program-specific. While all of the institutions participating share some of the same goals—the recruitment and retention of more diverse and better-prepared secondary mathematics teachers—the strategies that work in one context may not be effective in another. For this reason, the Networked Improvement Community (NIC) model is exceptionally well-designed to respond to this problem. The four features of a NIC provide explicit guidance for the varied programs in the RAC, not only to focus their work on specific needs, but to also contributes to the RAC's understanding of more, and less, effective efforts:

1. Focused on a well-specified aim;
2. Guided by a deep understanding of the problem, the system that produces it, and a theory of improvement relevant to it;
3. Disciplined by the rigor of improvement science; and
4. Coordinated to accelerate the development, testing, and refinement of interventions and their effective integration into practice across varied educational contexts (LeMahieu, 2015).

It is essential that programs hoping to learn and benefit from the work of the RAC find ways to contextualize the work to their own settings.

California State University East Bay (CSUEB). CSUEB is one of the 23 campuses that are part of the California State University (CSU) system. The Teacher Education Program is a fifth-year, post-baccalaureate program. CSUEB is one of three CSU campuses in the San Francisco Bay Area and is one of the most diverse public colleges on the U.S. mainland (35.8% Hispanic/Latino, 22.5% Asian, 14.3% White, 10.1 % African American/Black, 5.2% Multiple Ethnicity, 1% Hawaii/Other Pacific Islander, 0.2% American Indian/Alaskan Native, and 5% unknown). The demographics of the prospective teachers enrolled in the Department of Teacher Education's teacher preparation programs, however, do not match the diversity of the undergraduate population. Considering that the majority of both the undergraduate and the post-graduate population come from the local community, and the graduates of the teacher preparation programs go on to teach in local schools, this discrepancy is puzzling. For this reason, as well as the teacher shortages mentioned above, the CSU has created EduCorps as a means of identifying, recruiting, and supporting prospective teachers from current undergraduates, as well as local community college and high school students.

Two main features of EduCorps are faculty nominations and a Celebration of Teaching event. Faculty nominations are made by undergraduate faculty who are asked to identify students who demonstrate qualities that make them a good fit for teaching. The intention of the faculty nomination process is to reach out to students who may not otherwise be considering a career in teaching. Nominated students receive an email informing them that they have been nominated and inviting them to attend the Celebration of Teaching event hosted by the university. At this event, students hear from teacher education faculty, district personnel, and current teachers.

The first CSUEB Celebration of Teaching was well attended and received; the second event took place in March of 2019. It was attended by a mix of CSUEB undergraduates, career-changers, and community college students. Several of the attendees have since reached out to the department of teacher education to follow up and apply to the program. The program is currently determining how best to track the effectiveness of the events.

Mississippi State University (MSU). MSU is the largest land-grant institution in Mississippi. As a Carnegie designated "R1: Doctoral Universities—Very high research activity," it is also the leading research institution in the state. MSU is responsible for the education of close to half of the teachers in Mississippi. The College of Education, collaboratively with the University of Mississippi, houses the Mississippi Excellence in Teaching Scholarship program, awarding scholarships to top performing high school students wishing to enter the teaching profession. In addition to tuition, scholarship recipients are provided a study abroad opportunity as well as a trip to a national teaching conference and mentoring through their first years of teaching. While this scholarship is awarded across the high-needs areas, mathematics is specifically targeted. The successes and barriers of recruiting these students to this program have helped to inform the recruitment practices at MSU. In 2018, MSU was awarded a Noyce Capacity Building grant to further examine the recruitment and retention of teachers in mathematics and science.

MSU has been a member of the MTE-Partnership since its inception. The state networked community of mathematics educators supports the work of MTE-Partnership and has drawn on the work of MTE-Partnership and other partnerships to inform work within the state of Mississippi.

Florida International University (FIU) is a public research university in south Florida that is classified by Carnegie as a "R1: Doctoral Universities—Very high research activity" and is currently classified among the top 100 public universities in the country by US News and World Report. FIU and its School of Education and Human Development is the largest producer of teachers in south Florida. The university has an enrollment of over 56,000 undergraduate and graduate students including 68% Hispanic/Latino, 13% Black or African American, 9% White, 2% Asian and 6% Non-resident Alien. Secondary mathematics education students are part of FIUteach, a replication of UTeach at University of Texas-Austin (Brainard, 2007) for preparing both mathematics and science secondary school teachers. FIUteach has recently secured funding from the National Science Foundation (NSF) to provide Noyce Scholarships to FIUteach students with a focus on recruitment of diverse mathematics and science teacher candidates to teach in hard-to-staff schools.

FIU has participated in the MTE-Partnership since its inception. During FIU's participation in the Partnership, we have revised our approach to recruitment in order to involve faculty and staff throughout all aspects of the process. For example, we are in regular communication with university administrators and staff involved in recruitment and new student orientation efforts at FIU. We have found this communication to be of utmost importance to maintain common messaging about our FIUteach program, as well as coordinate recruitment efforts effectively by keeping abreast of changes the University is making to their efforts from year to year. Examples of recruitment strategies we have found successful include sending direct emails to incoming STEM students about the FIUteach program timed in conjunction with the new student orientations and providing a short pitch about the program at new student orientations during the overview introductory session that directs interested students to our table for more information during the Resource Fair event that follows. We also communicate regularly with the student advisors in order to coordinate the messaging about the FIUteach program that potential and current students are receiving. Additionally, program faculty join the advisors as part of new student orientations during the time designated for groups of STEM students to select and enroll in classes.

University of Hawai'i at Hilo & Manoa. University of Hawai'i at Hilo & Manoa serves as the hub of the MTE-P Hui team. As members of the PR^2 RAC, the MTE-P Hui team developed PDSA cycles to examine strategies for recruitment and retention of mathematics teachers. During the 2018–2019 academic year, their work continued to attract, prepare, and maintain highly qualified secondary mathematics teachers for Hawai'i schools. One effort involves working to continue to develop connections between the new-teacher efforts of the University of Hawai'i (UH) system and a campaign specific to recruiting mathematics teachers, as well as supporting other outreach efforts to prospective mathematics teachers. In addition, the members have worked with the UH system to develop another round of advertising targeting teacher recruitment as well as working with the university administration to submit a proposal to the state Legislature that would provide scholarships for teacher education.

LONG-TERM GOAL

There is little research presently available on the experiences of prospective teachers of color in secondary teacher preparation programs. As recruitment efforts continue, the PR^2 RAC hopes to contribute to the field to better understand the strategies that encourage diverse prospective teachers to pursue teaching mathematics as a career, as well as those that help them complete secondary education mathematics programs and enter the profession. The field needs both quantitative and qualitative data to understand both larger trends and specific needs of prospective teachers at different institutions.

Currently there is a large campaign to broaden the participation of diverse students in STEM fields. This effort needs to involve increasing the diversity of classroom teachers, including that of teachers of mathematics as underscored by AMTE (2017). In an initial review of the literature, the PR^2 RAC has found a lack of research into how to attract prospective teachers from a diverse variety of backgrounds. Instead, much of the work is based on research conducted around the turn of the century (McGee, 2015). This discrepancy necessitates the need for current research to reflect the changes in population demographics and the needs of prospective teachers. Initial research efforts of the RAC indicate that demographics and recruitment are dependent on the communities and populations served by each institution of higher learning, which highlights the complexity of increasing the diversity of the prospective teachers. The PR^2 RAC needs to develop a framework that addresses the overarching issue of prospective teacher recruitment and retention, but is also flexible enough to respond to the needs to specific institutions.

In addition, universities continually perfect recruitment strategies to attract more and more students to their institutions. Teacher preparation programs need to capitalize on and learn from these efforts with their own targeted recruitment. It is time for administrators of teacher preparation programs to understand that prospective teacher recruitment needs dedicated recruitment specialists; it cannot be one more responsibility of one or two faculty members. For most teacher preparation faculty, recruitment efforts are "on top of their regular school or university duties" (Dickey, 2017). To date, the RAC's efforts are focused on understanding the needs of our current demographics, investigating methods to attract more diverse prospective teachers, and begin to build a body of knowledge to support targeted recruitment efforts into teacher education.

FOCUS ON IMPROVEMENT

The work of the PR^2 RAC evolved from the efforts of the MATH RAC described in Chapter 14 of this book. A resource recommended in Table 15.1 as a starting point is the AMTE (2017) *Standards for Preparing Teachers of Mathematics*. This document provides standards for effective mathematics teacher preparation programs including recruiting, nurturing and graduating high-quality teachers from diverse communities and backgrounds. Another resource for those interested in increasing the number and diversity of students entering teacher preparation programs, particularly in high need areas such as mathematics, and providing support for these students, we recommend exploring the California State EduCorps website proposed in Table 15.1. This website is a model of how a university state system can collaborate to increase numbers of teachers in hard to staff fields in their state. A key undertaking of this collaboration is the faculty nomination of students who show predisposition toward teaching. These nominees are invited to a recruitment event. In this chapter, the section about California State University, East Bay, provides further information about this venture. An additional resource that is not directly related to recruitment but serves as a platform for discussing secondary preparation programs is NCTM's (2018) *Catalyzing Change*. While this book focuses on needed changes in the secondary mathematics classroom, its recommendations provide guidance for those involved in mathematics teacher recruitment and retention. A substantive outcome of the MATH RAC was the *Secondary Mathematics Teacher Recruitment Campaign Implementation Guide* recommended in Table 15.1. The content of each of the nine modules of this recruitment guide are summarized in Table 14.1 of Chapter 14 in this book. These modules will be available through OpenCanvas, the online platform adopted by the MTE-Partnership to provide access of resources to the broader community. Finally, in Table 15.1, we recommend reading the USDOE (2016) report, *The State of Racial Diversity in the Educator Workforce*. This report provides valuable foundational knowledge about the present racial diversity of students, teachers, and principals in schools. The report underscores the importance of efforts to increase number of diverse teacher candidates. It also provides examples of programs and efforts producing successful results in this area.

Reading List

The following annotated bibliography provides more information about these recommended readings.

1. Association of Mathematics Teacher Educators. (2017). *Standards for preparing teachers of mathematics*. Retrieved from: http://amte.net/standards
   – AMTE released the first-ever standards for preparation of mathematics teachers in PK–12 classrooms. This comprehensive document provides a framework for the development of exemplary preparation programs, including addressing the recruitment and retention of teacher candidates from diverse communities. Further, guidance is provided for how the work within the larger teacher education community, including accreditation groups, policy makers, and other stakeholders.

2. California State University System. (n.d.). *California State University EduCorps*. Retrieved from: https://www2.calstate.edu/educorps
   – Built upon the framework of the national EduCorp, this site provides a networked support for students entering programs that lead to teaching licensure. This site serves as a model of how a university system can work together to increase the numbers of teachers in especially critical needs areas across the state.

3. National Council of Teachers of Mathematics. (2018). *Catalyzing change in high school mathematics: Initiating critical conversations.* Reston, VA: National Council of Teachers of Mathematics.

    – *Catalyzing Change* is designed to provide a framework for discussing the state of high school mathematics. Although not specifically written for the preparation of teachers, this book discusses the changes needed to make our nation's' students competent in numbers and statistics. This book should be referenced as we consider reforms to our teacher preparation programs.

4. Ranta, J., & Dickey, E. (2015). *Secondary mathematics teacher recruitment campaign implementation guide.* Washington, DC: Association of Public and Land-grant Universities. Retrieved from: http://bit.ly/MATHImplGuide

    – The *Secondary Mathematics Teacher Recruitment Campaign Implementation Guide* is a result of the initial work of the MATH RAC (see Chapter 14). This guide consists of nine modules designed to lead you through a recruitment campaign. Each module contains specific guidance to aid even a novice through recruitment of secondary mathematics teacher candidates.

5. U.S. Department of Education, Office of Planning, Evaluation and Policy Development, Policy and Program Studies Service. (2016). *The state of racial diversity in the educator workforce.* Washington, DC: Author. Retrieved from http://www2.ed.gov/rschstat/eval/highered/racial-diversity/state-racial-diversity-workforce.pdf

    – This report discusses the current racial diversity of students, teachers, and principals in schools, as well as the diversity of teacher candidates in teacher preparation programs and other facets of the educator pipeline. Across the different sections of the report, selected efforts are presented as a spotlight to describe endeavors with successful results for increasing the diversity of teachers, as well as school and district leaders.

For current information on the PR[2] RAC, please visit mtep.info/PR2.

CONCLUSION

Recruitment and retention to mathematics teacher education programs are complex endeavors. Institutions of higher learning and other types of credentialing programs are diverse in many ways: the types of institution, the diversity of the students served, the diversity of the communities served by the teachers prepared, and the policies that govern the programs. Recruitment is often an added responsibility to faculty already carrying a full teaching and research load. There is still much to learn about effective recruitment and retention.

ENDNOTE

1.   In addition to the lead authors Dana Franz, Maria L. Fernandez, and Julie McNamara, Diane Barrett also contributed to the writing of this chapter.

REFERENCES

ACT. (2018). *ACT national profile report.* Retrieved from http://www.act.org/content/dam/act/unsecured/documents/cccr2018/P_99_999999_N_S_N00_ACT-GCPR_National.pdf

ACT, & United Negro College Fund. (2016). *The condition of college and career readiness 2015: African American students.* Retrieved from: http://www.act.org/content/dam/act/unsecured/documents/6201-CCCR-African-American-2015.pdf

Association of Mathematics Teacher Educators. (2017). *Standards for preparing teachers of mathematics.* Retrieved from http://amte.net/standards

Brainard, J. (2007, December 21). Texas offers a model for training math and science teachers. *Chronicle of Higher Education, 54*(17), A8–A10.

Bryk, A., Gomez, L. M., Grunow, A., & LeMahieu, P. (2015). *Learning to improve: How America's schools can get better at getting better.* Cambridge, MA: Harvard Education Press.

Council for the Accreditation of Educator Preparation. (2013). *The CAEP Standards.* Retrieved from http://caepnet.org/standards

D'Amico, D., Pawlewicz, R. J., Earley, P. M., & Mcgeehan, A. P. (2017). Where are all the black teachers? Discrimination in the teacher labor market. *Harvard Educational Review, 87*(1), 26–49.

Dickey, E. (2017). Reflections on the partnership. In W. M. Smith, B. R. Lawler, J. Bowers, & L. Augustyn (Eds.), *Proceedings of the sixth annual Mathematics Teacher Education Partnership conference* (pp. 162–171). Washington, DC: Association of Public and Land-grant Universities.

Gershenson, S., Holt, S. B., & Papageorge, N. W. (2016). Who believes in me? The effect of student-teacher demographic match on teacher expectations. *Economics of Education Review, 52*, 209–224.

Gist, C. D. (2016). Voices of aspiring teachers of color: Unraveling the double bind in teacher education. *Urban Education, 52*(8), 927–956.

LeMahieu, P. (2015). (2015, August 18). *Why a NIC?* [Blog post]. Retrieved from https://www.carnegiefoundation.org/blog/why-a-nic/

Mathematics Teacher Education Partnership. (2014). *Guiding principles for secondary mathematics teacher preparation.* Washington, DC: Association of Public and Land-grant Universities. Retrieved from mtep.info/guidingprinciples

McGee, E. O. (2015). Robust and fragile mathematical identities: A framework for examining racialized experiences and high achievement among black college students. *Journal for Research in Mathematics Education, 46*(5), 599–625.

McKown, C., & Weinstein, R. S. (2008). Teacher expectations, classroom context, and the achievement gap. *Journal of School Psychology, 46*, 235–261.

Muller, C., Riegle-Crumb, C., Schiller, K. S., Wilkinson, L., & Frank, K. A. (2010). Race and academic achievement in racially diverse high schools: Opportunity and stratification. *Teachers College Record, 112*(4), 1038–1063.

National Research Council. (2010). *Preparing teachers: Building evidence for sound policy*. Washington, DC: The National Academies Press.

Pringle, B. E., Lyons, J. E., & Booker, K. C. (2010). Perceptions of teacher expectations by African American high school students. *The Journal of Negro Education, 79*(1), 33–40.

Ranta, J., & Dickey, E. (2015). *Secondary mathematics teacher recruitment campaign implementation guide*. Washington, DC: Association of Public and Land-grant Universities. Retrieved from: http://bit.ly/MATHImplGuide

U.S. Department of Education, Office of Planning, Evaluation and Policy Development, Policy and Program Studies Service. (2016). *The state of racial diversity in the educator workforce*. Washington, DC: Author. Retrieved from http://www2.ed.gov/rschstat/eval/highered/racial-diversity/state-racial-diversity-workforce.pdf

CHAPTER 16

RETAINING BEGINNING SECONDARY MATHEMATICS TEACHERS THROUGH INDUCTION AND LEADERSHIP SUPPORT

Lisa Amick, James Martinez,
Megan W. Taylor, and Frederick Uy

Transforming the preparation of secondary mathematics teachers is at the core of the Mathematics Teacher Education Partnership (MTE-Partnership). Since its inception in 2012, this initiative continues to improve mathematics teacher education across the United States. The MTE-Partnership has established the *Guiding Principles for Secondary Mathematics Teacher Preparation Programs* (2014) and five Research Action Clusters (RAC) to carry out these principles. In fact, two of the *Guiding Principles* are focused on recruitment, selection, and support; principle 8 is centered on teacher candidates, and principle 9 is focused on in-service and beginning teachers. These are the principles upon which the Secondary Teacher Retention and Induction in Diverse Educational Settings (STRIDES) RAC was built. As the name suggests, its main goal is attracting and maintaining an adequate supply of secondary mathematics teachers. This RAC works with

The Mathematics Teacher Education Partnership: The Power of a Networked Improvement Community to Transform Secondary Mathematics Teacher Preparation, pages 351–370.
Copyright © 2020 by Information Age Publishing
351

teacher preparation programs in higher education institutions as well as in K–12 partner districts to actively recruit and retain high-quality and diverse teachers. Local educational agencies (LEAs) support this effort by monitoring and supporting STRIDES efforts in the field. Finally, Standard P.5. of the *Standards for Preparing Teachers of Mathematics* (Association of Mathematics Teacher Educators [AMTE], 2017), specifically delineates teacher recruitment/retention as one of the "five important standards that are essential aspects of effective mathematics teacher preparation programs" (p. 43).

CONTEXT

Every student in the U.S. deserves to have a highly qualified teacher. That said, at least 50% of all teachers leave the profession within the first five years (Foster, 2010), creating a revolving door of beginning teachers into and out of the profession. The possibility of a student not having at least one qualified teacher during their elementary and secondary schooling is a reality, and mathematics instruction is a specific area of concern. Additionally, the rate of departure for mathematics teachers is highest in high poverty schools (e.g., Goldring, Taie, & Riddles, 2014). This crisis of teacher retention has led to an increase in underprepared and unprepared mathematics teachers entering the classroom, as teacher preparation efforts have prioritized recruitment over preparation and retention. According to the Learning Policy Institute, 40% of newly hired mathematics or science teachers are underprepared, and underprepared teachers are far more likely to teach in schools serving students of color and low-income students (Carver-Thomas, 2018). Many teacher education experts agree that addressing the mathematics teaching crisis meaningfully will require building a more cohesive system of teacher preparation, support, and development that prioritizes recruitment and retention efforts equally (Mehta, Theisen-Homer, Braslow, & Lopatin, 2015).

TABLE 16.1. Strategies to Building a Strong and Stable Teacher Workforce

Key Strategies	Descriptors
High-retention pathways into teaching	• Common, clear vision of good teaching that permeates all experiences • Rigorous criteria for recruiting, selecting, and assessing a diverse, effective pool of teachers • Development of teachers' expertise in child development and content-specific and culturally relevant pedagogical knowledge • Sustained clinical practice • Sustained, high-quality mentoring and induction • Service scholarships, student loan forgiveness, and competitive compensation
Deep partnerships across teacher preparation programs, districts, and schools	• Explicit support for long-term teacher learning and leadership • High-quality school principals • Cost-sharing and strategic reallocation of existing funds

Fortunately, the field knows a great deal about how to build a more cohesive system of support for preparing and retaining high-quality mathematics teachers. The features of a system that is designed to build a strong and stable teacher workforce overtime have been identified (Espinoza, Saunders, Kini, & Darling-Hammond, 2018), detailed in Table 16.1. Not surprisingly, they emphasize a robust commitment to providing more cohesive, long-term support for teachers. All of the most promising teacher preparation and support programs in the field attend to these features to some degree, and of those, programs like UTeach (https://uteach.utexas.edu), the Woodrow Wilson Academy for Teaching and Learning (https://woodrowacademy.org/), and Trellis Education (http://www.trelliseducation.org), focus on more effective systems for preparing and retaining mathematics teachers, in particular. Additionally, there is national advocacy for existing and emerging teacher residency models to hold these key features as non-negotiable (The Sustainable Funding Project, 2016).

THEORY OF CHANGE

In the fall of 2014, a group of mathematics education researchers and teacher educators were brought together with the specific purpose of engaging in a review of literature on teacher attrition and retention, and to suggest a course of action for how the MTE-Partnership might positively impact teacher retention across its partners. The working group worked to operationalize the research by defining key strategies that develop a more systematic approach to teacher preparation, support, and retention, and including these strategies in a driver diagram with an improvement target specific to MTE-Partnership partner universities and school districts (see Figure 16.1). The defined improvement target addresses the fact that almost a third of America's teachers leave the field before their third year of teaching (National Commission on Teaching and America's Future, 2002), and that this attrition is more acute for mathematics teachers and for teachers serving "high-poverty, high-minority, urban, and rural public schools" (Ingersoll & Perda, 2010, p. 588).

The identified primary drivers—long-term support, importance of professional communities, cohesiveness, role of administrators, and professional pathways—came from this investigation of key strategies for building a strong and stable teacher workforce, with particular respect to secondary mathematics teaching. First, the quality of professional support—both in a teacher's preparation to teach and during the first few years in the classroom—is critical to retention (Jacob & McGovern, 2015) and should be focused specifically on the teaching and learning of mathematics (e.g., Loucks-Horsley, Stiles, Mundry, Love, & Hewson, 2010). As important is the cohesiveness of that support over time and the coherence of various support activities (Primary Driver 3; Fantilli & McDougall, 2009), especially when research shows that teachers' improvement tends to stagnate after year 2 (Kane, Rockoff, & Staiger, 2008). The top three primary drivers are addressed through the use of formal, professional learning communities (PLCs).

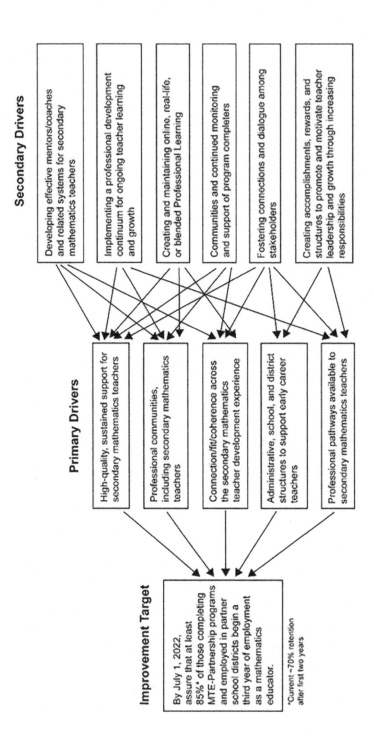

FIGURE 16.1. Driver Diagram for STRIDES. The driver diagram for STRIDES guides the work of the STRIDES RAC. The overarching goal is to ensure early career mathematics teachers make it to their third year of teaching and the drivers are focus areas that will help researchers attain that goal.

If incorporated at the site at which they teach, early career teachers who benefit from effective PLCs are more likely to continue to teach and are more likely to improve their teaching over time (Donovan, 2015). As noted in the driver diagram (see Figure 16.1), administrators at the university, district, and school level play a critical role in this kind of enactment of high-quality, sustained preparation and support for new teachers (Espinoza et al., 2018). Administrators must carve out professional pathways for teachers that not only ensure the critical components of a teacher's pathway into the profession are protected and prioritized (a primary driver, see Figure 16.1), but also incentivize teachers to continue to teach.

The secondary drivers were defined from the highest-leverage support mechanisms for teachers linked to retention that all MTE-Partnership partner teacher preparation programs would be able to operationalize in context-specific ways. It felt important to the planning team that the driver diagram be a unifying theory of change for the potentially varied innovations of individual MTE-Partnership programs, and thus the secondary drivers provided the working groups' suggestions for access points.

· THE WORK OF STRIDES

As noted earlier, the STRIDES RAC is focused on addressing the MTE-Partnership's guiding principle on student recruitment, selection, and support. Teacher preparation programs actively recruit high-quality and diverse teacher candidates and monitor/support them as they complete their programs. Since the inception of the MTE-Partnership, the national problem of retaining secondary mathematics teachers within the profession has been a priority. A RAC focused on both recruitment and retention was proposed at the 2013 annual MTE-Partnership Conference, but as the details of that RAC were fleshed out, the workload seemed to be too cumbersome for one research team to tackle. Recruitment was determined to be a higher priority at the time, and this led to the formation and implementation of the RAC named Marketing for Attracting Teacher Hopefuls (MATH; see Chapter 14 of this book). As the MTE-Partnership grew and gained more members, and teacher retention continued to be a concern, the need for a separate RAC that focused on retention was again discussed.

Understanding the Nature of Support for Early Career Mathematics Teachers

Members of the STRIDES RAC decided early on that the work of the team must focus on understanding and providing support for both prospective and early in-service teachers, given the role of a cohesive, continuum of professional learning on teacher growth and retention. Thus, to launch early initiatives aimed at improving teacher retention rates, STRIDES members designed an initial survey in the summer of 2015 to gather preliminary data on the nature and quality of professional support for prospective, first-, second-, and third-year teachers. Collecting, analyzing and understanding data about support within school structures is connected to the primary driver of administrative, school, and district structures to support early career teachers. Early on, STRIDES determined that the support and advocacy of administrators were needed in getting new teachers to stay in the profession and build their careers. Specific research questions guiding this work were: What is the perceived scope, nature and impact of professional support for early career mathematics teachers, and how does this: (a) change as teachers progress in their teaching career, and (b) relate to how likely it is a teacher will remain teaching?

Initial Survey. Researchers from 13 institutions and secondary mathematics teachers from four school districts designed a 19-question pilot survey, Reflection on Professional Activities. This survey was created through an iterative design and vetting process, having stemmed from a discussion centered on research-based reasons that teachers leave the teaching field. It was important to the RAC leaders that survey questions were connected to past research and reflective of current issues regarding job satisfaction and support of early career mathematics teachers. In addition to an online search of related questions from other instruments used in past research studies, the survey went through careful examination among the diverse group of STRIDES members to ensure the questions captured all the essential components of being a professional. These components included in-school and out-of-school support, mentoring, professional learning, and becoming a part of a learning community, to name a few.

Aside from demographic questions, the participants were provided a list of professional development activities and asked in which of those listed did they participate, and how worthwhile were those identified. The survey also allowed these prospective and early career teachers to respond to the open-response prompt, "please describe the most meaningful activity you participated in (that keeps) you inspired as a mathematics teacher... (including) information about the name, location, participants, and nature of the work, your role, and whether it was required or not." A follow-up question allowed the participants to state why they thought this "most meaningful activity" was inspiring. Other questions on the pilot survey asked the participants to describe "the most meaningful activity you participated in that improved your teaching practice in some way" and describe and rank the importance of "professional learning communities (they were) currently participating in." Finally, the survey gauged the importance of administrators at their schools by asking the participants to "describe any professional support you received from administrators."

Second-Round Survey. The data collected from this first pilot survey were used by members of the STRIDES RAC to create a more detailed, second-round pilot survey that delved deeper into the attitudes early career mathematics teachers specifically had with regard to: (a) professional learning opportunities, (b) professional learning communities, and (c) administrators. This second-round survey, administered in the spring of 2016, allowed a second group of participants to specify activities that have helped them grow professionally, and the degree to which these activities were worthwhile to them. In addition, collected were data about instructional context (i.e., public, private, etc.), and whether the early career teachers served students from special populations (i.e., special education, English language learner, gifted). Participant feelings regarding the degree that specific professional development activities changed these teachers' practices, as well as the level of "inspiration" these activities invoked were surveyed; thus, researchers could discern connections between these two measures. Open-ended responses allowed survey participants to provide additional details regarding their support systems. Finally, the degree that the participants felt that their administrators supported them professionally was measured, including specific relevant dimensions (e.g., assessment, instruction, curriculum, classroom management, collegial collaboration, course assignments/loads).

Early results of the second-round pilot survey indicated that early career teachers needed specific support and intervention with their teaching of mathematics and cooperation with colleagues and administrative persons at their schools. However, even after two rounds of pilot surveys, what specific support, by whom, and how often were still unclear. Further analysis of the data of the second-round pilot survey was accomplished by STRIDES RAC members at the 2016 annual MTE-Partnership Conference in Atlanta, Georgia. In adherence with principles related to Plan-Do-Study-Act (PDSA) cycles, the second-round pilot survey provided motivation for the development of a final, third survey, which was sent to all MTE-Partnership member institutions in the fall of 2016.

Third survey. This newly designed third survey incorporated revisions of questions posed in both pilot surveys as well as additional questions that evoked more specific responses than what was gained previously. The full-scale survey gathered information from participants from a wide geographic area and included responses from 141 early career mathematics teachers across the U.S. Figure 16.2 provides detail on how far the participants were into their tenure.

For further analysis of this survey, four questions were chosen by the committee. Results from these four select questions from this final survey would exemplify the degree that the prospective and early career teachers surveyed felt supported, by whom, and in what areas. Additionally, the data reveal participant attitudes about the support they felt they received, the importance they placed on professional learning activities, their understanding of their career choice, and how long they planned to remain in the teaching profession.

The teachers of the study reported, when looking back, would they still choose to become a teacher. The results of this question are depicted in Figure 16.3. This illustration provides important insight into what these teachers think of the profession. Eighty-one percent of the teachers who responded say that they certainly/probably would become a teacher again if they had it to do over. This percentage is higher than what is generally provided in similar studies (National Math + Science Initiative, 2013), possibly explained by the fact that these teachers all were

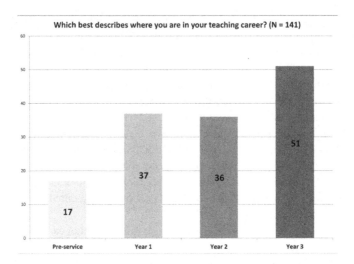

FIGURE 16.2. Years of Service. A Fall 2016 survey aimed to reach teachers in their first few years of teaching, and this data shows that the target population was reached. Seventeen of the respondents were currently student teaching, 37 were in their first year, 36 in their second year, and 51 were in their third year of teaching.

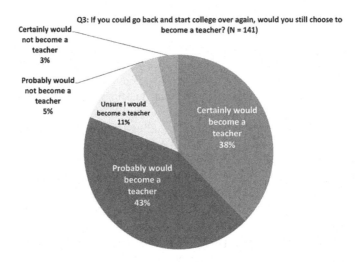

FIGURE 16.3. Repeat Career Choice. Fall 2016 survey results (n=141) asking prospective and early career mathematics teachers whether they would still choose to become a teacher if they were to go back and start college over again.

in MTE-Partnership schools, closely affiliated with, and supported by, mathematics faculty at higher education institutions. Therefore, this number may be higher than the general population of teachers. The rest of the data should be looked at through a lens of teachers who have high job satisfaction. Only 8 percent of teachers responding to this survey were fairly confident or confident that they would not choose teaching as a profession if they had it to do over, and with teacher turnover rates being so high, it can be assumed that this is much lower than the general population of teachers who are currently teaching.

The information in Figure 16.4 revealed to the researchers what types of professional learning activities that new teachers were utilizing and how influential those activities were in relation to their enthusiasm for teaching mathematics. Researchers found that a majority of the survey respondents did not feel that online resources, professional courses, or professional conferences were moderately or very influential forms of professional development. In addition, the most utilized forms of professional learning came from working with a mentor/coach and collaboration/planning with colleagues. Another finding depicted in Figure 16.4 is that school/department meetings and professional development workshops were

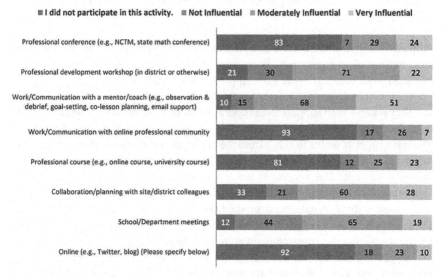

FIGURE 16.4. Professional Learning Activities Related to Enthusiasm. Fall 2016 survey data of prospective and early career mathematics teachers, asking about the relative influences of various professional development opportunities across the past five months.

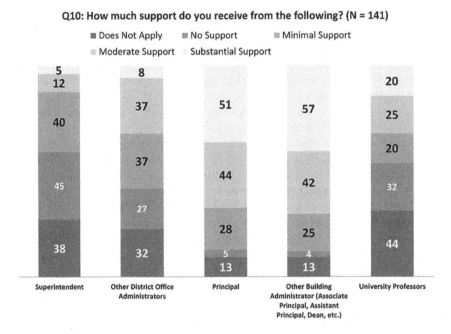

FIGURE 16.5. Current Support. Fall 2016 survey of prospective and early career mathematics teachers (n=141), asking them about forms of support they receive.

utilized by approximately 60% to 65% of the teachers surveyed and were moderately influential on enthusiasm for teaching. Local efforts seem to be more utilized and influential than any other form of professional learning.

As shown in Figure 16.5, approximately 50% to 60% of the teachers reported that they receive zero support from superintendents or university professors, either because none is provided or because it is not available at their current position. The most support seems to come from the principal and other building administrators such as assistant/associate principals and others in similar roles. Therefore, in-house administrator support is being utilized and is more impactful than the support provided through the central/district office or through local universities.

To further disaggregate the data, the team asked another question with regard to the details of who were the support providers for these teachers. The data in Figure 16.6 gave the research team some insight regarding how specific people are supporting survey respondents in different areas of teaching. For example, principals and assistant principals seem to be providing the most affirmation, while also providing support in a plethora of different areas; district office administrators are supporting more with curriculum materials than any other area. Deans and superintendents are not providing many of the support areas that were given as choices. Assistant principals were reported as providing more support

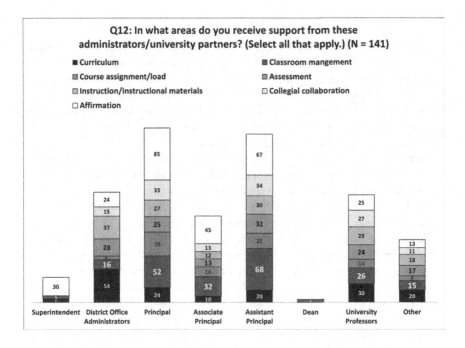

FIGURE 16.6. Forms of Support. Fall 2016 survey of prospective and early career mathematics teachers (n=141), asking them who provides them with support. Votes that were fewer than five in a category for the superintendent and the dean are not shown in the chart.

with classroom management than anyone else. It is essential to know what types of support are being provided by whom, so that when interventions are created and implemented, they can be targeted and specific in hopes to be as impactful as possible.

Summary of Select Data. Figures 16.3–16.6 revealed to the STRIDES RAC members that prospective and early career mathematics teachers from MTE-Partnership-affiliated schools, whose general dispositions about remaining in the teaching profession were positive, expressed strong feelings about the degree that specific learning activities increased their enthusiasm for teaching and the degree that they felt supported by a range of administrators in a variety of forms. One surprising result is that, compared to other support options, the novice teachers surveyed did not feel that online resources were as influential in increasing their enthusiasm for teaching mathematics. Planned professional development and collaborating with colleagues were more influential, in their opinion. Also significant was the role of the site principal, relative to other administrative supporters considered.

Development of Support Interventions During Induction

Results of this final pilot survey, aided by results of previous surveys, motivated the STRIDES RAC members to focus on developing interventions that increase professional support for early career teachers. These interventions aspired to increase, both in the short and long term: (a) professional learning activities, (b) participation in professional communities, and (c) support from administrators. The interventions were meant to further support systems that were working and also meant to fill in gaps where support was lacking.

Initial Steps. The 2016 annual meeting attendees ended their work at the meeting by splitting into sub-groups (supporting the professional learning of the early career mathematics teachers and supporting administrators) to develop these interventions to launch later that fall. Each group examined qualitative and quantitative responses from early career teachers who participated in prior surveys, summarized key themes and questions for the group, and presented key inferences for all to access. Based on each inference, individual STRIDES members brainstormed ways members might revise the survey again and investigate ways to ensure that early career mathematics teachers' professional learning and support are more effective. In effect, members identified three change ideas to press forward the RAC's secondary drivers (see Figure 16.1). STRIDES members self-selected into one of three groups to create PDSA cycles that would guide early interventions in fall of 2016. These groups included attention to developing long-term collaborative groups for early career mathematics teachers, examining the role of administrators and site-based colleagues, and designing training and support for teacher mentors. Each group focused on a particular change idea that correlated to interventions that would address the most pressing needs identified by early career teachers in the initial surveys (see Table 16.2).

Progress made between June 2016 and June 2017 was primarily accomplished using online (i.e. Zoom platform) meetings by STRIDES RAC members. The data from the final pilot survey was accomplished during this time and online conversations about next steps was considered. In preparation for the 2017 annual MTE-Partnership Conference held in New Orleans, co-leaders of the STRIDES RAC prepared overview materials to assist all newcomers to the RAC, recognizing that these persons could contribute substantively, given an objective look at the previous work performed by STRIDES.

Move to Interventions. During the 2017 annual MTE-Partnership Conference, STRIDES RAC members (veteran and new) revisited the aims set forth one year earlier, identified progress and roadblocks, revisited the RAC driver diagram, and established goals for collaborative interactions. Persons with a variety of backgrounds/skill sets were present (mathematicians, mathematics educators, and school district representatives), many of which were new to STRIDES having been recruited during open session time in the first morning (second day) of the annual meeting. After some debate, goals for the working meeting of the STRIDES RAC during the conference were established to: (1) develop specific

TABLE 16.2.   STRIDES Key Change Ideas, Summer 2016

Change Idea	Description of Change Idea	What We Want to Learn
Long-Term Collaborative Groups for Early Career Mathematics Teachers	Teacher collaborative groups that focus sustained attention on one teaching practice for early career mathematics teachers can support a sense of professionalism and professional improvement for early career mathematics teachers.	How can collaborative teacher groups be supported to engage in sustained attention to a particular teaching practice that early career mathematics teachers can grow in? How does participation in a collaborative group develop a sense of accomplishment in relation to the selected practice? What structure of collaborative group creates a positive perception about their profession and future professional trajectory?
Role of Administrators and Site-Based Colleagues	Support administrators by creating a common vision and using strategies to reinforce retention of early career mathematics teachers.	What targeted supports for administrators impact teacher retention?
Training and Supporting Teacher Mentors	Provide training for mentor teachers in order to improve teacher retention.	How can mentor teachers use the learning cycle to facilitate early career (prospective through third year) mathematics teachers enactment of the eight core teaching practices defined by the National Council of Teachers of Mathematics (NCTM, 2014)?

interventions based on collected data; and (2) determine which STRIDES members would commit to future tasks.

To achieve the second goal, the group re-analyzed data from the final pilot survey with two foci: to set the stage to develop intervention(s) targeting professional learning and support for early career teachers and ensure that resulting interventions would be consistent with data collected. The group distilled themes and questions from the survey results to identify possible interventions for bolstering professional learning and support of early career teachers. In addition to analyzing the survey data, a number of topics were discussed which further defined potential STRIDES interventions, including: compensation for participating educators for participating in interventions, how retention connects to improved student learning and can this be measured, and the potential of using (online) school district level informational meetings with administrators and/or other teachers to highlight retention issues (e.g. placement).

Additionally, the group reviewed a mock 10-minute intervention meeting (devised by a STRIDES member), which was piloted in May with an early career mathematics teacher and site principal from Pacifica High School, a public comprehensive high school in Oxnard, California. After a discussion of feedback from the participants, the overall consensus that this interactive collaboration, which had the early career mathematics teacher and site principal engage in a discussion about research-based best practices in teaching mathematics, had merit and could be used to support the STRIDES goals. In addition, the STRIDES RAC connected the interventions to the secondary drivers in the RAC driver diagram, specifically ones related to mentoring/coaching and administrative support.

As a result of this analysis and related discussions, a research question was identified that would help to define future efforts—how does targeted, interpersonal support by on-site instructional leaders impact the retention of early career secondary mathematics teachers? The following interventions were proposed: mentoring of early career mathematics teachers, support for administrators to support early career mathematics teachers, training and supporting mentors, and determination of possible funding for continued work. Survey results suggested early career secondary mathematics teachers found these first three support systems to be impactful or are supports they wish they had. Note the first two interventions reflect the modifications to the change ideas identified in 2016 (see Table 16.2). And while the third intervention remained similar, the fourth was added to respond to the need to consider financial elements of the project.

The group also agreed that due to time constraints, the administration of future surveys would not be needed to further define PDSA cycle interventions. The work STRIDES set forth for the 2017–2018 academic year is summarized in Table 16.3, organized around the four proposed change ideas.

Building Momentum. While analyzing the final pilot survey results, the members realized that, of all the input related to teacher support communicated in the survey by participants, two key focus areas arose: mentoring and administration. Therefore, during the 2018 annual MTE-Partnership Conference, STRIDES divided itself into two sub-groups aligned with these focus areas. Each sub-RAC identified research teams to examine additional interventions during the 2018–2019 academic year. For example, some members of the administration sub-RAC will design a protocol for the early career secondary mathematics teacher to view a mathematics classroom video together with the school principal and then reflect on it jointly. The mentoring sub-RAC is developing a protocol for a short five-minute check-up meeting between the early career mathematics teacher and a mentor on a mutually agreed upon frequency. Each is described in greater detail below.

The administrative sub-RAC examined the impact of a brief meeting between an early career mathematics teacher and administrator around a video of mathematics instructional practice on the teacher's feelings of support. Two groups of teachers were used in this study, a comparison group that will operate "business as usual" and an intervention group that would experience a targeted experience. Participants completed an online survey that included questions on demographics, current feelings about support, personal connections at their current schools,

TABLE 16.3. STRIDES Key Change Ideas, Summer 2017

Change Idea	Description of Change Idea	What We Want to Learn
Long-Term Collaborative Groups for Early Career Mathematics Teachers	Teacher collaborative groups that focus sustained attention on one teaching practice for early career mathematics teachers can support a sense of professionalism and professional improvement for early career mathematics teachers.	How can collaborative teacher groups be supported to engage in sustained attention to a particular teaching practice that early career mathematics teachers can grow in? How does participation in a collaborative group develop a sense of accomplishment in relation to the selected practice? What structure of collaborative group creates a positive perception about their profession and future professional trajectory?
Role of Administrators and Site-Based Colleagues	Support administrators by creating a common vision and using strategies to reinforce retention of early career mathematics teachers.	What targeted supports for administrators impact teacher retention?
Training and Supporting Teacher Mentors	Provide training for mentor teachers in order to improve teacher retention.	Can mentor teachers use the learning cycle to facilitate early career (prospective through third year) mathematics teachers' enactment of the eight core teaching practices defined by NCTM (2014)?
Funding Sources	Determination of funding sources for continued work for STRIDES.	What sources of funding are available and applicable to the STRIDES work? What are applicable dates for submittals and which STRIDES member(s) will be working on this?

self-efficacy, best practice strategies, and job satisfaction. The comparison group of teachers participated in a 10-minute meeting at the school site with their site principals. These participants viewed a five-minute video on best mathematics teaching practices and spent five minutes afterward discussing the content of the video in the context of their school site. Within 45 to 60 days of the administration of the initial survey, all teachers in the study and their principals filled out another online survey for comparison to the one completed prior to the collaborative session. The study hoped to illuminate the effects that a brief meeting about best practices in mathematics between early career teachers and their site principals had on their feelings of support.

The mentoring sub-RAC worked with a group of teachers in their first year of teaching to understand effective mentoring. For example, previous STRIDES work identified that mentor teachers were quite impactful in regard to positive professional support to early career secondary mathematics teachers. Thus, the mentoring sub-RAC was attempting to make sure every first-year teacher they were working with had a mentor and aimed to strengthen that relationship. Some of the interventions included monthly targeted monthly emails for early career mathematics teachers to read and discuss with their mentor teachers. A second intervention was to create a virtual panel of teachers in years 3–5 who provided a sounding board for questions and comments for the early career mathematics teachers. The mentoring sub-RAC collected data throughout the year to capture the details of and impacts of these interventions and also submitted a grant to the CPM organization to assist in funding their intervention.

Reflecting on the Work of STRIDES

In summary, STRIDES has drawn upon the survey data collected from early career secondary mathematics teachers regarding the support they are receiving, to strengthen the systems that promote retention, and remove or modify the systems that are do not. The work of retaining well-prepared, beginning teachers must start as soon as they enter their teacher preparation programs (or in some cases, begin their internship, whichever comes earlier) and the aim of STRIDES is to establish support systems that are effective for teachers. It is the goal of STRIDES to begin these support systems with teachers while they are in their teacher preparation programs and continue them throughout the early years of their careers. STRIDES researchers hope that by impacting the support that these teachers are getting during such a critical time in their careers, teacher retention can be impacted within this group of teachers, and the intervention program will be disseminated so that teacher retention is impacted at-large.

FOCUS ON IMPROVEMENT

STRIDES has learned a lot from studying early career secondary mathematics teachers. The two major findings were that (1) they are greatly benefiting from in-house mentors that provide meaningful professional support, and (2) they could use some help to strengthen those relationships. And while bringing new teachers into the profession is important, keeping them in the profession by adequately supporting them is arguably as important. Hence, the induction process is the focus for improving teacher retention once in the career. Loosely defined, induction programs are strategic processes and assessments provided to inexperienced teachers. These programs help by giving new teachers the needed tools and practice while at the beginning of their careers. Inductions may be in the form of mentoring, planning, professional learning, observations, dialogues, and evaluations. STRIDES supports work to improve and sustain induction programs, with an aim to 85% of MTE-Partnership graduates remain in the profession after three years (see Figure 16.1).

In the induction process, STRIDES has found value in beginning with surveying the needs of the early career mathematics teachers. Such a survey provides a better idea of how to collaborate better with these early career teachers in mathematics.

Reading List

To better understand these and other needs of early career secondary mathematics teachers, and to devise an agenda for retaining them in the careers, consider the following list of recommended readings to initiate efforts for secondary mathematics teacher retention.

1. Bullough, R. (2012). Mentoring and new teacher induction in the United States: A review and analysis of current practices. *Mentoring & Tutoring: Partnership in Learning, 20*(1), 57–74.
   - The article reviewed practices in mentoring and induction in three states: California, New York, and Texas. Further, it discussed how views of mentoring were changing, what literature and studies in induction were lacking, and what the future of mentoring might be.
2. Nishimoto, M. C. (2012). *Secondary preservice teacher expectations of the principal's role in new teacher induction* (Doctoral dissertation). Retrieved from https://digitalscholarship.unlv.edu/cgi/viewcontent.cgi?referer=https://scholar.google.com/&httpsredir=1&article=3802&context=thesesdissertations
   - Nishimoto discussed the challenges faced by early career mathematics teachers during induction in the attempt of understanding and placating their expectations versus realities. He analyzed and clarified the roles of principals in the induction process and recommended how principals could understand the initial beliefs of beginning teachers and use this understanding to improve induction experiences.
3. Bastian, K. C., & Marks, J. T. (2017). Connecting teacher preparation to teacher induction: Outcomes for beginning teachers in a university-based support program in low-performing schools. *American Educational Research Journal, 54*(2), 360–394.
   - Bastian and Marks highlighted the induction model developed and implemented by North Carolina's public university system, with an emphasis on low-performing K–12 schools. The study revealed that overall, teachers in the induction program were more likely to go back to the same school. The findings contributed to efforts in retaining teachers.
4. SRI Education. (2017). *A comprehensive model of teacher induction: Implementation and impact on teachers and students.* Menlo Park, CA: Author.
   - The report detailed how the New Teacher Center (NTC) implemented an induction model at three sites: Broward County Public Schools, Florida; Chicago Public Schools, Illinois; and the Grant Wood Area

Education Agency—a consortium of rural districts in Iowa. In this study, NTC identified four key components:

1. Build the capacity of districts and school leaders to support the mentoring program,
2. Select and assign full-time release mentors to caseloads of no more than 15 teachers each,
3. Provide mentors more than 100 hours of intensive training through institutes and in-field support from lead coaches, and
4. Provide regular, high-quality mentoring to first- and second-year teachers using a system of NTC-developed online formative assessment tools.

The induction model was found to be feasible for implementation in the various district contexts. In addition, the induction model helped students in the beginning teachers' classrooms perform better in English language arts and mathematics.

5. Ronfeldt, M., & McQueen, K. (2017). Does new teacher induction really improve retention? *Journal of Teacher Education, 68*(4), 394–410.
   – While it has been suggested that the implementation of induction programs will reverse the trend of new teacher turnover, this claim is not always true. Ronfeldt and McQueen found that providing beginning teachers with induction supports in their first year of teaching resulted in fewer leaving the profession. They also found that different types of teachers from different types of schools received almost the same amount and quality of support, with the exception of teachers who were Black and who were working in schools with greater numbers of English language learners.

For current information on the STRIDES RAC, please visit mtep.info/STRIDES.

Data-Driven Decisions

The STRIDES RAC has gathered both qualitative and quantitative data on early career mathematics teachers' needs and perspectives, which has motivated the group to synthesize three main factors that contribute to retention in the profession. These factors include: (a) professional learning activities, (b) participation in professional communities, and (c) support from administrators. The RAC members felt that a pragmatic approach to address these factors is warranted, and therefore began the task of devising, and in some cases implementing two interventions, one addressing (a) and (b), and the other addressing (c). STRIDES encourages members of the mathematics education and K–12 communities to join the team in its efforts to further investigate, and ultimately implement, effective methods to increase support for these teachers. Joining these efforts at this time would include creating a partnership between the local K–12 district and a university partner and working with early career secondary mathematics teachers and providing them with target support strategies that were developed by the STRIDES research team.

CONCLUSION

Two primary conclusions resulting from STRIDES activities are: (a) supporting early career mathematics teacher retention will involve a cohesive effort across teachers' entrance into and early years in the profession, and (b) the preparation and the support provided to early career mathematics teachers by teacher educators, mentors, coaches, administrators, and colleagues must be more cohesive. The change ideas distilled from the survey data and brainstorm of ways we can better support early career mathematics teachers are meant to be symbiotic; early career mathematics teachers need to participate in professional communities of other new teachers, and need targeted support from administrators, colleagues, and mentors to become more effective over time and to have the highest likelihood of remaining in the profession. STRIDES members currently continue to meet in small research groups based on these change ideas to implement PDSA cycles at their respective institutions, as well as meet virtually throughout the year in sub-groups and as a whole STRIDES research team. A next round of the latest revision of the STRIDES survey will inform the research teams as they continue to engage in the PDSA cycles that target retention efforts for secondary mathematics teachers. The long-term aspirational goal of STRIDES is to create a sustainable and cohesive system of professional support to retain high-quality mathematics teachers in the field.

REFERENCES

Association of Mathematics Teacher Educators. (2017). *Standards for preparing teachers of mathematics*. Raleigh, NC: Author. Retrieved from: http://amte.net/standards

Carver-Thomas, D. (2018). *California districts report another year of teacher shortages*. Learning Policy Institute Blog [Blog post]. Retrieved from https://learningpolicyinstitute.org/blog/california-districts-report-another-year-teacher-shortages

Donovan, M. S. (2015). *A proposal for integrating research and teacher professional preparation: Innovation and induction corridors*. White Paper. Washington, DC: Strategic Education Research Partnership.

Espinoza, D., Saunders, R., Kini, T., & Darling-Hammond, L. (2018). *Taking the long view: State efforts to solve teacher shortages by strengthening the profession*. Palo Alto, CA: Learning Policy Institute.

Fantilli, R. D., & McDougall, D. E. (2009). A study of novice teachers: Challenges and supports in the first years. *Teaching and Teacher Education, 25*(6), 814–825.

Foster, E. (2010). *How boomers can contribute to student success: Emerging encore career opportunities in K–12 education*. Washington, DC: National Commission on Teaching and America's Future.

Goldring, R., Taie, S., & Riddles, M. (2014). *Teacher attrition and mobility: Results from the 2012–13 teacher follow-up survey* (NCES 2014-077). U.S. Department of Education. Washington, DC: National Center for Education Statistics. Retrieved from http://nces.ed.gov/pubsearch

Ingersoll, R., & Perda, D. (2010). Is the supply of mathematics and science teachers sufficient? *American Educational Research Journal, 47*(3), 563–595.

Jacob, A., & McGovern, K. (2015). *The mirage. Confronting the hard truth about our quest for teacher development.* Brooklyn, NY: The New Teacher Project.

Kane, T. J., Rockoff, J. E., & Staiger, D. O. (2008). What does certification tell us about teacher effectiveness? Evidence from New York City. *Economics of Education Review, 27*(6), 615–631.

Loucks-Horsley, S., Stiles, K., Mundry, S., Love, N., & Hewson, P. (2010). *Designing professional development for teachers of science and mathematics* (3rd Edition). Thousand Oaks, CA: Corwin.

Mathematics Teacher Education Partnership. (2014). *Guiding principles for secondary mathematics teacher preparation.* Washington, DC: Association of Public and Land-grant Universities. Retrieved from mtep.info/guidingprinciples

Mehta, J., Theisen-Homer, V., Braslow, D., & Lopatin, A. (2015). *From quicksand to solid ground: Building a foundation to support quality teaching.* Cambridge, MA: The Transforming Teaching Project. Retrieved from http://www.totransformteaching.org/wp-content/uploads/2015/10/From-Quicksand-to-Solid-Ground-Building-a-Foundation-to-Support-Quality-Teaching.pdf

National Commission on Teaching and America's Future. (2002). *Unraveling the teacher shortage problem: Teacher retention is the key.* Washington, DC: Author.

National Council of Teachers of Mathematics. (2014). *Principles to actions: Ensuring mathematical success for all.* Reston, VA: Author.

National Math + Science Initiative (2013). *STEM education statistics.* Retrieved from http://www.nms.org/Portals/0/Docs/STATS%20fact%20sheet%2008_24.pdf

The Sustainable Funding Project. (2016). *For the public good: Quality preparation for every teacher.* New York, NY: Bank Street College of Education.

SECTION V

THE POWER OF A NETWORKED IMPROVEMENT COMMUNITY

This section provides a look back on the past seven years of work of the MTE-Partnership, as well as the prospects for its continued development. Chapter 17 outlines the major outcomes from the MTE-Partnership research efforts described across this book, followed by suggestions for how secondary mathematics teacher preparation programs might use these outcomes to support local improvement efforts, with particular attention to the AMTE Standards (2017). The chapter concludes with thoughts of the MTE-Partnership leadership about the continuing evolution of the MTE-Partnership over the coming years. Chapters 18 and 19 provide commentaries on the work of the MTE-Partnership from two leading scholars who have been involved in the development of the Networked Improvement Community design, which is being used by the MTE-Partnership.

CHAPTER 17

LOOKING BACK TO LOOK AHEAD

Transforming Secondary Mathematics Teacher Preparation

W. Gary Martin, Alyson E. Lischka, Wendy M. Smith, and Brian R. Lawler

The Mathematics Teacher Education Partnership (MTE-Partnership) was organized by the Association of Public and Land-grant Universities (APLU) in 2012 to
address problems with the quantity and quality of secondary mathematics teachers being produced by member institutions. The MTE-Partnership encompasses
more than 90 universities, colleges, and their K–12 school partners, working collaboratively using a Networked Improvement Community (NIC) design (Bryk,
Gomez, Brunow, & LeMahieu, 2015; Martin & Gobstein, 2015) to achieve the
aim of increasing the quantity of teacher candidates who meet a "gold standard"
(Association of Mathematics Teacher Educators [AMTE], 2017) of preparedness
by 2020. The target for the quantitative portion of this aim is an increase in the
number of graduating secondary mathematics teachers by 40% by 2020 across
the MTE-Partnership institutions, with an emphasis on increasing diversity. The
target for the latter, qualitative portion of the aim of meeting a gold standard
is documentation by programs that their graduates are capable of providing the

*The Mathematics Teacher Education Partnership: The Power of a Networked Improvement
Community to Transform Secondary Mathematics Teacher Preparation*, pages 373–389.
Copyright © 2020 by Information Age Publishing

ambitious instruction and deep learning compelled by rigorous state standards, based on benchmarks developed by the MTE-Partnership. Research Action Clusters (RACs) have worked on particular problems of practice that will support programs in meeting this ambitious aim.

This chapter presents a summary of what the MTE-Partnership has accomplished, as described in the preceding chapters of the book; suggestions that may support programs in their efforts to improve; and thoughts on the continuing development of the MTE-Partnership.

A LOOK BACK ON MTE-PARTNERSHIP PROGRESS

Toward the aim of supporting teacher preparation programs to produce a greater number of graduates, better prepared to provide the ambitious instruction and deep learning compelled by rigorous state standards, MTE-Partnership adopted the NIC model (Bryk et al., 2015; Martin & Gobstein, 2015). The NIC model is well suited to organize and support improvement research for several reasons (Bryk et al., 2015). First, NICs are a source of innovation. Second, NICs provide diverse contexts within which to test those ideas. Third, NICs provide the social connections that accelerate testing and diffusion of innovations. Fourth, NICs provide a safe environment in which to engage comparative analyses. Finally, NICs permit the identification of patterns that would otherwise look particular to each context in turn.

The broad MTE-Partnership community operates as a NIC. However, the NIC model also operates at a variety of levels of work across the partnership. In the previous chapters, the five aforementioned elements of NICs are demonstrated in the work of the individual RACs, and are also illustrated by the various included vignettes, serving to highlight the impact of the greater NIC on individual institutions. In this way, we see progress toward transforming teacher preparation programs as aligned with the MTE-Partnership *Guiding Principles* (MTE-Partnership, 2014) and summarized in the current MTE-Partnership driver diagram (see Figure 17.1). The following sections of this chapter review the progress that has been made, drawing from the previous chapters in this book, and are organized according to the area of the MTE-Partnership driver diagram that is addressed: Mathematical Preparation (including both the Active Learning Mathematics [ALM] and Mathematics of Doing, Understanding, Learning, and Educating for Secondary Schools [MODULE(S²)] RACs), Clinical Experiences (including each of its sub-RACs), and Recruitment and Retention (including both Program Recruitment and Retention [PR²] and Secondary Teacher Retention and Induction in Diverse Educational Settings [STRIDES] RACs). The review then closes with a discussion of the work of the Equity and Social Justice and the Program Transformation Working Groups, as these working groups signify the forward momentum of the MTE-Partnership.

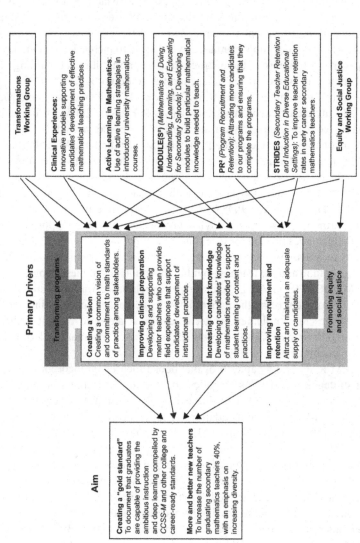

FIGURE 17.1. Revised MTE-Partnership Driver Diagram. Revised driver diagram depicting the current organization of the MTE-Partnership.

Mathematical Preparation of Secondary Mathematics Teachers

The MODULE(S²) and ALM RACs are extending research in the area of mathematical preparation of secondary teachers. Both RACs work to build on foundational research on the mathematical knowledge needed for teaching (see Chapter 4 of this book; Ball, Thames, & Phelps, 2008; Rowland, 2013) and envision the type of practices that programs are responsible for developing in prospective teachers, according to current standards and recommendations. These RACs attend to the MTE-Partnership's primary driver related to increasing content knowledge, by working to improve the instruction and opportunities for learning mathematical content in courses taken in teacher preparation programs.

As described in Chapter 5, the MODULE(S²) RAC is focused on the development of educative curriculum materials (Davis & Krajcik, 2005) for upper-level mathematics content courses, with an aim to incorporate activities that develop mathematical knowledge needed for teaching. Embracing MTE-Partnership Guiding Principle 1, Partnerships as the Foundation, the RAC involves mathematics and mathematics education faculty along with K–12 school partners in the development of the materials. The curriculum materials, developed and refined through Plan-Do-Study-Act (PDSA) Cycles, demonstrate progress achieved by the MTE-Partnership toward improving prospective teachers' opportunities for learning mathematics needed for teaching. The materials allow MTE-Partnership institutions to move closer to achieving the RAC aim to develop mathematical knowledge, both in and for teaching, within the graduates of these programs. Further, the MODULE(S²) RAC is generating research-based ways of understanding both the development of mathematical knowledge needed for teaching and how to support instructors implementing activities aimed at this development, through the creation of a framework to analyze development of mathematical knowledge needed for teaching.

The ALM RAC (see Chapter 6 of this book) focuses on incorporating active learning strategies in early mathematics courses so that learners in these courses have mathematical experiences that exemplify teaching practices set forth in current research-based recommendations (e.g., National Council of Teachers of Mathematics [NCTM], 2014). The ALM RAC, also addressing the MTE-Partnership primary driver of increasing mathematical content knowledge of program graduates, is implementing PDSA Cycles to better understand how to support a variety of faculty (e.g., tenure-track, transitional, non-tenure track, adjunct, full-time-temporary) in adopting active learning strategies in precalculus through calculus 2 courses. Through vignettes that provide descriptions of the work, the ALM RAC demonstrated several affordances of the NIC structure (Bryk et al., 2015): opportunities to test ideas in diverse contexts, providing a safe environment to engage in comparative analysis, and identification of patterns that arise in the adoption of new strategies (active learning strategies). By not only encouraging the use of active learning strategies among MTE-Partnership institutions and others, but also studying the support needed to sustain such practices among mostly transitional faculty, the work of the ALM RAC exemplifies a NIC structure and demonstrates the ways in which a NIC can accelerate reform.

Clinical Experiences

Improvement of clinical experiences provided by secondary mathematics teacher preparation programs was identified as a primary driver in the MTE-Partnership driver diagram (see Figure 17.1), as described in Chapter 7 of this book. Given the widely varying contexts for secondary mathematics teacher preparation (Taylor & Ronau, 2006; Yee, Otten, & Taylor, 2018), the Clinical Experiences RAC began with a focus on building effective partnerships in which all stakeholders are supported and has since grown to support clinical experiences in a variety of ways: modifying the structure of placements, modifying the ways in which traditional experiences are enacted through co-planning and co-teaching, and recently moving into building supports for methods courses that include clinical experiences. The work of this RAC is centered on developing teacher candidates' implementation of the Standards for Mathematical Practice in the *Common Core State Standards for Mathematics* (*CCSS-M*; National Governors Association Center for Best Practices & Council of Chief State School Officers, 2010) and the mathematics teaching practices (NCTM, 2014) in clinical experiences and provides a pathway for programs to reach the goals set forth in the *Standards for the Preparation of Teachers of Mathematics* (AMTE, 2017). The work of the Clinical Experiences RAC (overviewed in Chapter 8 of this book) addresses the primary driver in the MTE-Partnership driver diagram (see Figure 17.1) about improving clinical preparation. Their work is further divided into three sub-RACs, which focus on methods experiences, the co-planning and co-teaching model, and the paired placements model.

As described in Chapter 9, the Methods sub-RAC has conducted a series of PDSA Cycles through which they are designing materials for use in methods courses with co-requisite field experiences. This work attends to the MTE-Partnership Guiding Principle 1, Partnerships as the Foundation, by fostering mutually beneficial relationships between teacher preparation programs and mentor teachers. The materials, which are focused on the Standards for Mathematical Practice, lesson planning, and providing student feedback, are intended to engage prospective teachers and mentor teachers in conversations about the topics. By doing so, both mentor teachers and teacher candidates grow in their understanding of effective mathematics teaching.

The Co-Planning and Co-Teaching Clinical Experiences (CPCT) sub-RAC (see Chapter 10 of this book) has undertaken the research-based translation of co-planning and co-teaching models often found in special education preparation programs to the secondary mathematics preparation setting. Through PDSA Cycles, the CPCT team supported mentor teachers in enacting co-planning and co-teaching with prospective teachers and studied the effectiveness of the implementation. Chapter 10 of this book details the findings and demonstrates how this work moves the field forward in building bi-directional relationships between preparation programs and mentor teachers, which further attends to the MTE-Partnership's first guiding principle.

The Paired Placement sub-RAC identified a need to support the research-based practice of placing two teacher candidates with one mentor teacher in field experiences. Although existing research demonstrated the promise of this practice, supports were not available to encourage sustainability of paired placements. The Paired Placement sub-RAC has developed methods by which teacher preparation programs could adopt and support this practice. Through PDSA Cycles, the members refined syllabi, revised student teaching assignments, and prepared implementation guides to assist those adopting the paired placement practice. The vignettes shared in Chapter 11 of this book highlight the principle that the NIC model allows for diverse contexts in which innovations can be tested and refined. Together, the work of the entire Clinical Experiences RAC provides research-based practices to move the field of secondary mathematics teacher preparation closer to achieving the vision set forth by the MTE-Partnership *Guiding Principles* (see Chapter 12 of this book).

Recruitment and Retention

The nationwide shortage of highly qualified secondary mathematics teachers has been well-documented, persisted for decades, and continues at an acute level today (Cross, 2017). This shortage is even more acute for teachers of color in the field and in teacher preparation programs (Ingersoll, May, & Collins, 2017). To achieve its improvement aim—increase the number of graduating secondary mathematics teachers, with an emphasis on diversity—the MTE-Partnership established a primary driver related to improving recruitment and retention. An initial working group (see Chapter 13 of this book) addressing that driver blossomed into two RACs: the Marketing to Attract Teacher Hopefuls (MATH) and Secondary Teacher Retention and Induction in Diverse Educational Settings (STRIDES). The MATH RAC was later reorganized as the Program Recruitment and Retention (PR²) RAC.

The MATH RAC was the first to emerge from this working group, focusing exclusively on the recruitment of more—and more diverse—students into secondary mathematics teacher preparation programs (see Chapter 14 of this book). Research efforts were directed by four primary drivers: increasing awareness, converting awareness to interest, infrastructure for purposeful and ongoing recruitment, and specific attention to attracting and graduating traditionally underrepresented candidates. The MATH RAC's work culminated in a guide to support the development of a secondary mathematics teacher recruitment campaign (Ranta & Dickey, 2015). The recruitment guide is organized into nine modules, each with multiple examples from a variety of RAC institutions. The nine modules are overview, campaign planning, research, branding, social media, public relations, advertising through paid media, website identity, and lessons learned.

As this guide came to fruition, members of the RAC recognized that not enough had been learned yet about how to improve the diversity of candidates drawn to secondary mathematics teacher preparation programs and, further, that retention in the program was deserving of attention. To make this shift, the MATH RAC ended its work and most members, along with some new members, began the work to launch a new RAC in early 2017—allowing the group to revisit an analysis of the system that is producing the current, problematic outcomes. The new RAC, PR², set its aim on bringing more—and more diverse—secondary mathematics teachers into the workforce, beyond entry into the teacher preparation program. An analysis of the research literature, along with the expertise of RAC members, allowed the PR² RAC to build from the MATH RAC driver diagram, adding primary drivers: improving the perception of teaching; providing support for teacher candidates in the program; advocating with policy makers; and attending to issues of diversity, equity, and social justice. Chapter 15 includes reports of initial efforts of PR² members to understand local systems and begin to understand change ideas through PDSA Cycles.

A few years after the MATH RAC members began their work, the growth of the MTE-Partnership resulted in greater interest in the retention of secondary mathematics teachers during the first years of their career. The STRIDES RAC was launched to improve retention rates among early career secondary mathematics teachers (see Chapter 16 of this book). This group examined the support structures available to early career mathematics teachers, especially to learn which ones seemed to have significant impact, as reported by the teachers. These factors included (a) professional learning activities, (b) participation in professional communities, and (c) support from administrators. The results led members to organize two strands of work, focused on administrators and on mentors.

Equity and Social Justice

The MTE-Partnership Planning Committee decided that a working group was necessary to ensure explicit attention to issues of equity and social justice in the preparation of secondary mathematics teachers. Following its formation, the new Equity and Social Justice Working Group (ESJWG) began to identify and develop a shared understanding of the problem space, the system of teacher preparation as it is embedded in larger systems of schooling and society, and a working theory for improvement. The discussion of the work of ESJWG in Chapter 3 of this book emphasized the work to understand the problem and develop a working theory, the group's initial driver diagram. ESJWG functions differently from the RACs, in that its aim cannot be achieved without contribution from each of the RACs. And similarly, each of the RACs cannot achieve their aim without explicit attention to the issues raised by ESJWG. As a result, a primary role of the ESJWG within the MTE-Partnership is to serve as a hub for liaisons to each RAC, creating a bidirectional opportunity to learning. An example of this was the support the ESJWG provided to each RAC at the 2018 MTE-Partnership conference; RACs revisited driver diagrams in order to identify refinements to better examine underlying equity and justice elements of their improvement aims.

In addition to liaison efforts, the ESJWG has begun PDSA Cycles of its own, toward achieving elements of the work that serve the working group itself or lie outside the work of other RACs. One team is developing a tool to measure the aim of the working group to improve teacher candidates' equity-driven sociopolitical dispositions and knowledge and use of equitable teaching practices over the course of their program. The tool will focus on observation of teaching, where both aspects should be visible. A second team is learning how to help interested mathematicians and mathematics teacher educators have access to instructional resources that help them achieve an aim related to equity or justice. For example, when a mathematics teacher educator wishes to develop a lesson or module to prepare secondary mathematics teachers to "identify and implement practices that draw on students' mathematical, cultural, and linguistic resources/strengths and challenge policies and practices grounded in deficit-based thinking" (Indicator C.4.3; AMTE, 2017, p. 22), they will find a catalog or repository of readings, activities, and other supports.

Program Transformation

Finally, as described in Chapter 2, the Transformations Working Group has begun to explore how the work across all the RACs and working groups can be woven together into a coherent plan for secondary mathematics teacher preparation program transformation. Meeting the aspirational AMTE Standards (2017) is a complex endeavor that requires collaborations across stakeholders. The MTE-Partnership organized its work into RACs in order to accelerate progress in addressing particular problems of practice. Moving forward, local programs will need to integrate the findings of each of the RACs in order to achieve broader transformation. Each local program has unique contexts and may have differing priorities, but all need to attend to issues related to partnerships, equity, recruitment, retention, mathematics courses, methods courses, clinical experiences, and induction. Mobilizing a multi-faceted process focused on program transformation presents substantial challenges in garnering the necessary support and resources.

The previous section of the chapter has outlined the major areas of progress in the MTE-Partnership research efforts. The following section provides an overview of how programs might use this book to accelerate their efforts to improve what they are doing.

ACCELERATING PROGRAM IMPROVEMENT

The AMTE Standards "set forth an ambitious but achievable vision to prepare teachers who can effectively support the mathematics learning of each and every student" (2017, p. 163). Virtually all secondary mathematics teacher educators will recognize areas in which their programs need to improve in order to meet that vision. It is our belief that this book will provide useful guidance that may accelerate improvement efforts; it represents the compilation of over seven years of work conducted by hundreds of people working to improve secondary mathematics teacher preparation. As programs look to use that information in their improvement efforts, they may find it difficult to know where to start. Moreover, different programs may be facing vastly different challenges in improving secondary mathematics teacher preparation and therefore may be at different points in their journey toward program transformation, and the steps they need to take will vary. However, this section outlines general guidelines for how programs can accelerate their efforts to improve.

The focus of the MTE-Partnership, as well as this book, is on the *improvement* of secondary mathematics teacher preparation. The NIC design used by the MTE-Partnership has proven particularly valuable in promoting a focus on improvement, which means combining a focus on using evidence to guide improvement with a focus on working collaboratively, in order to enhance progress. Using the NIC approach, MTE-Partnership members are committed to the proposition that the quality of the teacher candidates they produce and of the programs they develop and maintain must be of the highest levels. Although external accountability through agencies like the Council for the Accreditation of Educator Preparation (CAEP) or state departments of education may address a baseline of quality, programs incorporating improvement science approaches seek continuous improvement beyond such external expectations.

The NIC approach often runs counter to the typical practices of those involved in educational research, who may be inclined to use more rigorous instrument design and longer-term data collection than is necessary or warranted to guide improvement efforts. Members of the MTE-Partnership have experienced (and may continue to experience) a learning curve in coming to embrace this approach. However, the immediacy of trying new ideas and then quickly making decisions based on evidence is truly empowering, allowing progress to be made at a quicker pace than is possible in traditional research paradigms. Rather than attempting to provide definitive, generalizable answers to research questions, the NIC design seeks solutions to problems of practice that can be adapted to fit a range of contexts. Note, however, that MTE-Partnership participants have produced many refereed articles and presentations about their work, and funding agencies seem increasingly open to proposals using a NIC design. In fact, MTE-Partnership groups have received four awards[1] from the National Science Foundation to support different areas of research incorporating the NIC design.

Becoming a part of the MTE-Partnership NIC will provide a broad coalition of support for other programs' improvement efforts; see http://mtep.info/get-involved for more information about how a program can become involved in the MTE-Partnership. However, the NIC model should also be enacted on other levels. The NIC-Transform project is focusing on organizing local secondary mathematics teacher preparation programs as NICs, in partnership with their allied stakeholders, as a means to promote program transformation. Many MTE-Partnership teams have built NICs that include programs across a state or region. For example, California State University has built a network including all campuses involved in secondary mathematics teacher preparation, and Nebraska's network includes all institutions involved in mathematics teacher preparation. Of course, the general advice of incorporating a NIC design applies to those working in teacher preparation in other content areas, to those working on related problems in mathematics education, and to those working in other education arenas. The book *Learning to Improve: How America's Schools Can Get Better at Getting Better* (Bryk et al., 2015) is an excellent starting point for becoming more familiar with the NIC design.

The AMTE Standards (2017) suggest initial steps in undertaking program improvement efforts. Engaging with stakeholders to build a common vision, assessing the current context to determine what is working and what needs to be improved, and selecting change strategies for initial exploration are all necessary steps to take. Without establishing a foundation of shared vision and priorities with their partners, programs are less likely to embark on a course of action that results in substantive change, and they are more likely to undertake actions of convenience without a clear target. Of course, this is consistent with the work of the Transformations Working Group (see Chapter 2) and with the NIC design, more generally. We particularly recommend the book *Increasing Student Success in STEM: A Guide to Systemic Institutional Change* (Elrod & Kezar, 2016) for the strong set of tools it provides to explore a context. Again, these initial actions are applicable to other problem areas in education.

This book includes a plethora of research and insights on particular problem areas in secondary mathematics teacher preparation that may become priorities for improvement efforts. The chapters describing the RACs and working groups include specific suggestions for starting points in *focus on improvement* sections, with the Clinical Experiences RAC including a final chapter in Section III exclusively focused on improvement. Consistent with the commitment to continual improvement, updated information for each RAC and working group, along with other ongoing research, is shared annually in the proceedings of the conferences, which are publicly available at http://mtep.info/conferences. Finally, each RAC and working group maintains a page of resources at http://mtep.info/rac-list; these pages also include links to relevant auxiliary websites they maintain. These resources are specific to secondary mathematics teacher preparation, but they may offer suggestions for starting points in other areas of teacher preparation as well.

WHERE THE MTE-PARTNERSHIP IS HEADED

The MTE-Partnership has made major strides since its founding in 2012, establishing a NIC focused on improving the preparation of secondary mathematics teachers. Five RACs are actively working on different aspects of the problem space, and two working groups are exploring cross-cutting themes related to program transformation and to equity and social justice. However, the work of the MTE-Partnership is far from complete, and a NIC must continue to develop, grow, and evolve, always recognizing that its theory of action is "possibly wrong and definitely incomplete" (Bryk et al., 2015, p. 79). In its most recent discussions, the MTE-Partnership planning team has identified two major areas on which the MTE-Partnership must focus as it continues to mature over the coming years: revisiting the framework on which the MTE-Partnership was founded, and taking a broader view of transformation that extends beyond the immediate goal of improving secondary mathematics teacher preparation programs. Each area is explored in turn.

Revisiting the MTE-Partnership Framework

MTE-Partnership's foundational elements are its *Guiding Principles for Secondary Mathematic Teacher Preparation* (2014) and its commitment to the NIC design. The organization's work as a NIC was discussed in Chapter 1. The early work involved mapping out the problem space, developing an aim, identifying major drivers for progress toward the aim, and organizing RACs to address those drivers. To maintain its vitality, an organization needs to regularly revisit its foundations, and the MTE-Partnership's foundations have been periodically revisited by its planning team over the past years. The *Guiding Principles* were revised several times, most recently to increase attention to equity, and the driver diagram has been expanded to include increased attention to equity and social justice and a new emphasis on program transformation. New RACs and working groups have been formed, as earlier RACs met their goals or ceased to be viable as new priorities emerged. However, the MTE-Partnership has now reached a pivotal moment where its foundations need to be reconsidered rather than tweaked to both incorporate what MTE-Partnership members have learned as a community over the past years and to address the changing circumstances that programs face.

The release of the AMTE Standards (2017) has provoked many new discussions about mathematics teacher education across the nation, as programs and projects consider how their efforts align with the document's recommendations. Within the MTE-Partnership, we have discussed the relationship of the AMTE Standards to the *Guiding Principles* document that has directed the MTE-Partnership. Although the two documents are generally consistent in their visions, they are framed somewhat differently and are not identical in the issues they address. At present, the RACs and working groups reference both documents: the *Guiding Principles*, because these principles were the organizational framework for the MTE-Partnership, and the AMTE Standards, because they are a common, guiding reference point for the field of mathematics teacher education. However, the work of the MTE-Partnership needs to be clearly aligned with the AMTE Standards. For example, the *Guiding Principles* might be updated to explicitly reference the AMTE Standards, or the AMTE Standards might be adopted as a replacement for the *Guiding Principles*, or the two might be combined in some manner. As with the current *Guiding Principles*, this work will be done in a collaborative manner that reflects the knowledge and values of the MTE-Partnership members.

The changing landscape for mathematics teacher preparation over the past few years also necessitates a reconsideration of the aims and drivers for the MTE-Partnership. Nationwide decreases in the number of teacher candidates being produced have made the first element of MTE-Partnership's aim—producing more secondary mathematics teachers—increasingly difficult to achieve. Thus, the MTE-Partnership's aim and its supporting metrics need to be revisited so that they can continue to serve as a worthy target for the NIC. Likewise, the drivers posited as major causal explanations leading to the aim need to be reconsidered. Particular areas of secondary mathematics teacher preparation were judged to be of the highest priority and, as a result, were included as primary drivers at the time the MTE-Partnership was launched. Given the progress made recently, other areas may emerge as greater priorities. In addition, new forces are impacting many programs, which may suggest the need for additional drivers that were not prioritized initially. For example, virtual learning and teaching environments are becoming increasingly prevalent; their impact may need to be considered in the experiences prospective teachers have, as well as in how teacher preparation programs are organized and delivered. The context of teacher preparation is also shifting, with a significant number of candidates being produced by post-baccalaureate and alternative certification programs, which suggests the need to forge new partnerships to promote nationwide change in mathematics teacher preparation.

As new drivers are proposed, new working groups and RACs will need to be considered to explore potential change ideas. In particular, to meet the needs of secondary school learners who have been marginalized or ill-served, equity and social justice must be increasingly infused across the work of the NIC and reflected in its drivers. As described in earlier chapters, the initial framing of the MTE-Partnership took more than two years to develop, involving an intensive process of garnering input and consensus. Likewise, a serious reconsideration of the framework also will require an extended period of time, especially given that the membership is now engaged in carrying out the work of the current RACs and working groups. However, this type of work is inherently part of an effective NIC.

A final area that any organization needs to consider is what it means to belong to that organization. With a NIC, membership is not based on paying dues; rather, membership means "accepting responsibility for something larger" (Bryk et al., 2015, p. 166), in this case the improvement of secondary mathematics teacher preparation. In the MTE-Partnership model, institutions join rather than individuals, given that change is programmatic rather than individual. Yet, individuals do the actual work of improvement, and an institution's participation depends on such individuals. Thus, ensuring a sense of identity and belonging by individuals has been a continuing challenge for the MTE-Partnership (Martin & Gobstein, 2015). Moreover, as individuals' responsibilities, roles, and interests shift, their involvement in the MTE-Partnership may change, with corresponding shifts in the institutional commitment to engagement. Thus, a process for revisiting the membership of the MTE-Partnership has been launched over the past few months, developing an explicit process for teams to renew their institutional commitment to participate, identifying the key individuals involved in the work. This process is vital to the continued success of the MTE-Partnership, particularly in ensuring that changes to the basic framework of the MTE-Partnership reflect the needs of its membership. Furthermore, in order to fulfill its mission of nationwide transformation of secondary mathematics teacher preparation, the MTE-Partnership must expand to encompass a broader range of programs. This expansion might include both encouraging existing teams to add other programs from their state and recruiting institutions in additional states in which the MTE-Partnership is not yet active to form new teams. As mentioned earlier, this future work many also include alliance with teacher preparation programs that are not university-based. An additional area of discussion involves how to engage with researchers who are working on projects related to the MTE-Partnership agenda but who are not at institutions involved with MTE-Partnership, which will require considering new models of membership or collaboration.

Broadening the Transformational Agenda

From its founding in 2012, the MTE-Partnership has had the overarching goal of transforming secondary mathematics teacher preparation, in alignment with the *CCSS-M* and other rigorous standards. More recently, the goal has expanded to encompass the AMTE Standards. In this section, we discuss attaining this goal at two levels: program transformation and system transformation.

The natural inclination for any program engaged in improvement may be to focus on one or two areas in which change is most needed. In fact, the initial organization of the MTE-Partnership, in which teams directly participated in one or two RACs or working groups, may have reinforced this tendency. However, as discussed in Chapter 2, program transformation becoming more aligned with the MTE-Partnership vision—including the AMTE Standards—must be the ultimate goal for institutions participating in the MTE-Partnership. As a result, having the Transformation Working Group and the NIC-Transform (NSF #1834551, 1834539) project create specific tools and protocols will be essential in supporting teams as they work toward a broader transformational agenda that fully achieves the vision of the MTE-Partnership and the AMTE Standards. To the degree possible, it will be useful to develop a general roadmap (or alternative roadmaps) for how teams might progress toward program transformation, accounting for the range of contexts that teams may encounter. Particular focus on how equity and social justice can be infused across programs will be a critical component of that roadmap.

Essential to transformation, the ability to effectively share resources across a change effort often proves challenging. In the MTE-Partnership, the RACs have produced valuable resources for members and the broader community of mathematics teacher educators that support programs' transformational efforts. Disseminating the resources and managing the quality and intellectual property rights are current challenges. As discussed in Chapter 2, the focus needs to be on creating a dynamic environment that supports the generation and sharing of resources that are created by the RACs and working groups, along with discussion of how the resources are used, that may guide both their implementation and continued development. The NIC-Transform project has been piloting possible models and is also coordinating discussions about approaches being taken by the RACs and working groups in the area of knowledge generation and management. In addition, other means of sharing information will be considered, such as webinars, asynchronous discussions, and print resources. Developing a database of community members who might provide expertise in particular areas may also be a valuable tool for programs, and providing contextual information will help those seeking assistance to identify relevant experts. We also intend to explore a system of badging to aid in the process of acknowledging expertise and leadership in particular areas.

In addition, system-level transformation will be an essential component of the MTE-Partnership's emerging agenda to support program transformation. For example, initial experiences of NIC-Transform participants suggest that joining forces with people working in teacher preparation in other disciplines or grade levels will be useful in leveraging resources and developing a broader coalition for change. For example, a program could partner with secondary science or STEM educators to work on strategies for recruiting students to their programs. Or, given a context in which mathematics teacher educators work across grades, programs could explore ways to improve partnerships with school districts more generally across K–12 mathematics teacher preparation, rather than solely focusing on the secondary grades. Although the MTE-Partnership will maintain focus on its goal of improving secondary mathematics teacher preparation, the MTE-Partnership will encourage the formation of NICs working in related areas of teacher preparation over the coming years in order to create a broader context for improving teacher preparation. Whereas the forms such alliances will take may vary, working with persons in other fields or areas to form NICs, building on the experiences of the MTE-Partnership, will prove fruitful for those fields or areas, and the MTE-Partnership will benefit from having such partners in other areas.

In many cases, broader policy factors impact the improvements that programs are able to incorporate. For example, many states demand increasingly higher scores on various tests required for certification, such as the *Praxis II* for secondary mathematics and the Teacher Performance Assessment (edTPA), which may exacerbate the challenges programs face in increasing their number of graduates. Although the focus of the MTE-Partnership has been on developing approaches to program improvement that teams can incorporate, efforts to address this broader policy landscape will support teams' efforts to improve. For example, the MTE-Partnership could develop a policy brief that addresses how high standards of quality can be maintained without creating artificial barriers, such as high scores on external assessments, that do little to ensure quality. Policy briefs on this and other relevant issues will be useful as teams advocate for supportive policies with state boards of education, certification bodies, and other entities impacting their ability to effectively prepare secondary mathematics teachers. The MTE-Partnership has the potential to further leverage its relationship with the Association of Public and Land-grant Universities to have a significant impact on policy on both state and national levels.

Additional support for change in secondary mathematics teacher preparation, particularly in the policy arena, may be found in developing coalitions with related organizations. The MTE-Partnership should seek stronger connections with other organizations that have related agendas in mathematics teacher education (such as the Association of Mathematics Teacher Educators), in teacher education more generally (such as the American Association of Colleges for Teacher Education or the Council of Academic Deans from Research Education Institutions), or in mathematics education as a whole (such as the Association of State Supervisors of Mathematics, the Conference Board of the Mathematical Sciences, or the National Council of Teachers of Mathematics). The nature of these collaborations may vary from more informal relationships to more formal affiliation. But whatever form they may take, the MTE-Partnership must increasingly seek collaboration with related organizations in order to accomplish its transformational agenda.

Finally, many organizations are engaged in defining or certifying program quality, which may have a profound impact on programs' improvement efforts. More immediately, the MTE-Partnership should be highly engaged in discussions about standards that affect secondary mathematics teacher preparation, such as the Specialized Program Area (SPA) standards collaboratively developed by NCTM (2012) for the Council for the Accreditation of Educator Preparation, or the next generation of the Conference Board of the Mathematical Sciences' *The Mathematical Education of Teachers II*. In the longer term, the MTE-Partnership planning team has discussed developing an internal process of certifying program quality, which might include badging of particular aspects of a program that meet certain criteria, or identification of an entire program, as meeting MTE-Partnership criteria of quality. Having such a process may become increasingly important as some states and schools are eliminating national program accreditation or approval processes. The MTE-Partnership, through its commitment to continuous improvement and accountability to its own professional community, could provide programs in such contexts with a valuable service.

CONCLUSION

The MTE-Partnership has emerged as an exemplar of a NIC working to address the "wicked problem" of secondary mathematics teacher preparation (Gomez, Russell, Bryk, Mejia, & LeMahieu, 2016). This community of professionals has a strong commitment to continued progress toward a shared vision of high-quality mathematics teacher preparation. The NIC model represents a different research paradigm than is typical, yet this model provides powerful tools to develop solutions for significant problems of practice. The use of PDSA Cycles and other improvement science tools has proved powerful in emerging from the mindset that improvements must be perfected before they are implemented to the mindset that improvement is a journey upon which must be embarked. Furthermore, working collaboratively with colleagues from across the nation provides opportunities to leverage expertise and produce solutions that are adaptable to the range of contexts in which they work.

The MTE-Partnership has made significant progress toward its goal of transforming secondary mathematics teacher preparation, and the journey continues. Regardless of the level of engagement a program currently has in the MTE-Partnership problem space or related problem spaces, the MTE-Partnership members encourage interested people and programs to begin the journey of improving their practice. We welcome new members to the MTE-Partnership, look to actively partner with others who are doing work parallel to ours, and support others engaged in the work of networked improvement. The current outcomes of secondary mathematics teacher preparation are the product of our current system—including its history—and it will take the coordinated involvement of all of us to effect significant and lasting improvements.

ENDNOTE

1. The work of the MTE-Partnership is supported in part by four grants from the National Science Foundation, including *Using Networked Improvement Communities to Design and Implement Program Transformation Tools for Secondary Mathematics Teacher Preparation* (NIC-Transform; NSF #1834551, 1834539), *Student Engagement in Mathematics through an Institutional Network for Active Learning* (SEMINAL; NSF #1624643, 1624610, 1624628, 1624639), *Clinical Experiences Research Action Cluster* (NSF #1726998, 1726853, 1726362), and *Mathematics of Doing, Understanding, Learning, and Educating for Secondary Schools* (MODULE(S²); NSF #1726744, 1726707, 1726098, 1726252, 1726723, 1726804). Earlier work of the MTE-Partnership was supported in part by a grant from the Helmsley Charitable Trust.

REFERENCES

Association of Mathematics Teacher Educators. (2017). *Standards for preparing teachers of mathematics*. Raleigh, NC: Author. Retrieved from http://amte.net/standards
Ball, D., Thames, M. H., & Phelps, G. (2008). Content knowledge for teaching: What makes it special? *Journal of Teacher Education, 59*, 389–407.

Bryk, A., Gomez, L. M., Grunow, A., & LeMahieu, P. (2015). *Learning to improve: How America's schools can get better at getting better*. Cambridge, MA: Harvard Education Press.

Cross, F. (2017). *Teacher shortage areas: Nationwide listing, 1990–1991 through 2017–2018*. Office of Postsecondary Education, U.S. Department of Education, Washington, DC. Retrieved from https://www2.ed.gov/about/offices/list/ope/pol/ateacher-shortageareasreport2017-18.pdf

Davis, E. A., & Krajcik, J. (2005). Designing educative curriculum materials to promote teacher learning. *Educational Researcher, 34*(3), 3–14.

Elrod, S., & Kezar, A. (2016). *Increasing student success in STEM: A guide to systemic institutional change*. Washington, DC: Association of American Colleges and Universities.

Gomez, L. M., Russell, J. L., Bryk, A. S., Mejia, E. M., & LeMahieu, P. G. (2016). The right network for the right problem. *Phi Delta Kappan, 98*(3), 8–15.

Ingersoll, R., May, H., & Collins, G. (2017). *Minority teacher recruitment, employment, and retention: 1987 to 2013*. Palo Alto, CA: Learning Policy Institute. Retrieved from https://learningpolicyinstitute.org/product/ minority-teacher-recruitment

Martin, W. G., & Gobstein, H. (2015). Generating a networked improvement community to improve secondary mathematics teacher preparation: Network leadership, organization, and operation. *Journal of Teacher Education, 66*(5), 482–493.

Mathematics Teacher Education Partnership. (2014). *Guiding principles for secondary mathematics teacher preparation*. Washington, DC: Association of Public and Land-grant Universities. Retrieved from http://mtep.info/guidingprinciples

National Council of Teachers of Mathematics. (2012). *NCTM CAEP standards (2012)— Secondary (initial preparation)*. Retrieved from: https://www.nctm.org/uploadedFiles/Standards_and_Positions/CAEP_Standards/NCTM%20CAEP%20Standards%202012%20-%20Secondary.pdf

National Council of Teachers of Mathematics. (2014). *Principles to actions: Ensuring mathematical success for all*. Reston, VA: Author.

National Governors Association Center for Best Practices, Council of Chief State School Officers. (2010). *Common core state standards: Mathematics*. Washington, DC: Author.

Ranta, J., & Dickey, E. (2015). *Secondary mathematics teacher recruitment campaign implementation guide*. Washington, DC: Association of Public and Land-grant Universities. Retrieved from http://bit.ly/MATHImplGuide

Rowland, T. (2013). The knowledge quartet: The genesis and application of a framework for analysing mathematics teaching and deepening teachers' mathematics knowledge. *Sisyphus-Journal of Education, 1*(3), 15–43.

Taylor, M., & Ronau, R. (2006). Syllabus study: A structured look at mathematics methods courses. *AMTE Connections, 16*(1), 12–15.

Yee, S., Otten, S., & Taylor, M. (2018). What do we value in secondary mathematics teaching methods? *Investigations in Mathematics Learning, 10*(4), 187–201.

CHAPTER 18

REFLECTIONS ON THE MTE-PARTNERSHIP

The Power of Networked Improvement Communities to Solve Complex Problems of Practice

Jennifer Lin Russell

Networked Improvement Communities (NICs) provide a social structure for organizing inter-organizational collaboration to address a specific problem of practice (Dolle et al., 2013; Russell et al., 2017). The concept was introduced to the educational field by Tony Bryk, Louis Gomez, and Alicia Grunow from the Carnegie Foundation for the Advancement of Teaching (Bryk, Gomez, & Grunow, 2011), and subsequently, there has been incredible interest in this way of organizing systems for improvement in the field of education. A community of educators, reformers, researchers, and policymakers has gravitated to the NIC concept as a model for solving complex educational problems, and the Mathematics Teacher Education Partnership (MTE-Partnership) has been at the vanguard of this movement providing a strong case for the power of this idea.

The MTE-Partnership illustrates how NICs are a promising approach for solving complex problems of practice in education (Gomez, Russell, Bryk, LeMahieu, & Mejia, 2016). When Howard Gobstein, Gary Martin, and colleagues set out to

The Mathematics Teacher Education Partnership: The Power of a Networked Improvement Community to Transform Secondary Mathematics Teacher Preparation, pages 391–399.

improve mathematics teaching and learning at scale, they thought the NIC concept provided a strong model for organizing the MTE-Partnership work given the complexity of the problem they were trying to solve. Even restricting the focus of the work to prospective teacher preparation proves to be a highly complex endeavor, as represented by the table of contents of this book. Improving the preparation of mathematics teachers includes redesigning university-based course curriculum, changing the pedagogy employed by mathematics professors, rethinking apprenticeship opportunities that happen in partnership with K–12 schools, and in order to do so, transforming teacher preparation programs. One of the challenges of working on a complex problem like improving mathematics teaching and learning is knowing how to divide it into manageable domains of action, and the MTE-Partnership is an excellent case from which to learn. By thoroughly examining the challenges and limitations within current teacher preparation programs, the network was able to specify a theory of improvement that pointed to concrete and high leverage components of the teacher education process that, if improved, would produce substantially better prepared teachers. This work to understand the problem and identify high leverage drivers for a change enabled the network to organize hundreds of programs into working groups, concurrently tackling different aspects of the problem.

In this commentary, I reflect on three key aspects of the MTE-Partnership's approach to tackling a complex problem that make it an important example of the NIC concept in action. First, I show how the MTE-Partnership is on its way to *becoming a scientific-professional learning community (SPLC)*, capable of solving a complex problem of practice. Second, I highlight how the MTE-Partnership illustrates the importance of networked improvement work that directly *targets the core technology of the education process*. Finally, I reflect on promising examples of *cross organizational and cross sector collaboration* (e.g., higher education and K–12 partnerships), embedded in the MTE-Partnership's work.

FORGING A SPLC TO IMPROVE SECONDARY MATHEMATICS TEACHER PREPARATION

The NIC concept aims to provide a social structure for catalyzing the development of scientific professional learning communities (Russell et al., 2019). They are *professional* in the sense that they are voluntary associates of professionals working together to solve pressing problems faced in their work practice. NICs are *learning* communities in the sense that they are organized to generate new insights and distribute knowledge and tools to support improved practice and outcomes. And they share characteristics of *scientific* communities in that the learning is driven by disciplined inquiry cycles anchored in evidence of practice.

My colleagues and I (Russell et al., 2019) posit three primary indicators that a NIC is operating as a scientific-professional learning community. First, NICs are communities that are held together by shared goals, language, norms, theories, and practices. We posit that a shared theory and aligned measures contribute to clarity in goals, common language, and normative expectations for practice in a scientific-professional learning community. Second, NICs support learning and problem solving through disciplined inquiry. NICs seek to break down traditional boundaries between producers and consumers of knowledge, by providing structures and routines for practitioners to produce knowledge about practical improvement. Specifically, practitioners engage in disciplined inquiry using established methods drawn from improvement science, such as Plan-Do-Study-Act (PDSA) Cycles. An aim of routines such as the PDSA is to add a more formalized, scientific process to the reflective practice that is characteristic of professional learning communities (Tichnor-Wagner, Wachen, Cannata, & Cohen-Vogel, 2017). Finally, NICs coordinate and accelerate learning through strategic knowledge management. Central to the production of knowledge in a network is the consolidation of diverse contributions into new knowledge products. In NICs, practitioners engage in small experiments within their professional practice and learn through successive inquiry cycles. But accelerating learning in such a community requires that a central hub harvests and manages this learning by making it visible to others in the network and facilitating the spread of the most promising change ideas (Wohlstetter & Lyle, 2018).

The MTE-Partnership is making promising steps toward catalyzing a scientific-professional learning community to support mathematics teaching and learning improvement. The Partnership has taken explicit actions to build a community held together by shared goals, language, norms, theories, and practices. The work is grounded in a shared vision for mathematics teacher preparation developed through an intensive process of engagement with its members, composed of school-university partnerships. The MTE-Partnership's *Guiding Principles for Secondary Mathematics Teacher Preparation Programs* (2014) capture the Partnership's growing understandings of teacher preparation and the institutional structures needed to support it. Specifically, working groups were established to better understand four major challenges: developing a common vision across stakeholders involved in teacher preparation, improving candidates' mathematical preparation, preparing and supporting mentor teachers, and recruitment and retention of teacher candidates. This work resulted in the specification of a working theory of improvement, represented as a driver diagram. The driver diagram helps members of the community stay focused on the high leverage aspects of the teacher education system that are expected to help it meet its aims. As the driver diagram is elaborated to represent the improvement work that people in the network are doing, it captures the shared goals, language, theories of improvement, and promising practices uncovered by the community.

The MTE-Partnership pursues learning and problem solving through disciplined inquiry, an additional indicator that they are becoming an SPLC. Driven by the guiding principles and the driver diagram, the MTE-Partnership formed five Research Action Clusters (RACs) to test program changes and interventions aligned with the network's working theory of improvement. The RACs employ

PDSA inquiry cycles to iteratively prototype, test, and refine interventions. For example, the Clinical Experiences RAC utilized a series of PDSAs to design, test, and refine a module consisting of three activities that can be used in methods courses to help teacher candidates and mentors gain a better understanding of the *Standards for Mathematical Practice* (National Governors Association Center for Best Practices, Council of Chief State School Officers, 2010). Other teams who engaged in the Clinical Experiences RAC tested and refined tools and routines to support teacher candidates and mentor teachers engaging in co-planning and co-teaching. Through a series of iterative PDSAs, they drew on a variety of measures such as mini-surveys and observations to refine the tools and routines that support effective co-teaching and co-planning. These examples, and others described in this book, show how the RACs are systematically designing and testing changes to teacher education programs that are helping participating programs make progress toward their aim of improved preparation for secondary mathematics teachers. While not always fully explicated, there is evidence that participating faculty are collaborating to collect and analyze data capturing teaching and learning practice in order to generate promising changes and interventions that can be spread through the network.

A final indicator the MTE-Partnership is becoming an SPLC is evident in its work to coordinate and accelerate learning through strategic knowledge management. For learning networks, this is a critical but elusive component of the work. While likely less developed than the other components of an SPLC, there are signs that the MTE-Partnership is beginning to take active steps toward developing this capacity. Routines embedded in the MTE-Partnership's annual conferences and working group meetings support collegial exchange of promising teacher education program changes. The writing of white papers, and ultimately this book, represent a commitment to documenting and spreading what the network is learning about how to improve teacher preparation. As the MTE-Partnership moves forward, members of the Transformation Working Group are actively seeking a viable design to support knowledge management, a system that would enable users to access tested strategies and then log outcomes and contextual factors important to consider when users further test the strategies and interventions.

IMPROVING THE TECHNICAL CORE IN HIGHER EDUCATION ORGANIZATIONS

The progress the MTE-Partnership is making toward becoming a scientific-professional learning community is admirable. In addition, the MTE-Partnership stands out as a reform initiative that aims to improve teaching and learning in higher education institutions at scale. While critical to improving learning outcomes, many reform initiatives shy away from directly intervening on the core technology of educational organizations: the interaction between instructor, students, and content (Ball & Forzani, 2007; Grossman et al., 2009). Improvement science was designed to help understand and optimize routine work processes, and so it has been easily taken up to improve school routines such as interventions to address student absences, getting students to complete financial aid forms to make them eligible for post-secondary funding, and providing regular feedback to beginning teachers. But improving classroom instruction is more complicated because it requires a number of interdependent processes such as the development or selection of new curricular materials, training instructors to use different pedagogical practices, and monitoring how changes in pedagogy and instructional materials influence student thinking and learning. Additionally, improving student learning, particularly at the secondary and post-secondary levels, requires intervening on teaching across many teachers/instructors and classrooms. Consequently, improving complex instructional processes requires adapting standard improvement science methods. Since improvement science methods focus on mapping processes and looking for inconsistent or ineffective use of standard practices (Gawande, 2010; Langley, Nolan, Nolan, Norman, & Provost, 2009), NICs must adjust this methodology to work within the complexity of the interdependent processes that make up the instructional core.

The MTE-Partnership has taken the complexity of this challenge head on in their work to improve teaching and student learning in secondary mathematics teacher education programs. The MTE-Partnership wisely started by mapping the potential points of leverage in this highly complex system of activity. They identified different instructional spaces that needed to be improved—post-secondary mathematics courses, teaching methods courses, and clinical experiences in secondary schools/classrooms. They further identified different components of the teacher education process in these spaces to target for improvement including course content, instructor pedagogy, and mentee-mentor interactions. Then each instructional space and component became a focus for improvement by one or more working group that identified and tested practice changes. For example, the two mathematical preparation RACs, Active Learning Mathematics (see Chapter 6 in the book) and Mathematics of Doing, Understanding, Learning, and Educating for Secondary Schools (MODULE(S²) (see Chapter 5 in this book) include working groups focused on building teacher candidates' mathematical content knowledge through improving the curriculum of post-secondary mathematics courses. Additionally, these RACs worked with mathematics instructors to promote their uptake of pedagogical practices consistent with the type of pedagogy teacher candidates learn in their teaching methods courses. This work is ambitious in many ways, in part because it is unusual to see organized efforts to improve teaching in the post-secondary sector.

Another key to accelerating the improvement of core educational processes is getting the considerable knowledge our field has accumulated about effective structures, routines, and practices to be used reliably by educators (Hiebert, Gallimore, & Stigler, 2002). The MTE-Partnership's RACs have been strategic in their work to ground improvement in existing research and practical frameworks such as the Association of Mathematics Teacher Educators' (2017) *Standards for the Preparation of Teachers of Mathematics.* The chapters in this book show how improvement work was deeply grounded in existing research and practical knowledge. This is an important feature of the work given that it is critical to accelerating improvement, and there is tendency for educators in NICs to experiment with changes that are not rooted in the best research knowledge about teaching and learning,

Relatedly, the network illustrates a third lesson for instructionally focused improvement work: focus on leveraging research and practice-based knowledge about teaching and learning *in a specific discipline.* There is an emerging consensus in the educational research community that efforts to improve core teaching and learning processes must focus on disciplinary-specific practices, rather than general instructional practices (McConachie & Petrosky, 2010; Resnick, 2010; Stein, Engle, Smith, & Hughes, 2008). The depth of knowledge we aim for students to gain in order to be college and career ready requires that students are exposed to authentic opportunities for disciplinary reasoning and practice. The MTE-Partnership is sharply focused on improving the preparation of secondary mathematics teachers, recognizing that teachers must have rigorous disciplinary specific training in order to create powerful learning opportunities for students.

For these three reasons—targeting multiple points of leverage in complex learning systems, grounding improvement in the field's existing knowledge base, and a focus on disciplinary specific teaching and learning processes—the MTE-Partnership is a useful case for informing how to design instructionally focused networked-improvement efforts more broadly.

TACKLING IMPROVEMENT THROUGH CROSS-ORGANIZATIONAL AND CROSS-SECTOR PARTNERSHIPS

I will conclude with a final reflection about what we can learn from the MTE-Partnership: the need to tackle problems of practice across organizations and sectors in order to achieve ambitious, instructionally focused improvement goals at scale. For many educational problems of practice, improvement requires coordinated effort across organizational and even sectoral boundaries. A student's learning trajectory spans multiple school organizations, out-of-school learning providers, and post-secondary institutions. The production of sufficient high-quality teachers requires attention to training experiences in higher education institutions and later within school-based pre-service and in-service placements. A NIC provides a social organization to coordinate improvement that spans organizations and sectors. Grounded by a shared theory of improvement represented as a driver diagram, teams in many organizations can be testing interventions in parallel and under varying conditions that all contribute to learning how to solve a shared problem of practice. With a strong central hub providing coordination, analytic support, and knowledge management, the network can identify a set of evidence-based interventions that drive practice improvement, which can then be spread through organizations in the network.

The MTE-Partnership is on the way to achieving this vision. The network has drawn in over 100 member organizations spanning institutions of higher education (IHEs), school districts, schools, state departments of education, and education reform organizations. Members from participating organizations join RACs that align with their interests, expertise, and improvement needs. The participating IHEs represent considerable context variation in the way that their teacher education programs are organized and operated. For example, they range from comprehensive four-year universities to community colleges. This range provides tremendous opportunities for learning as promising interventions developed in one context can be tested and refined in other contexts. Additionally, a large NIC with many participating organizations from one field has great potential to drive practical improvement at scale. As the network begins to identify a set of proven interventions and concrete practice changes that support teacher education program transformation, there is the potential to shift the focus of the network to the uptake of this new vision of mathematics teacher education. With widespread proof of concept for these changes, the network has the potential to influence practice beyond the participating programs and IHEs, as organizations tend to align with programs seen as innovative in a given field (DiMaggio & Powell, 1983; Lounsbury & Pollack, 2001).

Additionally, it is notable that the work of the MTE-Partnership spans the K–12 and post-secondary sectors. The problem of practice the MTE-Partnership is addressing naturally implicates both sectors: teacher candidates begin their training in post-secondary institutions but participate in clinical training in K–12 schools. This disconnect is well known in the teacher education process; there are often weak connections and coordination between teacher education coursework and clinical experiences (Leatham & Peterson, 2009; Wilson, Floden, & Ferrini-Mundy, 2001). Yet, it is relatively rare for educational reform efforts to work across sectors and doing so is challenging to coordinate (Russell et al., 2015). There are promising examples of cross-sector collaboration in the Clinical Experiences RAC, which is composed of three sub-RACs each focusing on different aspects of the clinical experiences continuum. For example, one of the sub-RACs, composed of several universities and partnering school districts, collaborated to refine and implement a paired-placement clinical experience for prospective teachers that fosters collaboration, reflection, and equitable teaching practices among the prospective teachers and their mentor teacher (see Chapter 11 in this book). Another sub-RAC focuses on co-planning and co-teaching approaches, and their work has informed the work of the paired placement sub-RAC and also has merit on its own for improving clinical experiences (see Chapter 10). The third sub-RAC focuses on developing modules to help university faculty to create courses in which mentor teachers and teacher candidates experience essential learning opportunities that link university coursework and clinical experiences in the field (see Chapter 9). Together mathematics teacher educators, mathematicians, and school partners work together across the three sub-RACs to improve the quality of secondary mathematics education field experiences. The MTE-Partnership is an important illustration of how NICs can enable productive collaboration among diverse organizations that span sectors of the educational field, in order to address complex problems of practice.

As educators, researchers, reformers, funders, and policymakers gravitate to the concept of networked improvement, it is important to have rich written cases that describe the design decisions involved in launching and operating improvement networks. This book makes a strong contribution to the field's understanding of the early uptake of this complex organizational form. Getting a window into the way MTE-Partnership was designed and refined in operation helps us to see the opportunities and challenges associated with catalyzing scientific professional learning communities in education that tackle important problems of practice.

REFERENCES

Association of Mathematics Teacher Educators. (2017). *Standards for preparing teachers of mathematics*. Raleigh, NC: Author. Retrieved from http://amte.net/standards

Ball, D. L., & Forzani, F. M. (2007). What makes education research "Educational"? *Educational Researcher, 36*(9), 529–540.

Bryk, A. S., Gomez, L. M., & Grunow, A. (2011). Getting ideas into action: Building networked improvement communities in education. In M. Hallinan (Ed.) *Frontiers in Sociology of Education* (pp. 127–162). New York, NY: Springer.

DiMaggio, P., & Powell, W. W. (1983). The iron cage revisited: Collective rationality and institutional isomorphism in organizational fields. *American Sociological Review, 48*, 147–160.

Dolle, J. R., Gomez, L. M., Russell, J. L., & Bryk, A. S. (2013). More than a network: Building professional communities for educational improvement. In B. J. Fishman, W. R. Penuel, A. R. Allen, & B. H. Cheng (Eds.), *Design-based implementation research: Theories, methods, and exemplars*. National Society for the Study of Education Yearbook (pp. 443–463). New York, NY: Teachers College Record.

Gawande, A. (2010). *The checklist manifesto: How to get things right*. New York, NY: Metropolitan Books.

Gomez, L. M., Russell, J. L., Bryk, A. S., LeMahieu, P. G., & Mejia, E. (2016). The right network for the right problem. *Phi Delta Kappan, 98*(3), 8–15.

Grossman, P., Compton, C., Igra, D., Ronfeldt, M., Shahan, E., & Williamson, P. W. (2009). Teaching practice: A cross-professional perspective. *Teachers College Record, 111*(9), 2055–2100.

Hiebert, J., Gallimore, R., & Stigler, J. W. (2002). A knowledge base for the teaching profession: What would it look like and how can we get one? *Educational Researcher, 31*(5), 3–15.

Langley, G. L., Nolan, K. M., Nolan, T. W., Norman, C. L. & Provost, L. P. (2009). *The improvement guide: A practical approach to enhancing organizational performance* (2nd ed.). San Francisco, CA: Jossey Bass.

Leatham, K. R., & Peterson, B. E. (2009). Secondary mathematics cooperating teachers' perceptions of the purpose of student teaching. *Journal of Mathematics Teacher Education, 13*, 99–119.

Lounsbury, M., & Pollack, S. (2001). Institutionalizing civic engagement: Shifting logics and the cultural repackaging of service-learning in US higher education. *Organization, 8*(2), 319–339.

Mathematics Teacher Education Partnership. (2014). *Guiding principles for secondary mathematics teacher preparation.* Washington, DC: Association of Public and Land-grant Universities. Retrieved from mtep.info/guidingprinciples

McConachie, S. M., & Petrosky, A. R. (2010). *Content matters: A disciplinary literacy approach to improving student learning.* San Francisco, CA: Jossey-Bass.

National Governors Association Center for Best Practices, Council of Chief State School Officers. (2010). *Common core state standards: Mathematics.* Washington, DC: Author.

Resnick, L. B. (2010). Nested learning systems for the thinking curriculum. *Educational Researcher, 39*(3), 183–197. Retrieved from https://doi.org/10.3102/0013189X10364671

Russell, J. L., Bryk, A. S., Khachatryan, E., LeMahieu, P., Peurach, D., Zoltners, J., & Hannan, M. (2019, April). *The social organization of networked improvement communities.* American Educational Research Association Annual Meeting, Toronto, Canada.

Russell, J. L., Meredith, J., Childs, J., Stein, M. K., & Prine, D. W. (2015). Designing interorganizational networks to implement education reform: An analysis of state Race to the Top applications. *Educational evaluation and policy analysis, 37*(1), 92–112.

Russell, J. L., Bryk, A. S., Dolle, J., Gomez, L. M., LeMahieu, P., & Grunow, A. (2017). A framework for initiation of Networked Improvement Communities. *Teachers College Record, 119*(5), 1–36.

Russell, J. L., Bryk, A. S., Khachatryan, E., LeMahieu, P., Peurach, D., Sherer, J. Z., & Hannan, M. (2019, April). *The social organization of networked improvement communities.* American Educational Research Association Annual Meeting, Toronto, ON.

Stein, M. K., Engle, R. A., Smith, M. S., & Hughes, E. K. (2008). Orchestrating productive mathematical discussions: Five practices for helping teachers move beyond show and tell. *Mathematical Thinking and Learning, 10*, 313–340.

Tichnor-Wagner, A., Wachen, J., Cannata, M., & Cohen-Vogel, L. (2017). Continuous improvement in the public school context: Understanding how educators respond to Plan-Do-Study-Act cycles. *Journal of Educational Change, 18*(4), 465–494.

Wilson, S. M., Floden, R. E., & Ferrini-Mundy, J. (2001). *Teacher preparation research: Current knowledge, gaps, and recommendations.* Seattle, WA: Center for the Study of Teaching and Policy.

Wohlstetter, P., & Lyle, A. G. (2018). Inter-organizational networks in education. In M. Connolly, D. H. Eddy-Spicer, C. James, & S. D. Kruse (Eds.), *The SAGE handbook of school organization* (pp. 210–227). Thousand Oaks, CA: Sage Publishing.

CHAPTER 19

REFLECTIONS ON THE MTE-PARTNERSHIP

The Power of Networked Improvement Communities to Support Transformational Change

Paul LeMahieu, and Wendy M. Smith[1]

Transformational change in education is one of today's most important challenges. Transformational change means thinking about and relating to things differently. It means positioning oneself in new ways with respect to important questions, including how people relate to persistent problems in education and how people think about addressing these problems. When someone wants to make a change that is highly visible, they might change or introduce programs. However, such change typically lasts only as long as there is explicit sponsorship (social, political, or material). If someone wants to make a change that represents doing something truly differently: deep, widespread, and enduring, then they would be well advised to change how people think about the problem —and make this new way of thinking and acting the "new normal" in their organizations. The biggest challenge to transformational change has to do with the necessary shifts in mindsets and perspectives of the often diverse stakeholders (LeMahieu, Bryk, Grunow, & Gomez, 2017).

The Mathematics Teacher Education Partnership: The Power of a Networked Improvement Community to Transform Secondary Mathematics Teacher Preparation, pages 401–409.
401

The essence of this challenge in improving education is helping people see new ways to think and to put the changes in perspective. Such *perspective shifts* lead to changes in position with respect to the problems that need to be solved. One reason perspective shifts are so challenging is that people typically think about transformational change in terms of behaviors—to act in one way or another, to follow this procedure or that process—but changing the way a person thinks about something means that person is in the position to invent a better process to figure out better behaviors.

Transformational change as an improvement-oriented approach has as its first step a deep study of the problem, so that stakeholders understand not only the problem, but also the system that gives rise to and tolerates the problem as well as within which solutions have to succeed (Bryk, Gomez, Grunow, & LeMahieu, 2015; LeMahieu, Edwards, A.R., & Gomez, 2015). Transformational change is also highly situational: asking first what the problem is, and what the context of the problem is. What one person needs to do in order to meaningfully address a given problem might be different from someone else addressing the same problem in a different context. At the very least, it can mean modifications to or adaptations of similar ideas. Being told to use a given program, without consideration of context within which it must succeed, is unlikely to be associated with transformational change. Programs developed in one context cannot be dropped into a new system and reasonably expected to produce transformational change. Unless stakeholders have thought deeply about place and context, they will not understand the system well enough to address the problem meaningfully.

NETWORKED IMPROVEMENT COMMUNITIES AS A VEHICLE FOR TRANSFORMATIONAL CHANGE

For the Carnegie Foundation for the Advancement of Teaching, Networked Improvement Communities (NICs) are the integration of two sets of important and powerful ideas. The first are the ideas associated with improvement science, and the second are the ideas associated with networks. Neither of these are new ideas, but their combination and subsequent adaptation and application to educational settings has come about only recently (Bryk, Gomez, Grunow, & LeMahieu, 2015).

Improvement science has been around for close to 100 years, but most of that time, it was developed and used in business and industry, particularly manufacturing. In the past 25 years, improvement science has been increasingly and successfully used in health care. On the basis of that experience, the Carnegie Foundation has found it encouraging to apply similar improvement science to other professional contexts and people-serving sectors. Most recently, the Carnegie Foundation has been adapting improvement science ideas to work in educational contexts, thinking about developing methods (e.g., processes, tools, and routines) for employing these ideas in educational settings (Bryk, Gomez, Brunow, & LeMahieu, 2015).

Networks are even older than improvement science and go back to historical collectives of people evolving to clans, tribes, city states, and nations. Although ideas of collectives have been around for a long time, only recently have researchers developed the knowledge and tools to intentionally and systematically support communities so they are productive in new ways. People have figured out how to organize, coordinate, guide, and structure communities in ways in which they are able to get meaningful collective action on a shared problem. For these reasons, as well as others, networks provide a unique human environment in which to do improvement work.

By integrating these two sets of big and powerful ideas related to improvement science and networks, it becomes possible to apply improvement science as executed in the context of carefully structured activities. Having existing communities provides an architecture in which to do improvement work. This integration is powerful in a transformative sense in education, because such networks equip people to engage in meaningful collective work on specific problems and increase their capacity to solve them. People working on a specific aim and applying improvement science come away not only with solutions to the targeted problem, but also with the tools to know how to solve other problems that come along. It is the combination of the two ideas that generates power in the transformational sense: not just implementing a program or doing one thing differently but learning how to change the system by making the fundamental principles, methods, and processes of networked improvement the new normal within organizations.

A NIC does not just happen; an effective NIC incorporates *will-building, improvement processes and tools, capacity development, integration of content and process knowledge,* and *a network structure to support social learning at scale* (LeMahieu, Grunow, Nordstrum, & Baker, 2017). Without a doubt, the greatest challenge is to bring all these elements together in adequate amounts. *Will-building* refers to the idea of working on a problem that people care enough about that they will invest significant time and resources to engage in working on the problem. Will-building also includes getting people to agree that and understand why a particular state of affairs is a problem, and, further, to generate a sense of urgency that compels people to want to join with others to work on and solve that problem. This kind of will-building is the first challenge in getting an NIC to operate effectively, properly, and productively.

The second challenge to effective NICs involves learning to apply *improvement methods, processes, and tools.* No matter the approach taken, certain fundamental principles define the family of practices that constitute improvement science (Bryk, Gomez, Grunow, & LeMahieu, 2015). Developing the capability of those engaged in the enterprise is a necessary condition to ensure widespread use of the practices. Stakeholders need knowledge of the processes that guide how to do improvement science. Additionally, leaders must have particular and specialized knowledge, skills, and dispositions to encourage people to work together toward a common goal.

The third challenge is *integrating scholarly and practical knowledge and expertise relevant to the problem and its solution.* Effective improvement work respects in equal measure the knowledge of researchers and scholars as well as practitioners. People working together, particularly leaders of change efforts, need knowledge and skills to understand the problem at hand and potential solution strategies. Change efforts tend to be ineffective if the effort has access to research-based content knowledge but lacks the knowledge of practitioners who work close to, often in direct contact, with the problem, or it has access to practical knowledge but lacks scholarly knowledge. In education, teachers often have deep content knowledge, such as knowledge of mathematics, the teaching of mathematics, and uses of mathematics in teaching. It is a big mistake to undertake change efforts that do not include enough people with content knowledge. It can be equally challenging within a NIC to gather and balance community members to bring together their knowledge of improvement processes and content knowledge in all its necessary forms.

The fourth key challenge for developing effective NICs is *the development of network structures that support coordinated collective effort and widespread social learning.* As people start to engage in improvement science in a network—to build, launch and maintain a NIC—they engage with a big social learning challenge. In all kinds of NICs, even with a common problem to address, it is difficult to determine how to structure all the components and people to solve that problem. In a NIC, different people and groups work on different dimensions of a complex problem—this is precisely how networks can accelerate learning as well as the spread, uptake, and use of the new knowledge. A key promise of working as a NIC is that learning will be accelerated as members learn from one another (Bryk, Gomez, Brunow, & LeMahieu, 2015). However, effectively sharing knowledge generated by a NIC in ways that others can take up and apply that knowledge is a challenge with which the field continues to struggle. The NIC needs ways for members to learn what others are doing to contribute to meaningful collective action.

Together, the challenges of will-building, capacity development, integration of improvement and content knowledge, and network structure are all vital for an effective NIC. These dimensions are hard to enact, particularly when different people have different types of knowledge, skills, and dispositions. Yet, to address complex problems in education, NICs provide a promising way to organize improvement efforts to yield transformational changes.

THE MATHEMATICS TEACHER EDUCATION PARTNERSHIP AS A NETWORKED IMPROVEMENT COMMUNITY

The Mathematics Teacher Education Partnership (MTE-Partnership) is one of several NICs working on these challenges. The MTE-Partnership is focused on teacher preparation as its targeted problem, and it is using improvement tools to make transformative changes within the institutions that provide professional preparation programs. The MTE-Partnership has developed and maintains collective processes and builds in regular opportunities to meet together, including annual workshops. MTE-Partnership also represents a NIC whose network is geographically dispersed, adding to the challenge of working collectively.

The MTE-Partnership's work is making substantial and fundamental changes in the area of teacher preparation. The group is not just tinkering, preserving old forms of behavior or programs, while trying to make them work better, but instead is developing new ways of approaching the problem and the systems within which the problem is embedded.

The MTE-Partnership has grown in scope over the past several years, while remaining focused on its primary challenge of increasing the quality and quantity of secondary mathematics teachers. Often improvement efforts are challenged by a sort of *mission creep*, taking on additional problems as they grow in scope; the MTE-Partnership has avoided this pitfall. One of the MTE-Partnership's strengths has been the hard work and thoughtfulness the group put into the initiation of its network. The MTE-Partnership leaders took time to understand and clarify the problem they wanted to work on, establishing and securing widespread commitment to a common aim that the MTE-Partnership seeks to realize—along with their ways of going about working on that problem and achieving that aim (Martin & Gobstein, 2015). As new ideas have come in, they are intellectually consistent with what they have been doing, as opposed to distracting or diluting to what they have been doing.

The MTE-Partnership followed the core principles of networked improvement science, beginning with chartering its efforts. A charter is defined as a deep study of the problem, the system that tolerates (or even produces) the problem, and the system in which changes have to succeed. People study root causes to identify factors that produce or influence the problem, which culminates in two foundational documents to outline the planned changes: design principles and a theory of practice improvement. A theory of improvement is typically expressed in the form of a driver diagram; there are numerous examples of such diagrams throughout this volume. A set of design principles serve as a bridge to what is known from scholarship and best practice and articulates what kinds of solutions the NIC will adapt or invent and pursue. Having such principles is important to avoid chaos when convening a community. Such guiding principles allow the networked community to be very clear what is and is not within the intellectual scope of their effort. Then, the NIC can invite innovation, while avoiding the unconstrained efforts typical of groups in which individuals each pursue their own agendas and strategies. Design principles also help the group to collectively select viable change strategies, and not just follow the whims of particular individuals.

The MTE-Partnership leaders put in the effort at the beginning of forming their NIC to develop sound design principles (MTE-Partnership, 2014), which has helped the effort remain intellectually coherent and focused. The MTE-Partnership leaders understand and have remained committed to the processes of improvement science; even as new ideas came up, the design principles allowed the MTE-Partnership leaders to evaluate if the new ideas were important new directions or distractions. Design principles can give leaders the courage to avoid pursuing ideas that might otherwise have merit, but do not fit what the NIC is doing. The MTE-Partnership's *Guiding Principles for Secondary Mathematics Teacher Preparation Programs* (2014), have allowed the NIC to grow while remaining focused on the original problem.

In addition to a strong foundation with design principles, an effective NIC has a driver diagram (Bryk, Gomez, Grunow, & LeMahieu, 2015). Driver diagrams serve two main purposes. The first is intellectual: to articulate the intellectual landscape in a way that makes it possible to work meaningfully in addressing the most significant, high leverage factors in addressing a problem or achieving an aim. Understanding and mapping the intellectual landscape allows a NIC to discover the factors it ought to be concerned with that might make for improvements. Second, a driver diagram provides organization to a NIC's efforts. When a NIC is widely distributed all around the country in very different kinds of places (as the MTE-Partnership is)—with numerous host institutions that are politically and otherwise independent—a driver diagram can help them all focus on similar change strategies, adapted to their own contexts. Effective NICs view member diversity as powerful, yet one of the biggest challenges for a NIC is how to stimulate and guide a genuinely coherent collective and collaborative community and not have individuals each doing their own thing. Driver diagrams list the high leverage factors that a NIC wants to address and enables a network to collectively engage in meaningful work.

In the MTE-Partnership, there is an overall driver diagram to guide their collective efforts. The MTE-Partnership has taken that driver diagram and broken it into component parts (fractal like), creating Research Action Clusters (RACs) around each of the different drivers in the larger driver diagram. As a result, the MTE-Partnership built natural subgroups within the omnibus NIC that it was developing, with each subgroup guided by its own driver diagram such that each is engaged in meaningful collective action and effort. Many similar large-scale efforts have not been as successful in organizing subgroups and keeping their efforts focused. The MTE-Partnership saw those primary drivers in its theory of practice improvement and decided to build a "sub-NIC" around each of those drivers.

The MTE-Partnership has done far better than most NICs to stimulate organized, coherent, and collective action across diverse, independent entities. The RAC structure is a big part of this success, with each subgroup working on different dimensions and change strategies, even as all our working on the same overall problem. The true promise and biggest challenge of a NIC lives in this organization of work processes; the MTE-Partnership's RACs are helping the NIC make significant progress toward its aim.

ADVICE FOR THE MTE-PARTNERSHIP

The MTE-Partnership has been operating as a NIC long enough that it's time for high-level stock-taking. A NIC operates on *continuous improvement* because an effective educational system sees transformation as a process, not a destination. One must have constancy of purpose until realizing some gains, perhaps the achievement of initially stated aims. Then the leadership of the NIC must look and ask: "Do we have the same problem?" and "In light of how we answer this first question, do we need to work in the same way?" Such an examination might find that the driver diagram that was helping the NIC to make improvements is now holding the NIC back.

The MTE-Partnership should pause to consider what new insights they have about where improvement is needed, what they have learned about their theory of practice improvement over the years, and how that should influence the current working version of that theory and consequent action. Then the challenge becomes to determine how to hold onto the gains the NIC has made while possibly moving on to address new drivers of improvement. It is important to not lose the progress made thus far. Changes that have been experimental in nature can now become institutionalized as part of the new normal. It is then time to look to the driver diagram, or a next version of it, to ask, "Where do we direct new effort, in order to extend the gains even further?" As NICs mature, this revisiting of aims and their drivers is part of a healthy process and the nature of transformational change.

LEARNING FROM THE MTE-PARTNERSHIP

In addition to the MTE-Partnership's contribution to the field—bringing about transformational change to the preparation of secondary mathematics teachers—other institutions should also learn from the MTE-Partnership's process and adherence to improvement science principles. The MTE-Partnership really stands out in three ways: its use of improvement processes, its partnerships, and its commitment to evidence. First, the MTE-Partnership has been truly exemplary in the seriousness and thoughtfulness with which it prepared and launched its work as a NIC. As a result, their change efforts have been sustained over quite a period of time—longer than most change efforts in education pay attention to any one idea. It is important for other education change efforts to have similar long-term commitments to improvement processes.

Second, the MTE-Partnership's authentic and rich partnerships stand out. Higher education folks tend to be ferociously independent, particularly around intellectual independence. The culture of higher education has a deeply-rooted independent and individual perspective on how work gets done. What the MTE-Partnership has managed to do in terms of genuinely coherent and collaborative effort in developing and nurturing its active partnerships, not only among institutions of higher education, but together with K–12 school districts, is truly extraordinary. While collaboration is often built in at the K–12 level (and particularly so at the building level), this is not always the case in higher education. The MTE-Partnership's leaders have accomplished the development and maintenance of authentic collaborative partnerships in higher education better than that which is usually seen.

Finally, a challenge that the MTE-Partnership has faced is traditional thinking about roles, responsibilities, the nature of research, and the evidence that validates practices and improvements. Short-cycle improvement efforts, via the Plan-Do-Study-Act Cycle, are pretty far from the typical structure of longitudinal research projects in education. Collaboration, particularly across institutions, is not typically the norm. The methodological requirements necessary to warrant the attribution of affect to identified causes typically confound efforts to produce the knowledge necessary to replicate warranted practices across contexts. Moreover, well-established norms about roles and responsibilities tend to identify some as producers of knowledge and others as users (with and often manifest hierarchy attached to the distinction). Improvement science challenges many preconceptions about the nature of epistemic practices as well as roles and responsibilities of those who engage in them. People everywhere have conventional wisdoms about how such work is done. Higher education is not unique in having its own conventions, although the conventions for research methodologies in higher education are not universal to other contexts. The MTE-Partnership and the individuals who make it up have managed to negotiate these preconceptions in ways that redound to the benefit of their production of useful improvement knowledge. Others seeking to make improvements in educational systems can learn from how the MTE-Partnership NIC has formed and continues to operate as an exceptional partnership: leadership and membership, researchers and practitioners, together as improvers.

ENDNOTE

1. Thanks to Alli Davis and Lindsay Augustyn of the University of Nebraska–Lincoln who contributed to the writing of this chapter.

REFERENCES

Bryk, A., Gomez, L. M., Grunow, A., & LeMahieu, P. (2015). *Learning to improve: How America's schools can get better at getting better.* Cambridge, MA: Harvard Education Press.

LeMahieu, P. G., Bryk, A. S., Grunow, A., & Gomez, L. M. (2017). Working to improve: Seven approaches to quality improvement in education. *Quality Assurance in Education, 25*(1), 2–4.

LeMahieu, P. G., Edwards, A. R., & Gomez, L. M. (2015). At the nexus of improvement science and teaching. *Journal of Teacher Education, 66*(5), 1–4.

LeMahieu, P. G., Grunow, A., Nordstrum, L.E., & Baker, L. (2017). Networked improvement communities: The discipline of improvement science meets the power of Networks. *Quality Assurance in Education, 25*(1), 5–25.

Martin, W. G., & Gobstein, H. (2015). Generating a networked improvement community to improve secondary mathematics teacher preparation: Network leadership, organization, and operation. *Journal of Teacher Education, 66*(5), 482–493.

Mathematics Teacher Education Partnership. (2014). *Guiding principles for secondary mathematics teacher preparation.* Washington, DC: Association of Public and Land-grant Universities. Retrieved from mtep.info/guidingprinciples

EDITOR BIOGRAPHIES

W. Gary Martin is an Emily R. and Gerald S. Leischuck Endowed Professor at Auburn University, where he teaches undergraduate and graduate courses in mathematics education. His research interests center on high school mathematics, including trends in curriculum and standards, effective teacher preparation, and leadership development. He has participated on numerous national initiatives, including the writing teams for *Principles to Actions: Ensuring Mathematical Success for All* (National Council of Teachers of Mathematics, 2014) and *Standards for the Preparation of Teachers of Mathematics* (Association of Mathematics Teacher Educators, 2017). He is currently the co-director of the Mathematics Teacher Education Partnership.

Brian R. Lawler is an Associate Professor of Mathematics Education in the Bagwell College of Education at Kennesaw State University. His scholarship focuses on equity issues in mathematics education, in particular the ways in which power and knowledge intertwine to govern the learner's mathematical identity. This focus underscores his current research on how schools and districts transform their mathematics instruction to teach in ways that foreground student thinking in light of the pressures of high-stakes education. Through this work, he supports teachers, schools, and districts to de-track mathematics instruction and transform teaching practices. Lawler is currently leader of the Equity and Social Justice

The Mathematics Teacher Education Partnership: The Power of a Networked Improvement Community to Transform Secondary Mathematics Teacher Preparation, pages 411–412.

Working Group and serves on the Mathematics Teacher Education Partnership Planning Team.

Alyson E. Lischka is an Associate Professor of Mathematics Education in the Department of Mathematical Sciences at Middle Tennessee State University. Her research focuses on the implementation of ambitious teaching practices among prospective and practicing teachers, specifically through explorations of effective feedback and questioning practices in the teaching of mathematics and mathematics methods courses. She teaches undergraduate courses in content and methods for prospective secondary teachers along with courses in the Mathematics and Science Education doctoral program. Lischka is currently the leader of the Mathematics of Doing, Understanding, Learning, and Educating for Secondary Schools Research Action Cluster and serves on the Mathematics Teacher Education Partnership Planning Team.

Wendy M. Smith is the Associate Director of the Center for Science, Mathematics and Computer Education at the University of Nebraska–Lincoln. Her research interests include PK–20 mathematics education, institutional change, rural education, teacher change, teacher professional development, teacher leadership, professional networks, action research, and estimating teacher professional development effects on student achievement. As a former middle school mathematics teacher, she seeks to support more equitable student outcomes in mathematics and science. She is currently the leader of the Active Learning Mathematics Research Action Cluster and serves on the overall planning committee within the Mathematics Teacher Education Partnership.

Printed in the United States
By Bookmasters